Abraham Kandel, Horst Bunke, Mark Last (Eds.)

Applied Graph Theory in Computer Vision and Pattern Recognition

Studies in Computational Intelligence, Volume 52

Editor-in-chief
Prof. Janusz Kacprzyk
Systems Research Institute
Polish Academy of Sciences
ul. Newelska 6
01-447 Warsaw
Poland
E-mail: kacprzyk@ibspan.waw.pl

Further volumes of this series
can be found on our homepage:
springer.com

Vol. 33. Martin Pelikan, Kumara Sastry, Erick
Cantú-Paz (Eds.)
Scalable Optimization via Probabilistic Modeling, 2006
ISBN 978-3-540-34953-2

Vol. 34. Ajith Abraham, Crina Grosan, Vitorino
Ramos (Eds.)
Swarm Intelligence in Data Mining, 2006
ISBN 978-3-540-34955-6

Vol. 35. Ke Chen, Lipo Wang (Eds.)
Trends in Neural Computation, 2007
ISBN 978-3-540-36121-3

Vol. 36. Ildar Batyrshin, Janusz Kacprzyk, Leonid
Sheremetor, Lotfi A. Zadeh (Eds.)
Preception-based Data Mining and Decision Making in Economics and Finance, 2006
ISBN 978-3-540-36244-9

Vol. 37. Jie Lu, Da Ruan, Guangquan Zhang (Eds.)
E-Service Intelligence, 2007
ISBN 978-3-540-37015-4

Vol. 38. Art Lew, Holger Mauch
Dynamic Programming, 2007
ISBN 978-3-540-37013-0

Vol. 39. Gregory Levitin (Ed.)
Computational Intelligence in Reliability Engineering, 2007
ISBN 978-3-540-37367-4

Vol. 40. Gregory Levitin (Ed.)
Computational Intelligence in Reliability Engineering, 2007
ISBN 978-3-540-37371-1

Vol. 41. Mukesh Khare, S.M. Shiva Nagendra (Eds.)
Artificial Neural Networks in Vehicular Pollution Modelling, 2007
ISBN 978-3-540-37417-6

Vol. 42. Bernd J. Krämer, Wolfgang A. Halang (Eds.)
Contributions to Ubiquitous Computing, 2007
ISBN 978-3-540-44909-6

Vol. 43. Fabrice Guillet, Howard J. Hamilton (Eds.)
Quality Measures in Data Mining, 2007
ISBN 978-3-540-44911-9

Vol. 44. Nadia Nedjah, Luiza de Macedo
Mourelle, Mario Neto Borges,
Nival Nunes de Almeida (Eds.)
Intelligent Educational Machines, 2007
ISBN 978-3-540-44920-1

Vol. 45. Vladimir G. Ivancevic, Tijana T. Ivancevic
Neuro-Fuzzy Associative Machinery for Comprehensive Brain and Cognition Modeling, 2007
ISBN 978-3-540-47463-0

Vol. 46. Valentina Zharkova, Lakhmi C. Jain
Artificial Intelligence in Recognition and Classification of Astrophysical and Medical Images, 2007
ISBN 978-3-540-47511-8

Vol. 47. S. Sumathi, S. Esakkirajan
Fundamentals of Relational Database Management Systems, 2007
ISBN 978-3-540-48397-7

Vol. 48. H. Yoshida (Ed.)
Advanced Computational Intelligence Paradigms in Healthcare, 2007
ISBN 978-3-540-47523-1

Vol. 49. Keshav P. Dahal, Kay Chen Tan, Peter I. Cowling (Eds.)
Evolutionary Scheduling, 2007
ISBN 978-3-540-48582-7

Vol. 50. Nadia Nedjah, Leandro dos Santos Coelho,
Luiza de Macedo Mourelle (Eds.)
Mobile Robots: The Evolutionary Approach, 2007
ISBN 978-3-540-49719-6

Vol. 51. Shengxiang Yang, Yew-Soon Ong, Yaochu Jin (Eds.)
Evolutionary Computation in Dynamic and Uncertain Environments, 2007
ISBN 978-3-540-49772-1

Vol. 52. Abraham Kandel, Horst Bunke, Mark Last (Eds.)
Applied Graph Theory in Computer Vision and Pattern Recognition, 2007
ISBN 978-3-540-68019-2

Abraham Kandel
Horst Bunke
Mark Last
(Eds.)

Applied Graph Theory in Computer Vision and Pattern Recognition

With 85 Figures and 17 Tables

Springer

Prof. Abraham Kandel
National Institute for Applied
Computational Intelligence
Computer Science & Engineering Department
University of South Florida
4202 E. Fowler Ave.,
ENB 118
Tampa, FL 33620
USA
E-mail: kandel@csee.usf.edu

Prof. Dr. Horst Bunke
Institute of Computer Science
and Applied Mathematics (IAM)
Neubrückstrasse 10
CH-3012 Bern
Switzerland
E-mail: bunke@iam.unibe.ch

Dr. Mark Last
Department of Information Systems Engineering
Ben-Gurion University of the Negev
Beer-Sheva 84105
Israel
E-mail: mlast@bgu.ac.il

Library of Congress Control Number: 2006939143

ISSN print edition: 1860-949X
ISSN electronic edition: 1860-9503
ISBN-10   3-540-68019-5 Springer Berlin Heidelberg New York
ISBN-13   978-3-540-68019-2 Springer Berlin Heidelberg New York

This work is subject to copyright. All rights are reserved, whether the whole or part of the material is concerned, specifically the rights of translation, reprinting, reuse of illustrations, recitation, broadcasting, reproduction on microfilm or in any other way, and storage in data banks. Duplication of this publication or parts thereof is permitted only under the provisions of the German Copyright Law of September 9, 1965, in its current version, and permission for use must always be obtained from Springer-Verlag. Violations are liable to prosecution under the German Copyright Law.

Springer is a part of Springer Science+Business Media
springer.com
© Springer-Verlag Berlin Heidelberg 2007

The use of general descriptive names, registered names, trademarks, etc. in this publication does not imply, even in the absence of a specific statement, that such names are exempt from the relevant protective laws and regulations and therefore free for general use.

Cover design: deblik, Berlin
Typesetting by the SPi using a Springer LaTeX macro package
Printed on acid-free paper   SPIN: 11946359   89/SPi   5 4 3 2 1 0

Preface

Graph theory has strong historical roots in mathematics, especially in topology. Its birth is usually associated with the "four-color problem" posed by Francis Guthrie in 1852,[1] but its real origin probably goes back to the Seven Bridges of Königsberg problem proved by Leonhard Euler in 1736.[2] A computational solution to these two completely different problems could be found after each problem was abstracted to the level of a *graph model* while ignoring such irrelevant details as country shapes or cross-river distances. In general, a graph is a nonempty set of points (*vertices*) and the most basic information preserved by any graph structure refers to adjacency relationships (*edges*) between some pairs of points. In the simplest graphs, edges do not have to hold any attributes, except their endpoints, but in more sophisticated graph structures, edges can be associated with a direction or assigned a label. Graph vertices can be labeled as well. A graph can be represented graphically as a drawing (vertex = dot, edge = arc), but, as long as every pair of adjacent points stays connected by the same edge, the graph vertices can be moved around on a drawing without changing the underlying graph structure.

The expressive power of the graph models placing a special emphasis on connectivity between objects has made them the models of choice in chemistry, physics, biology, and other fields. Their increasing popularity in the areas of computer vision and pattern recognition can be easily explained by the graphs' ability to represent complex visual patterns on one hand and to keep important structural information, which may be relevant for pattern recognition tasks, on the other hand. This is in sharp contrast with the more conventional feature vector or attribute-value representation of patterns where only unary measurements – the features, or equivalently, the attribute values – are used for object representation. Graph representations also have a number of invariance properties that may be very convenient for certain tasks.

---

[1] Is it possible to color, using only four colors, any map of countries in such a way as to prevent two bordering countries from having the same color?

[2] Given the location of seven bridges in the city of Königsberg, Prussia, Euler has proved that it was not possible to walk with a route that crosses each bridge exactly once, and return to the starting point.

As already mentioned, we can rotate or translate the drawing of a graph arbitrarily in the two-dimensional plane, and it will still represent the same graph. Moreover, we can stretch out or shrink its edges without changing the underlying graph. Hence graph representations have an inherent invariance with respect to translation, rotation and scaling – a property that is desirable in many applications of image analysis. On the other hand, we have to pay a price for the enhanced representational capabilities of graphs, viz. the increased computational complexity of many operations on graphs. For example, while it takes only linear time to test two feature vectors or two tuples of attribute-value pairs, for identity, all available algorithms for the equivalent operation on general graphs, i.e., graph isomorphism, are of exponential complexity. Nevertheless, there are numerous applications where the underlying graphs are relatively small, such that algorithms of exponential complexity are applicable. In other problem domains, heuristics can be found that cut significant amounts of the search space, thus rendering algorithms with a reasonably high speed. Last but not least, for more or less all common graph operations needed in pattern recognition and machine vision, approximate algorithms have become available meanwhile, which can be substituted for their exact versions. As a matter of experience, often the performance of the overall task is not compromised by using an approximate algorithm rather than an optimal one.

This book intends to cover a representative, but in no way exclusive, set of novel graph-theoretic methods for complex computer vision and pattern recognition tasks. The book is divided into three parts, which are briefly described below.

Part I includes three chapters applying graph theory to low-level processing of digital images. The first chapter by Walter G. Kroptasch, Yll Haxhimusa, and Adrian Ion presents a new method for partitioning a given image into a hierarchy of homogeneous areas ("segments") using graph pyramids. A graphical model framework for image segmentation based on the integration of Markov random fields (MRFs) and deformable models is introduced in the chapter by Rui Huang, Vladimir Pavlovic, and Dimitris N. Metaxas. In the third chapter, Alain Bretto studies the relationship between graph theory and digital topology, which deals with topological properties of 2D and 3D digital images.

Part II presents four chapters on graph-theoretic learning algorithms for high-level computer vision and pattern recognition applications. First, a survey of graph based methodologies for pattern recognition and computer vision is presented by D. Conte, P. Foggia, C. Sansone, and M. Vento. Then Gabriel Valiente introduces a series of computationally efficient algorithms for testing graph isomorphism and related graph matching tasks in pattern recognition. Sebastien Sorlin, Christine Solnon, and Jean-Michel Jolion propose a new graph distance measure to be used for solving graph matching problems. Joseph Potts, Diane J. Cook, and Lawrence B. Holder describe an approach, implemented in a system called Subdue, to learning patterns in relational data represented as a graph.

Finally, Part III provides detailed descriptions of several applications of graph-based methods to real-world pattern recognition tasks. Thus, Gian Luca Marcialis, Fabio Roli, and Alessandra Serrau present a critical review of the main graph-based and structural methods for fingerprint classification while comparing them with the classical statistical methods. Horst Bunke et al. present a new method to visualize a time series of graphs, and show potential applications in computer network monitoring and abnormal event detection. In the last chapter, A. Schenker, H. Bunke, M. Last, and A. Kandel describe a clustering method that allows the use of graph-based representations of data instead of the traditional vector-based representations.

We believe that the chapters included in our volume will serve as a foundation for a variety of useful applications of the graph theory to computer vision, pattern recognition, and related areas. Our additional goal is to encourage more research studies that will deal with the methodological challenges in applied graph theory outlined by this book authors.

October 2006
Abraham Kandel
Horst Bunke
Mark Last

Contents

Part I Applied Graph Theory for Low Level Image Processing and Segmentation

Multiresolution Image Segmentations in Graph Pyramids
Walter G. Kropatsch, Yll Haxhimusa and Adrian Ion 3

A Graphical Model Framework for Image Segmentation
Rui Huang, Vladimir Pavlovic and Dimitris N. Metaxas 43

Digital Topologies on Graphs
Alain Bretto 65

Part II Graph Similarity, Matching, and Learning for High Level Computer Vision and Pattern Recognition

How and Why Pattern Recognition and Computer Vision Applications Use Graphs
Donatello Conte, Pasquale Foggia, Carlo Sansone and Mario Vento 85

Efficient Algorithms on Trees and Graphs with Unique Node Labels
Gabriel Valiente 137

A Generic Graph Distance Measure Based on Multivalent Matchings
Sébastien Sorlin, Christine Solnon and Jean-Michel Jolion 151

Learning from Supervised Graphs
Joseph Potts, Diane J. Cook and Lawrence B. Holder 183

Part III Special Applications

Graph-Based and Structural Methods for Fingerprint Classification
Gian Luca Marcialis, Fabio Roli and Alessandra Serrau 205

Graph Sequence Visualisation and its Application to Computer Network Monitoring and Abnormal Event Detection
H. Bunke, P. Dickinson, A. Humm, Ch. Irniger and M. Kraetzl 227

Clustering of Web Documents Using Graph Representations
Adam Schenker, Horst Bunke, Mark Last and Abraham Kandel 247

Part I

Applied Graph Theory for Low Level Image Processing and Segmentation

Multiresolution Image Segmentations in Graph Pyramids

Walter G. Kropatsch, Yll Haxhimusa and Adrian Ion

1 Introduction

"How do we bridge the representational gap between image features and coarse model features?" is the question asked by the authors of [1] when referring to several contemporary research issues. They identify the one-to-one correspondence between salient image features (pixels, edges, corners, etc.) and salient model features (generalized cylinders, polyhedrons, invariant models, etc.) as a limiting assumption that makes prototypical or generic object recognition impossible. They suggested to bridge and not to eliminate the representational gap, as it is done in the computer vision community for quite long, and to focus efforts on (1) *region segmentation*, (2) *perceptual grouping*, and (3) *image abstraction*. Let us take these goals as a guideline to consider multiresolution representations under the special viewpoint of segmentation and grouping. In [2] multiresolution representation is considered under the abstraction viewpoint.

Wertheimer [3] has formulated the importance of wholes (Ganzen) and not of its individual elements and introduced the importance of perceptual grouping and organization in visual perception. Regions as aggregations of primitive pixels play an extremely important role in nearly every image analysis task. Their internal properties (color, texture, shape, etc.) help to identify them, and their external relations (adjacency, inclusion, similarity of properties) are used to build groups of regions having a particular meaning in a more abstract context. The union of regions forming the group is again a region with both internal and external properties and relations.

Low-level cue image segmentation cannot and should not produce a complete final "good" segmentation, because there is no general "good" segmentation. Without prior knowledge, segmentation based on low-level cues will not be able to extract semantics in generic images. Using some similarity measures, the segmentation process results in "homogeneity" regions with respect to the low-level cues. Problems emerge because (1) homogeneity of low-level cues will not map to the semantics [4] and (2) the degree of homogeneity of a region is in general quantified by threshold(s) for a given measure [5]. Even though segmentation methods (including ours) that do not take the context of the image into consideration cannot produce a "good" segmentation, they can be valuable tools in image analysis in the same sense as efficient edge detectors are. Note that efficient edge detectors do not consider the context of the image, too. Thus, the low-level coherence of brightness, color, texture, or motion attributes should be used to sequentially come up with hierarchical partitions [6]. Mid and high-level knowledge can be used to either confirm these groups or select some further attention. A wide range of computational vision problems could make use of segmented images, were such segmentation rely on efficient computation, e.g., motion estimation requires an appropriate region of support for finding correspondences; higher-level problems such as recognition and image indexing can also make use of segmentation results in the problem of matching.

It is important for a grouping method to have the following properties [7]:

- Capture perceptually important groupings or regions, which reflect global aspects of the image
- Be highly efficient running in time linear in the number of image pixels, and
- Create hierarchical partitions [6]

To find region borders quickly and effortlessly in a bottom-up "stimulus-driven" way based on local differences in a specific feature, we propose a hierarchy of extended region adjacency graphs (RAG+) to achieve partitioning of the image by using a minimum weight spanning tree (MST). A RAG+ is a region adjacency graph (RAG) enhanced by nonredundant self-loops or parallel edges. Rather than trying to have just one "good" segmentation the method produces a stack of (dual) graphs (a graph pyramid), which down projected onto the base level gives a multilevel segmentation i.e., a labeled spanning tree. The MST of an image is built by combining the advantage of regular pyramids (logarithmic tapering) with the advantages of irregular graph pyramids (their purely local construction and shift invariance). The aim is reached by using the selection method for contraction kernels proposed in [8]. Borůvka's minimum spanning tree algorithm [9] with the dual-graph contraction algorithm [10] build in a hierarchical way an MST, while preserving the proper topology. For vision tasks, in natural systems, topological relations seem to play an even more important role than precise geometrical positions.

1.1 Overview of the Chapter

The plan of the chapter is as follows. In order to make the reading of this chapter easy, in Sect. 2 we recall some of the basic notions of graph theory. After a short introduction into image pyramids (Sect. 3) a detailed presentation of dual-graph contraction is given (Sect. 5). Using the dual-graph contraction algorithm from Sect. 5, Borůvka's algorithm is redefined in Sect. 6.3, so that we can construct an image graph pyramid, and at the same time, the minimum spanning tree. In Sect. 6 we give the definition of internal and external contrast and the merge decision criteria based on these definitions. In addition, the algorithm for building the hierarchy of partitions is introduced in this section. Also Sect. 6.5 reports on experimental results. Evaluation of the quality of the segmentation results is reported in Sect. 7. Parts of this chapter has been previously published in [11].

2 Basics of Graph Theory

In 1736, Leonard Euler was puzzled whether it is possible to walk across all the bridges on the river Pregel in Königsberg[1] only once and return to the starting point (see Fig. 1a). In order to solve this problem, Euler in an ingenious way, abstracted the bridges and the landmasses. He replaced each landmass by a dot (called vertex) and each bridge by an arch (called edge or line) (Fig. 1b). Euler proved that there is no solution to this problem. The Königsberg bridge problem was the first problem studied in what is nowadays called graph theory. This problem was a starting point also for another branch in mathematics, the topology. The definitions given later are compiled from the books [12–14], therefore the citations are not repeated. The interested reader can find all these definitions and more in the earlier mentioned literature.

Formally, one can define graph G on sets V and E as:

Definition 1 (Graph). *A graph $G = (V(G), E(G), \iota_G(\cdot))$ is a pair of sets $V(G)$ and $E(G)$ and an incidence relation $\iota_G(\cdot)$ that maps pairs of elements of $V(G)$ (not necessarily distinct) to elements of $E(G)$.*

The elements v_i of the set $V(G)$ are called vertices (or nodes, or points) of the graph G, and the elements e_j of $E(G)$ are its edges (or lines). Let an example be used to clarify the incidence relations $\iota_G(\cdot)$. Let the set of vertices of the graph G in Fig. 1b) be given by $V(G) = \{v_A, v_B, v_C, v_D\}$ and the edge set by $E(G) = \{e_a, e_b, e_c, e_d, e_f, e_g\}$. The incidence relation is defined as:

$$\iota_G(e_a) = (v_A, v_B), \iota_G(e_b) = (v_A, v_B), \iota_G(e_c) = (v_A, v_C), \iota_G(e_d) = (v_A, v_C),$$
$$\iota_G(e_e) = (v_A, v_D), \iota_G(e_f) = (v_B, v_D), \iota_G(e_g) = (v_C, v_D). \quad (1)$$

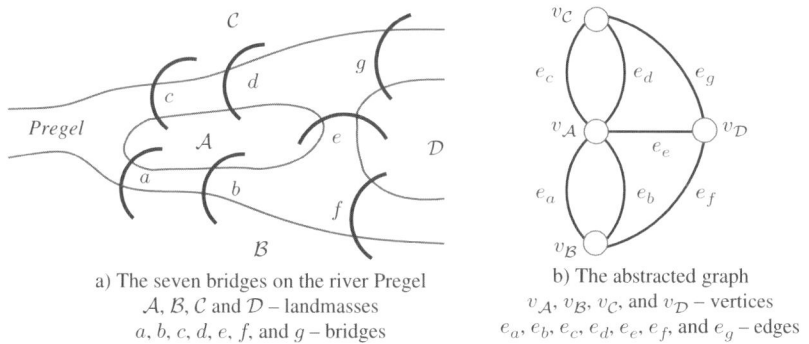

a) The seven bridges on the river Pregel
$\mathcal{A}, \mathcal{B}, \mathcal{C}$ and $\mathcal{D}$ – landmasses
$a, b, c, d, e, f,$ and g – bridges

b) The abstracted graph
$v_A, v_B, v_C,$ and v_D – vertices
$e_a, e_b, e_c, e_d, e_e, e_f,$ and e_g – edges

Fig. 1. The seven bridges problem and the abstracted graph

---

[1] Nowadays Pregoyla in Kaliningrad.

For the sake of simplicity of the notation, the incidence relation will be omitted, therefore one can write, without the fear of confusion:

$$e_a = (v_A, v_B), \ e_b = (v_A, v_B), \ e_c = (v_A, v_C), \ e_d = (v_A, v_C),$$
$$e_e = (v_A, v_D), \ e_f = (v_B, v_D), \ e_g = (v_C, v_D). \qquad (2)$$

i.e., the graph is defined as $G = (V, E)$ without explicit mentioning of the incidence relation. The vertex set $V(G)$ and the edge set $E(G)$ are simply written as V and E. There will be no distinction between a graph and its sets, one may write a vertex $v \in G$ or $v \in V$ instead of $v \in V(G)$, an edge $e \in G$ or $e \in E$, and so on. Vertices and edges are usually represented with symbols like $v_1, v_2, \ldots$ and $e_1, e_2, \ldots$, respectively. Note that in (2), each edge is identified with a pair of vertices. If the edges are represented with ordered pairs of vertices, then the graph G is called *directed* or *oriented*, otherwise if the pairs are not ordered, it is called *undirected* or *nonoriented*. Two vertices connected by an edge $e_k = (v_i, v_j)$ are called *end vertices* or *ends* of e_k. In the directed graph the vertex v_i is called the *source*, and v_j the *target* vertex of edge e_k. The elements of the edge set E are distinct i.e., more than one edge can join the same vertices. Edges having the same end vertices are called *parallel edges*.[2] If $e_k = (v_i, v_i)$, i.e., the end vertices are the same, then e_k is called a *self-loop*. A graph G containing parallel edges and/or self-loops is a multigraph. A graph having no parallel edges and self-loops is called a *simple graph*. The number of vertices in G is called its *order*, written as $|V|$; its number of edges is given as $|E|$. A graph of order 0 is called an *empty graph*,[3] and of order 1 is simply called *trivial graph*.[4] A graph is *finite* or *infinite* based on its order. If not otherwise stated all the graphs used in this chapter are finite and not empty.

Two vertices v_i and v_j are neighbors or *adjacent* if they are the end vertices of the same edge $e_k = (v_i, v_j)$. Two edges e_i and e_j are *adjacent* if they have an end vertex in common, say v_k, i.e., $e_i = (v_k, v_l)$ and $e_j = (v_k, v_m)$. If all vertices of G are pairwise neighbors, then G is *complete*. A complete graph on m vertices is written as K^m. An edge is called *incident* on its end vertices. The degree (or valency) $deg(v)$ of a vertex v is the number of edges incident on it. A vertex of degree 0 is called *isolated*; of degree 1 is called *pendant*. Note that a self-loop at a vertex v contributes twice in $deg(v)$.

Let $G = (V, E)$ and $G' = (V', E')$ be two graphs. $G' = (V', E')$ is a subgraph of G ($G' \subseteq G$) if $V' \subseteq V$ and $E' \subseteq E$, i.e., the graph G contains graph G'. Graph G is called also a supergraph of G' ($G \supseteq G'$). If either $V' \subset V$ or $E' \subset E$, the graph G' is called a proper subgraph of G. If $G' \subseteq G$ and G' contains all the edges $e = (v_i, v_j) \in E$ such that $v_i, v_j \in V'$, G' is the *(vertex) induced subgraph* of G and V' induces (spans) G' in G. It is written as $G' = G[V']$, i.e., since $V' \subset G(V)$, then $G[V']$ denotes the graph on V' whose edges are the edges of G with both ends in V'. If not otherwise stated, by induced subgraph, the vertex-induced subgraph is meant. If there are no isolated vertices in G', then G' is called the *induced subgraph of G on the edge set E'* or simply *edge induced subgraph of G*. If $G' \subseteq G$ and V' spans all

---
[2] Also called double edges.
[3] A graph with no vertices and hence no edges.
[4] A graph with one vertex and possibly with self-loops.

of G, i.e., $V' = V$ then G' is a spanning subgraph of G. A subgraph G' of a graph G is a maximal (minimal) subgraph of G with respect to some property Π if G' has the property Π and G' is not a proper subgraph of any other subgraph of G having the property Π. The minimal and maximal subsets with respect to some property are defined analogously. This definition will be used later to define a component of G as a maximal connected subgraph of G, and a spanning tree of a connected G is a minimal connected spanning subgraph of G.

Let $G = (V, E)$ be a graph with sets $V = \{v_1, v_2, \cdots\}$ and $E = \{e_1, e_2, \cdots\}$. A walk in a graph G is a finite nonempty alternating sequence $v_0, e_1, v_1, \ldots, v_{k-1}, e_k, v_k$ of vertices and edges in G such that $e_i = (v_i, v_{i+1})$ for all $1 \leq i \leq k$. This walk is called a $v_0 - v_k$ walk with v_0 and v_k as the terminal vertices and all other vertices are internal vertices of this walk. In a walk, edges and vertices can appear more than once. If $v_0 = v_k$, the walk is *closed*, otherwise it is *open*. A walk is a trail if all its edges are distinct. A trail is closed if its end vertices are the same, otherwise it is opened. By definition the walk can contain the same vertex many times. A path P is a trail where all vertices are distinct. A simple path is written as $P = v_0, v_1, v_2, \cdots, v_k$, where edges are not explicitly depicted since in a path all vertices are distinct and therefore in a simple graph all the edges are distinct too. Note that in a multigraph a path is not uniquely defined by this nomenclature, because of possible multiple edges between two vertices. Vertices v_0 and v_k are linked by the path P, also P is called a path from v_0 to v_k (as well as between v_0 and v_k). The number of edges in the path is called the path length. The path length is denoted with P^k, where k is the number of edges in the path. Note that by definition it is not necessary that a path contains all the vertices of the graph. Cycles, like paths, are denoted by the cyclic sequence of vertices $C = v_0, v_1, \cdots, v_k, v_0$. The length of the cycle is the number of edges in it is called k-cycle written as C^k. The minimum length of a cycle in a graph G is the girth $g(G)$ of G, and the maximum length of a cycle is its circumference. The distance between two vertices v and w in G denoted by $d(u, w)$, is the length of the shortest path between these vertices. The *diameter* of G, $diam(G)$ is the maximum distance between any two vertices of G.

Connectivity is an important concept in graph theory and it is one of the basic concepts used in this presentation. Two vertices v_i and v_j are connected in a graph $G = (V, E)$ if there is a path $v_i - v_j$ in G. A vertex is connected to itself. A nonempty graph is connected if any two vertices are joint by a path in G. Let graph $G = (V, E)$ be a nonconnected graph. The set V is partitioned into subsets $V_1, V_2, \cdots, V_p$ if $V_1 \cup V_2 \cup \cdots \cup V_p = V$ and for all i and j, $i \neq j$ $V_i \cap V_j = \emptyset$. $\{V_1, V_2, \cdots, V_p\}$ is called a partition of V. Since the graph G is nonconnected, the vertex set V can be partitioned into subsets $V_1, V_2, \cdots, V_p$, such that each vertex induced subgraph $G[V_i]$ is connected, and there exists no path between a vertex in subset V_i and a vertex in V_j, $j \neq i$. A maximally connected subgraph of G is called a component of graph G. A component of G is not a proper subgraph of any other connected subgraph of G. An isolated vertex is considered to be a component, since by definition it is connected to itself. Note that a component is always nonempty, and that if a graph G is connected then it has only one component, i.e., itself.

The following theorem is used in the Sect. 5 to show that after the edge removal from the cycle the graph stays connected.

Theorem 1. *If a graph $G = (V, E)$ is connected, then the graph remains connected after the removal of an edge e of a cycle $C \in E$, i.e., $G' = (V, E - \{e\})$ is connected.*

Proof. The proof can be found in [12].

From the earlier theorem one can conclude that edges that if removed disconnect a graph, do not lie on any cycle.

The definition of cut and cut-set are as follows. Let $\{V_1, V_2\}$ be partitions of the vertex set V of a graph $G = (V, E)$. The set $\mathcal{K}(V_1, V_2)$ of all edges having one end in one vertex partition (V_1) and the other end on the second vertex partition (V_2) is called a cut. A cut-set $\mathcal{K}_S$ of a connected graph G is a minimal set of edges such that its removal from G disconnects G, i.e., $G - \mathcal{K}_S$ is disconnected. If the induced subgraphs of G on vertex set V_1 and V_2 are connected then $\mathcal{K} = \mathcal{K}_S$. If the vertex set $V_1 = \{v\}$, the cut is denoted by $\mathcal{K}(v)$.

Trees are simple graph structures, and are extensively used in the rest of the discussion. A graph G is acyclic if it has no cycles. A tree of graph G is a connected acyclic subgraph of G. Vertices of degree 1 in a tree are called *leaves*, and all edges are called branches. A nontrivial tree has at least two leaves and a branch, for example the simplest tree consists of two vertices joined by an edge. Note that an isolated vertex is by definition an acyclic connected graph, and therefore a tree.

A spanning tree of graph G is a tree of G containing all the vertices of G. Edges of the spanning tree are called *branches*. The tree containing all vertices, and only those edges not in the spanning tree, is called *cospanning tree*, and its edges are called *cords*. An acyclic graph with k components is called a k-tree. If the k-tree is a spanning subgraph of G, then it is called a spanning k-tree of G. A forest F of a graph G is a spanning k-tree of G, where k is the number of component of G. A forest is simply a set of trees, spanning all the vertices of G. A connected subgraph of a tree T is called a subtree of T. If T is a tree then there is exactly one unique path between any two vertices of T.

And finally some basic binary and unary operations on graphs are described. Let $G = (V, E)$ and $G' = (V', E')$ be two graphs. Three basic binary operations on two graphs are as follows:

Union and Intersection. The *union* of G and G' is the graph $G''' = G \cup G' = (V \cup V', E \cup E')$, i.e., the vertex set of G''' is the union of V and V', and the edge set is the union of E and E', respectively. The *intersection* of G and G' is the graph $G''' = G \cap G' = (V \cap V', E \cap E')$, i.e., the vertex set of G''' has only those vertices present in both V and V', and the edge set contains only those edges present in both E and E', respectively.

Symmetric Difference. The *symmetric difference*[5] between two graphs G and G', written as $G \oplus G'$, is the induced graph G''' on the edge set $E \boxplus E' = (E \setminus E') \cup (E' \setminus E)$,[6] i.e., this graph has no isolated vertices and contains edges present either in G or in G' but not in both.

---

[5] Called also ring sum.
[6] Where $\setminus$ is the set minus operation and is interpreted as removing elements from X that are in Y.

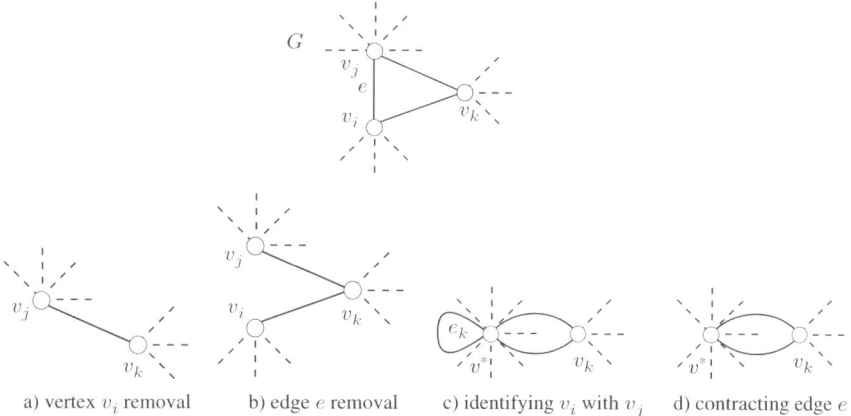

Fig. 2. Operations on graph

Four unary operations on a graph are as follows:

Vertex Removal. Let $v_i \in G$, then $G - v_i$ is the induced subgraph of G on the vertex set $V - v_i$; i.e., $G - v_i$ is the graph obtained after removing the vertex v_i and all the edges $e_j = (v_i, v_j)$ incident on v_i. The removal of a set of vertices from a graph is done as the removal of single vertex in succession. An example of vertex removal is shown in Fig. 2a.

Edge Removal. Let $e \in G$, then $G - e$ is the subgraph of G obtained after removing the edge e from E. The end vertices of the edge $e = (v_i, v_j)$ are not removed. The removal of a set of edges from a graph is done as the removal of single edge in succession. An example of edge removal is shown in Fig. 2b.

Vertex Identifying. Let v_i and v_j be two distinct vertices of graph G joined by the edge $e = (v_i, v_j)$. Two vertices v_i and v_j are identified if they are replaced by a new vertex v^* such that all the edges incident on v_i and v_j are now incident on the new vertex v^*. An example of vertex identifying is given in Fig. 2c.

Edge Contraction. Let $e = (v_i, v_j) \in G$ be the edge with distinct end points $v_i \neq v_j$ to be contracted. The operation of edge contraction denotes removal of the edge e and identifying its end vertices v_i and v_j into a new vertex v^*. If the graph G' results from G after contracting a sequence of edges, than G is said to be *contractible* to a graph G'. Note the difference between vertex identifying and edge contraction, in Fig. 2c and d. Vertex identifying preserves the edge e_k, whereas edge contraction first removes this edge. In Sect. 5 a detailed treatment of edge contraction and edge removal in the dual graphs context is presented.

3 Image Pyramids

Visual data is characterized by large amount of data and high redundancy with relevant information clustered in space and time. All this indicates a need of organization and aggregation principles, in order to cope with computational complexity

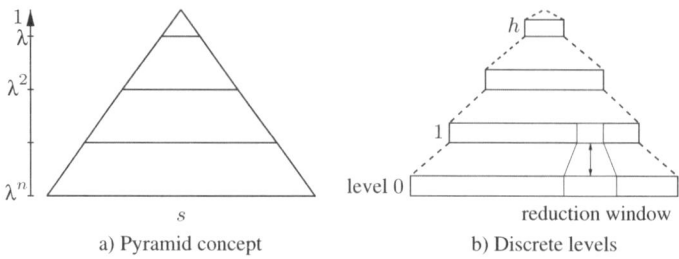

Fig. 3. Multiresolution pyramid

and to bridge the gap between raw data and symbolic description. Local processing is important in early vision, since operations like convolution, thresholding, mathematical morphology, etc. belong to this class. However, using them is not efficient for high- or intermediate-level vision, such as symbolic manipulation, feature extraction, etc., because these processes need both local and global information. Therefore a data structure must allow the transformation of *local* information (based on subimages) into *global* information (based on the whole image), and be able to handle both local (distributed) and global (centralized) information. Such a data structure, the pyramid, is known as *hierarchical architecture* [15], and it allows distribution of the global information to be used by local processes. The pyramid is a trade-off between parallel architecture and the need for a hierarchical representation of an image, i.e., at several resolutions [15].

An image pyramid (Fig. 3a,b) describes the contents of an image at multiple levels of resolution. High-resolution input image is at the base level. Successive levels reduce the size of the data by a *reduction factor* $\lambda > 1.0$. *Reduction windows* relate one cell at the reduced level with a set of cells in the level directly below. Thus, local independent (and parallel) processes propagate information up and down and laterally in the pyramid. The contents of a lower resolution cell are computed by means of a *reduction function* the input of which are the descriptions of the cells in the reduction window. Sometimes the description of the lower resolution needs to be extrapolated to the higher resolution. This function is called the *refinement* or *expansion function*. It is used in Laplacian pyramids [16] and wavelets [17] to identify redundant information in the higher resolution and to reconstruct the original data. Two successive levels of a pyramid are related by the reduction window and the reduction factor. Higher-level description should be related to the original input data in the base of the pyramid. This is identified by the *receptive field* (RF) of a given pyramidal cell c_i. The $RF(c_i)$ aggregates all cells (pixels) in the base level of which c_i is the ancestor.

Based on how the cells in subsequent levels are joint, two types of pyramids exist:

– Regular
– Irregular pyramids

These concepts are strongly related to the ability of the pyramid to represent the regular and irregular tessellation of the image plane.

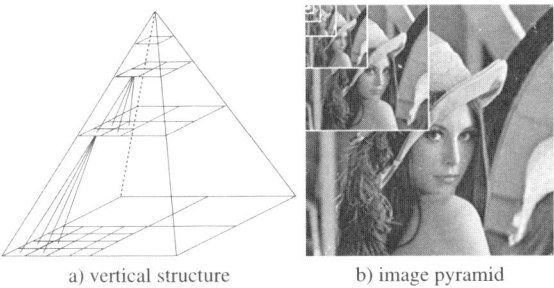

Fig. 4. $2 \times 2/4$ regular pyramid

3.1 Regular Pyramids

The *constant reduction factor* and *constant size reduction window* completely define the structure of the regular pyramid. The decrease rate of cells from level to level is determined by the reduction factor. The number of levels h is limited by the reduction factor $\lambda > 1.0$: $h \leq \log(image\_size)/\log(\lambda)$. The *main computational advantage* of regular image pyramids is due to their *logarithmic complexity*. Usually regular pyramids are employed in a regular grid tessellated image plane, therefore the reduction window is usually a square of $n \times n$, i.e., the $n \times n$ cells are associated to a cell on a higher level directly above. Regular pyramids are denoted using notation $n \times n/\lambda$. The vertical structure of a classical $2 \times 2/4$ is given in Fig. 4a. In this regular pyramid $2 \times 2 = 4$ cells are related to only one cell in the level directly above. Since the children have only one parent this class of pyramids is also called nonoverlapping regular pyramids. Therefore the reduction factor is $\lambda = 4$. An example of $2 \times 2/4$ regular image pyramid is given in Fig. 4b. The image size is $512 \times 512 = 2^9 \times 2^9$ therefore the image pyramid consist of $1 + 2 \cdot 2 + 4 \cdot 4 + ... + 2^8 \times 2^8 + 2^9 \times 2^9$ cells, and the height of this pyramid is 9. The pyramid levels are shown by a white border on the left upper corner of image. See [18] for extensive overview of other pyramid structures with overlapping reduction windows, e.g., $3 \times 3/2, 5 \times 5/4$. It is possible to define pyramids on other plane tessellation, e.g., triangular tessellation [15].

Thus, because of the rigid vertical structure, the regular image pyramid is an efficient structure for fast grouping and access to image objects across the input image. The regular pyramid representation of a shifted, rotated, and/or scaled image is not unique, and moreover it does not preserve the connectivity. Thus, [19] concludes that regular image pyramids have to be rejected as general-purpose segmentation algorithms. This major drawback of the regular pyramid motivated a search for a structure that is able to adapt on the image data. It means, that the regularity of the structure is to be abandoned.

3.2 Irregular Pyramids

Abandoning the regularity of the structure means that the horizontal and vertical neighborhood have to be explicitly represented, usually by using graph formalisms.

These irregular structures are usually called *irregular pyramids*. One of the main goals of irregular pyramids is to achieve the shift invariance, and to overcome this major drawback of their regular counterparts. Other motivations why one has to use irregular structures are [20]: arrangement of biological vision sensors is not completely regular; the CCD cameras cannot be produced without failure, resulting in an irregular sensor geometry; perturbation may destroy the regularity of regular pyramids; and image processing to arbitrary pixels arrangement (e.g., log-polar geometries [21]).

Two main processing characteristics of the regular pyramids should be preserved by building irregular ones [22]:

1. Operation are local, i.e., the result is computed independently of the order, this allows parallelization.
2. Bottom-up building of the irregular pyramid, with an exponential decimation of the number of cells.

The structure of the regular pyramid as well as the reduction process is determined by the type of the pyramid (e.g., $2 \times 2/4$). After removing this regularity constraint one has to define a procedure to derive the structure of the reduced graph G_{k+1} from G_k, i.e., a graph contraction method has to be defined. Irregular pyramids can be build by parallel graph contraction [23], or graph decimation [24]. Parallel graph contraction has been developed only for special graph structures, like trees, and is not discussed in this chapter. The graph decimation procedure is described in Sect. 5. An efficient random decimation algorithm for building regular pyramids, called *stochastic pyramids* (MIS) is introduced in [24]. A detailed discussion of this and similar methods is done in [25]. It is shown that MIS in some cases is not logarithmically tapered, i.e., the decimation process does not successively reduce the number of cells exponentially. The main reason for this behavior is that the cell's neighborhood is not bounded, for some cases the degree of the cell increases exponentially. In [25], two new methods based on maximal independent edge set (MIES and MIDES) that overcome this drawback are presented. An overview of the properties of regular and irregular pyramids is found in [26]. In irregular pyramids the flexibility is paid by less efficient data access.

Most information in vision today is in the form of array representation. This is advantageous and easily manageable for situations having the same resolution, size, and other typical properties equivalent. Various demands are appearing upon more flexibility and performance, which makes the use of array representations less attractive [27]. The increasing use of actively controlled and multiple sensors requires a more flexible processing and representation structure [2, 20]. Cheaper CCD sensors could be produced if defective pixels would be allowed, which yields in the resulting irregular sensor geometry [21, 28]. Image processing functions should be generalized to arbitrary pixel geometries [21, 29]. The conventional array form of images is impractical as it has to be searched and processed every time if some action is to be performed and (1) features of interest may be very sparse over parts of an array, leaving a large number of unused positions in the array; and (2) a description of additional detail cannot be easily added to a particular part of an array.

In order to express the connectivity or other geometric or topological properties, the image representation must be enhanced by a neighborhood relation. In the regular square grid arrangement of sampling points, it is implicitly encoded as 4- or 8-neighborhood with the well known paradox in conjunction with Jordan's curve theorem. The neighborhood of sampling points can be represented explicitly, too: in this case the sampling grid is represented by a *graph* consisting of vertices corresponding to the sampling points and of edges connecting neighboring vertices. Although this data structure consumes more memory space it has several advantages, as follows [20]: the sampling points need not be arranged in a regular grid; the edges can receive additional attributes too; and the edges may be determined either automatically or depending on the data. In irregular pyramids, each level represents a partition of the pixel set into cells, i.e., connected subsets of pixels. The construction of an irregular image pyramid is iteratively local [8, 24]:

- The cells have no information about their global position
- The cells are connected only to (direct) neighbors
- The cells cannot distinguish the spatial positions of the neighbors

This means that we use only local properties to build the hierarchy of the pyramid. Usually, on the base level (level 0) of an irregular image pyramid the cells represent single pixels and the neighborhood of the cells is defined by the 4-connectivity of the pixels. A cell on level $k + 1$ (parent) is a union of neighboring cells on level k (children). As shown in Sect. 5 this union is controlled by *contraction kernels* (*decimation parameters*). Every parent computes its values independently of other cells on the same level. This implies that an image pyramid is built in $O[log(image\_diameter)]$ parallel steps. Neighborhoods on level $k+1$ are derived from neighborhoods on level k. Two cells c_1 and c_2 are neighbors if there exist pixels p_1 in c_1 and p_2 in c_2 such that p_1 and p_2 are 4-neighbors.

Before we continue with the presentation of graph pyramids, a concept of planar graphs is needed. A planar graph separates the plane into regions called faces. This idea of separating the plane into regions is helpful in defining the dual graphs. Duality of a graph brings together two important concepts in graph theory: cycles and cut-sets. This concept of duality is also encountered in the graph-theoretical approach of image region and edge extraction. The definition of dual graphs representing the partitioning of the plane, allows one to apply transformations on these graphs, like edge contraction and/or removal to simplify them in the sense of less vertices and edges. Edge contraction and removal introduces naturally a hierarchy of dual graphs, the so-called *dual-graph pyramid*.

4 Planar and Dual Graphs

A graph $\widetilde{G}$ of finite sets of vertices V and edges E is called *plane graph* if it can be drawn in a plane in $\mathbb{R}^2$ such that [12]:

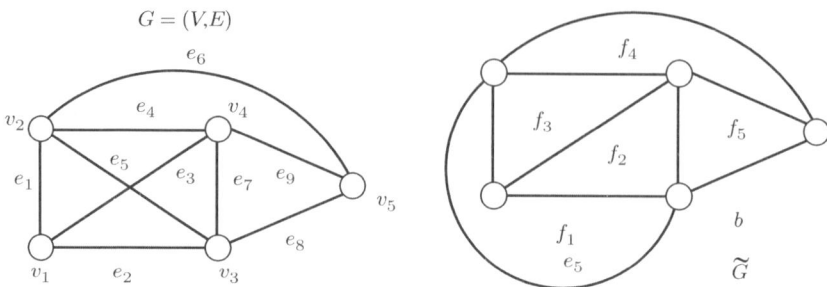

Fig. 5. A planar graph G and its embedding in a plane, the plane graph $\widetilde{G}$

- All $V \subset \mathbb{R}^2$
- Every edge is an arc[7] between two vertices
- No two edges are crossed

Note that $\mathbb{R} \setminus \widetilde{G}$ is an open set and its connected regions are faces f of $\widetilde{G}$. It is said that the plane graph divides the plane into regions. Since $\widetilde{G}$ is bordered, one of its faces is an unbounded one (infinite area). This face is called the *background face*.[8] The other faces enclose finite areas, and are called interior faces. Edges and vertices incident to a face are called the boundary elements of that face. A planar embedding of a graph G is an isomorphism between G and a plane graph $\widetilde{G}$. $\widetilde{G}$ is called a drawing of G. Similar to $\widetilde{G}$, G is drawn so that its edges intersect only on vertices.

A graph G is planar if it can be embedded on the plane. The concept of embeddings can be extended to any surface. A graph G is embeddable in surface S if it can be drawn in S so that its edges intersect only on their end vertices. A graph embeddable on the plane is embeddable on the sphere too. It can be shown by using the stereoscopic projection of the sphere onto a plane [14]. Note that the concept of faces is also applicable to spherical embeddings.

Let G in Fig. 5 represent a planar graph, in general with parallel edges and self-loops. Since the graph is embedded onto a plane, it divides the plane into faces. Let each of these faces be denoted by a new vertex say f, and let these vertices be put inside the faces, as shown in Fig. 5. From this point on the notion of face vertices and face are synonymous. Let the faces that are neighbors, i.e., that share the same edge e_2 (they are incident on the same edge), be connected by the edge, say $\bar{e}_2$, so that edges e_2 and $\bar{e}_2$ are crossed. At the end, for each edge $e_2 \in G$ there is an edge $\bar{e}_2$ of the newly created graph $\overline{G}$, which is called the dual graph of G. If e_2 is incident only with one face a self-loop edge $\bar{e}_2$ is attached to the vertex on the face in which the edge e_2 lays, of course e_2 and the self-loop edge $\bar{e}_2$ have to cross each other. The adjacency of faces is expressed by the graph $\overline{G}$. More formally one can define dual graphs for a given plane graph $G = (V, E)$ [14]:

---

[7] An arc is a finite union of straight line segments, and a straight line segment in the Euclidean plane is a subset of $\mathbb{R}^2$ of the form $\{x + \lambda(y - x) | 0 \leq \lambda \leq 1\} \forall x \neq y \in \mathbb{R}^2$.
[8] Called also exterior face.

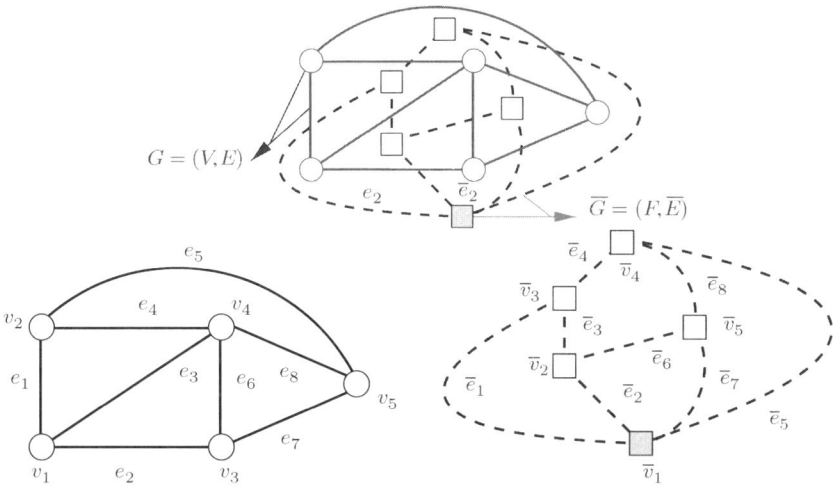

Fig. 6. A plane graph G and it dual $\overline{G}$

Definition 2 (Dual graphs). *A graph $\overline{G} = (\overline{V}, \overline{E})$ is a dual of $G = (V, E)$ if there is a bijection between the edges of G and $\overline{G}$, such that a set of edges in $\overline{G}$ is a cycle vector if and only if the corresponding set of edges in G is a cut vector.*

There is a one-to-one correspondence between the vertex set $\overline{V}$ of $\overline{G}$ and the face set F of G, therefore sometimes graph $\overline{G} = (\overline{V}, \overline{E})$ is written as $\overline{G} = (F, \overline{E})$ instead, without fear of confusion. In order to show that $\overline{G}$ is a dual of G, one has to prove that vectors forming a basis of the cycle subspace of $\overline{G}$ correspond to the vectors forming a basis of the cut subspace of G. The edges e_i of graph G in Fig. 6 correspond to edges $\overline{e}_i$ in graph $\overline{G}$. The cycles $\{e_1, e_3, e_4\}$, $\{e_2, e_3, e_6\}$, $\{e_4, e_5, e_8\}$, and $\{e_6, e_7, e_8\}$ form a basis of the cycle subspace of G. These cycles correspond to the set of edges $\{\overline{e}_1, \overline{e}_3, \overline{e}_4\}$, $\{\overline{e}_2, \overline{e}_3, \overline{e}_6\}$, $\{\overline{e}_4, \overline{e}_5, \overline{e}_8\}$, and $\{\overline{e}_6, \overline{e}_7, \overline{e}_8\}$, which form a basis of the cut subspace of $\overline{G}$. It follows according to the definition of the duality, that graph $\overline{G}$ is a dual of G. The graph G is called the *primal graph* and $\overline{G}$ the *dual graph*. Dual graphs are denoted by a line above the big letter. If a planar graph G' is a dual of G, then a planar G is a dual of G' as well, and every planar graph has a dual [12, 13].

In the following, two important properties of dual graphs with respect to the edge contraction and removal operations are given, the proofs are due to [14]. These properties are required to prove that during the process of dual-graph contraction graphs stay planar and are duals (Sect. 5). Let G and its dual $\overline{G}$ be two graphs. Let edge $\overline{e} \in \overline{G}$ correspond to edge $e \in G$. Note that a cycle in G corresponds to a cut in $\overline{G}$ and vice versa [14]. Let $\overline{G'}$ denote the graph $\overline{G}$ after the *contraction* of the edge $\overline{e}$, and G' the graph after the *removal* of the corresponding edge e from G.

Theorem 2. *A graph and its dual are duals also after the removal of an edge e in the primal graph G and the contraction of the corresponding edge $\overline{e}$ in the dual graph $\overline{G}$.*

Corollary 1. *If a graph G has a dual, then every edge-induced subgraph of G has also a dual.*

Theorem 3 (Whitney 1933). *A graph is planar if and only if it has a dual.*

Proof. The proofs can be found in [14] and [12].

4.1 Dual Image Graphs

An image is transformed into a graph such that, to each pixel a vertex is associated, and pixels that are neighbors in the sampling grid are joint by an edge. Note that no restriction on the sampling grid is made, therefore an image of regular as well as nonregular sampling grid can be transformed into a graph. The gray value or any other feature is simply considered as an attribute of a vertex (and/or an edge). Since the image is finite and connected, the graph is finite and connected as well. The graph which represents the pixels is denoted by $G = (V, E)$ and is called *primal graph*.[9] Note that pixels represent finite regions, and the graph G is representing in fact a graph with faces as vertices. The dual of a face graph (see Sect. 4) is the graph representing borders of the faces, which in fact are interpixel edges and interpixel vertices. This graph is denoted by $\overline{G}$ and is called simply *dual graph*. Based on Theorem 3, dual graphs are planar, therefore images with square grid are transformed into 4 – connected square grid graphs, since 8 – connected square grid graphs are in general not planar. [10]

The same formalism as done for the pixels can be used at intermediate levels in image analysis i.e., RAGs. RAGs can be the results of image segmentation processes. Regions are connected sets of pixels, and are separated by region borders. Their geometric dual though causes problems [10]. This section is concluded by a formal definition of the dual image graphs:

Definition 3 (Dual image graphs [30]). *The pair of graphs $(G, \overline{G})$, where $G = (V, E)$ and $\overline{G} = (\overline{V}, \overline{E})$ are called dual image graphs if both graphs $(G, \overline{G})$ are finite, planar, connected, not simple in general and duals of each other.*

Dual graphs can be seen as an extension of the well know region adjacency graphs (RAG) representation. Note that this representation is capable to encode not only adjacency relations but inclusion relations as well [10].

5 Dual-Graph Contraction

Irregular (dual graph) pyramids are constructed in a bottom-up way such that a subsequent level (say $k + 1$) results by (dually) contracting the precedent level (say k). In this section a short exposition of the dual-graph contraction is given, following the work of Kropatsch [10]. Building dual-graph pyramids using this algorithm is presented in Sect. 5.3. Dual-graph contraction (DGC) [10] proceeds in two steps:

---
[9] Also called neighborhood graph.
[10] This holds for square grid graphs of grid size $\geq 4 \times 4$.

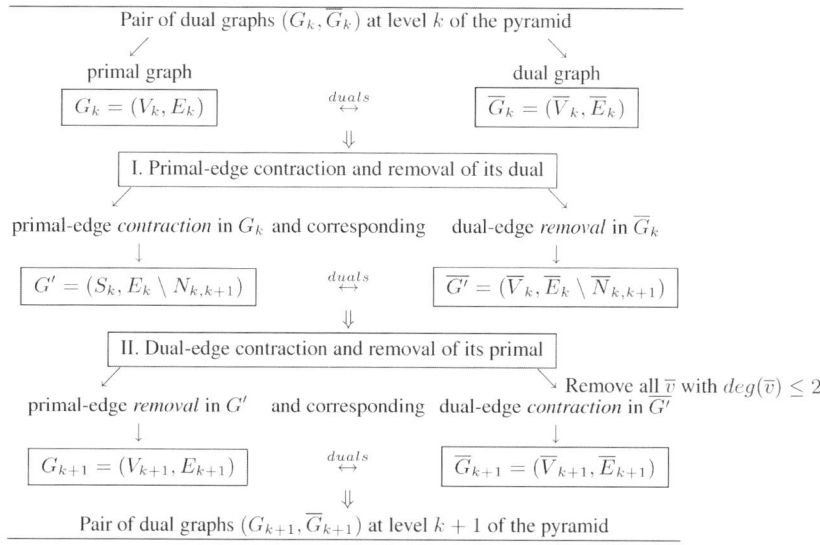

Fig. 7. Dual-graph contraction procedure (DGC)

1. Primal-edge contraction and removal of its dual
2. Dual-edge contraction and removal of its primal

In Fig. 7 examples of these two steps are shown in three possible cases. Note that these two steps correspond in [10] to the steps (1) dual-edge contraction, and (2) dual face contraction.

The base of the pyramid consists of the pair of dual image graphs $(G_0, \overline{G_0})$. In order to proceed with the dual-graph contraction a set of so-called contraction kernels (decimation parameters) must be defined. The formal definition is postponed until the Sect. 5.1. Let the set of contraction kernels be $\langle S_k, N_{k,k+1} \rangle$. This set consists of a subset of surviving vertices $S_k = V_{k+1} \subset V_k$, and a subset of nonsurviving primal edges $N_{k,k+1} \subset E_k$ (where index $k, k+1$ refer to contraction from level k to $k+1$). Surviving vertices in $v \in S_k$ are vertices not to be touched by the contraction, i.e., after contraction these vertices make up the set V_{k+1} of the graph G_{k+1}; and every nonsurviving vertex $v \in V_k \setminus S_k$ must be paired to one surviving vertex in a unique way, by nonsurviving primal edges (Fig. 8a). In this Figure, the shadowed vertex s is the survivor and this vertex is connected with arrow edges (ns) with nonsurviving vertices. Note that a contraction kernel is a tree of depth one, i.e., there is only one edge between a survivor and a nonsurvivor, or analogously one can say that the diameter of this tree is two.

The contraction of a nonsurviving primal edge consists in the identification of its endpoints (vertices) and the removal of both the contracted primal edge and its dual edge (see Sect. 2 for details on these operations). Figure 9a shows the normal situation, Fig. 9b the situation where the primal-edge contraction creates multiple edges, and Fig. 9c self-loops. In Fig. 9c, redundancies (lower part) are decided through the corresponding dual graphs and removed by dual-graph contraction. In Fig. 9, the

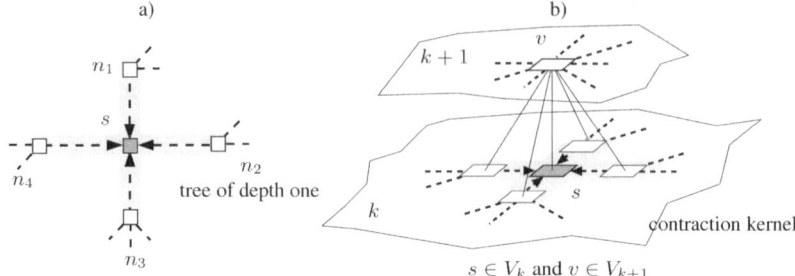

$s \in S_k$ and $n_1, n_2, n_3, n_4 \in N_{k,k+1}$ Creation of a new vertex v in the level $k+1$ of the pyramid

Fig. 8. (a) Contraction kernel and (b) parent–child relation

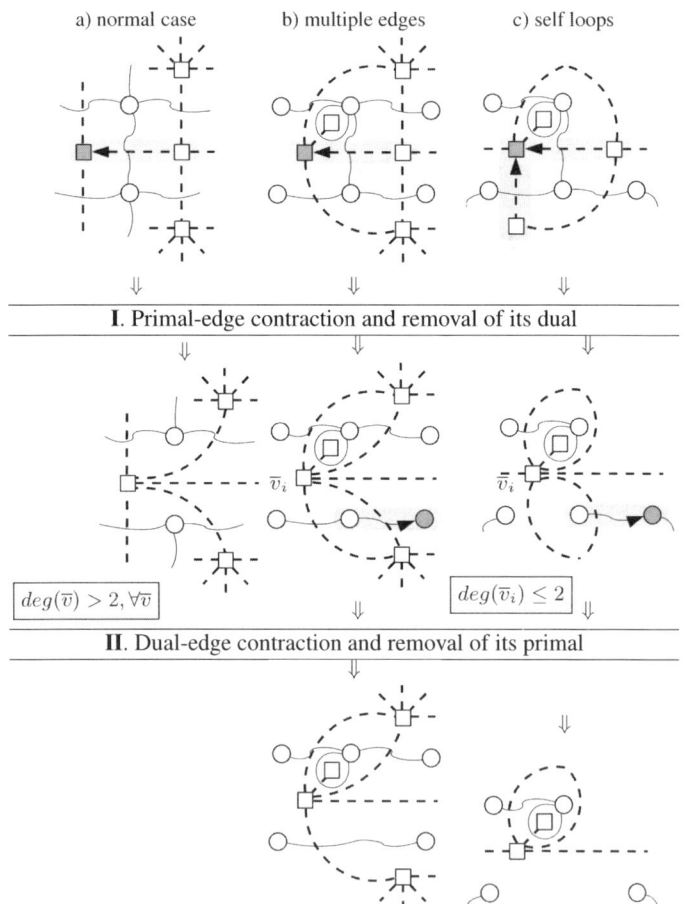

Fig. 9. Dual-graph contraction of a part of a graph

primal graph is shown with square, vertices with broken lines, and its dual with circle vertices and full lines.

In [10] it is shown that $\langle S_k, N_{k,k+1} \rangle$ determine the structure of an irregular pyramid. The relation between two pairs of dual graphs,$(G_k, \overline{G_k})$ and $(G_{k+1}, \overline{G_{k+1}})$, is established by dual-graph contraction with the set of contraction kernels $\langle S_k, N_{k,k+1} \rangle$ as:

$$(G_{k+1}, \overline{G_{k+1}}) = C[(G_k, \overline{G_k}), \langle S_k, N_{k,k+1} \rangle]. \qquad (3)$$

Dual-edge contraction and removal of its primal (second step) has a role of cleaning the primal graph by simplifying most of the multiple edges and self-loops,[11] but not those enclosing any surviving parts of the graph. They are necessary to preserve correct structure [10]. Dual-graph contraction reduces the number of vertices and edges of a pair of dual graphs, while preserving the topological relations among surviving parts of the graph. In [30,31] a detailed presentation of dual-graph contraction is given.

5.1 Contraction Kernels

Let S be the set of surviving vertices, and N the set of nonsurviving primal edges. The connected components[12] $CC(s), s \in S$, of subgraph (S, N) form a set of rooted tree structures $T(s)$ that, if contracted, each of them would collapse into the vertex s of the contracted graph. The number of these trees is $|S|$. The union of trees $T(s)$ contains the nonsurviving primal edges N. $T(s)$ is a *spanning tree* of the connected component $CC(s)$, or equivalently, (V, N) is a spanning forest of the graph $G = (V, E)$. In order to decimate the graph $G = (V, E)$ the set of *surviving vertices* $S \subset V$ and the set of *nonsurviving primal edges* $N \subset E$ must be selected, such that the following conditions are satisfied (1) graph (V, N) is a spanning forest of graph $G = (V, E)$, and (2) the surviving vertices $s \in S \subset V$ are the roots of the forest (V, N).

Definition 4 (Contraction kernels). *A set of disjoint rooted trees with length two of path going through the root is called a set of contraction kernels.*

Analogously, the trees $T(v)$ of the forest (V, N) with roots $v \in V$ are *contraction kernels*. After applying the dual-graph contraction algorithm on a graph, one has to establish a path connecting two surviving vertices on the resulted new graph. Let $G = (V, E)$ be a graph with decimation parameters (S, N).

Definition 5 (Connecting path [30]). *A path in $G = (V, E)$ is called a connecting path between two surviving vertices $s, s' \in S$ if it consists of three subsets of edges:*

- *The first part is a possibly empty branch of contraction kernel $T(s)$.*
- *The middle part is an edge $e \in E \setminus N$ that bridges the gap between (connects) the two contraction kernels $T(s)$ and $T(s')$.*
- *The third part is a possibly empty branch of contraction kernel $T(s')$.*

---

[11] Called also redundant edges.
[12] Neglected level indexes refer to contraction from level k to level $k+1$.

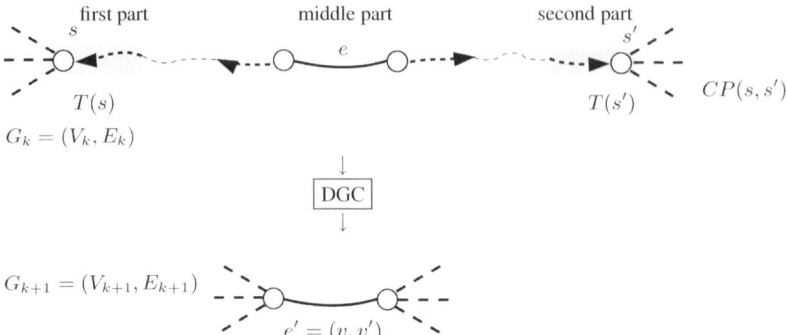

Fig. 10. Connecting path $CP(v, v')$, e is the bridge of this path

See Fig. 10 for explanation. The connecting path is denoted by $CP(s, s')$. Edge e is called the *bridge* of the connecting path $CP(s, s')$. Each edge $e' = (v, v') \in E_{k+1}$ has a corresponding connecting path $CP_k(s, s')$, where $s, s' \in S \subset V_k$ are survivors in the graph $G_k = (V_k, E_k)$. This means that two surviving vertices s and s', $s \neq s'$, that can be connected by a path[13] $CP_k(s, s')$ in G_k are connected by an edge in E_{k+1}. If the graph G_k is connected, after dual-graph contraction the connectivity of the graph G_{k+1} is preserved [30].

Dual-edge contraction can be implemented by (1) simply renaming all the non-surviving vertices to their surviving parent vertex (e.g., by using a find union set algorithm [32]), (2) deleting all nonsurviving edges N, and (3) their duals $\overline{N}$. We use different (MIS, MIES, and D3P) stochastic methods to build contraction kernels [25].

5.2 Equivalent Contraction Kernels

Reference [16] combines two or more successive reductions in one equivalent weighting function in order to compute any level of any regular pyramid directly from the base level. Similarly, [31] combines two (or more) dual-graph contractions (as shown in Fig. 11) of graph $G_k = (V_k, E_k)$ with decimation parameters $\langle S_k, N_{k,k+1} \rangle$ and $\langle S_{k+1}, N_{k+1,k+2} \rangle$ into one single equivalent contraction kernel (ECK) $N_{k,k+2} = N_{k,k+1} \circ N_{k+1,k+2}$:[14]

$$C[C[G_k, \langle S_k, N_{k,k+1} \rangle], \langle S_{k+1}, N_{k+1,k+2} \rangle] = C[G_k, \langle S_{k+1}, N_{k,k+2} \rangle] = G_{k+2} \tag{4}$$

The structure of G_{k+1} is determined by G_k and the decimation parameters $\langle S_k, N_{k,k+1} \rangle$. Simply overlaying the two sets of contraction kernels, $\langle S_k, N_{k,k+1} \rangle$ (the one from level k to $k + 1$) and $\langle S_{k+1}, N_{k+1,k+2} \rangle$ (the one from level $k + 1$

---

[13] By definition of the connectivity of a graph, there exists always a path between any two vertices of graph.
[14] Only G_k is shown instead of $(G_k, \overline{G_k})$ for simplicity.

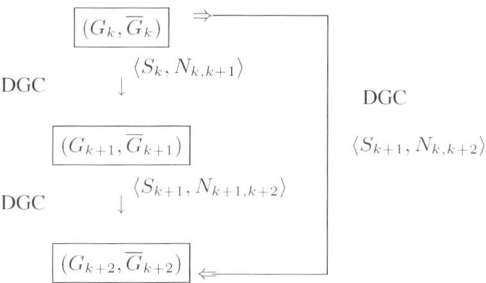

Fig. 11. Equivalent contraction kernel

to $k+2$) will not yield a proper equivalent contraction kernel $\langle S_{k+1}, N_{k,k+2} \rangle$. The surviving vertices from G_k to G_{k+2} are $S_{k+1} = V_{k+2}$. The edges of the searched contraction kernels must be formed by edges $N_{k,k+2} \subset E_k$. An edge $e_{k+1} = (v_{k+1}, v'_{k+1}) \in N_{k+1,k+2}$ corresponds to a connecting path $CP_k(v_{k+1}, v'_{k+1})$ in G_k.[15] By Definition 5, $CP_k(v_{k+1}, v'_{k+1})$ consists of one branch of $T_k(v_{k+1})$, one branch of $T_k(v'_{k+1})$, and one surviving edge $e_k \in E_k$ connecting the two contraction kernels $T_k(v_{k+1})$, and $T_k(v'_{k+1})$.

Definition 6 (Bridge [30]). *Function bridge*: $E_{k+1} \mapsto E_k$ *assigns to each edge* $e_{k+1} = (v_{k+1}, w_{k+1}) \in E_{k+1}$ *one of the bridges* $e_k \in E_k$ *of the connecting paths* $CP_k(v_{k+1}, w_{k+1})$:

$$bridge(e_{k+1}) = e_k. \quad (5)$$

Connecting two disjoint tree structures by a single edge results in a new tree structure. Now, $N_{k,k+2}$ can be defined as the result of connecting all contraction kernels T_k by bridges as:

$$N_{k,k+2} = N_{k,k+1} \cup \bigcup_{e_{k+1} \in N_{k+1,k+2}} bridge(e_{k+1}) \quad (6)$$

This definition satisfies the requirements of a contraction kernel [30]. Analogously, the earlier process can be repeated for any pair of levels k and k' such that $k < k'$. If $k = 0$ and $k' = h$, where h is the level index of the top of the pyramid, with the resulting equivalent contraction kernel ($N_{0,h}$), the base level (0) is contracted in one step into an apex $V_h = \{v_h\}$. ECKs are able to compute any level of the pyramid directly from the base.

5.3 Dual-Graph Pyramid

A graph pyramid is a pyramid where each level is a graph $G(V, E)$ consisting of vertices V and of edges E relating two vertices. In order to correctly represent the embedding of the graph in the image plane [33], we additionally store the dual graph $\overline{G}(V, \overline{E})$ at each level. The levels are represented as pairs $(G_k, \overline{G}_k)$ of *dual plane graphs* G_k and $\overline{G}_k$. See Sect. 4.1 for more details on this representation.

---
[15] If there are more than one connecting paths, one is selected.

The sequence $(G_k, \overline{G_k})$, $0 \leq k \leq h$ is called dual *graph pyramid*, where 0 is the base level index and h is the top level index, also called the height of the pyramid. Moreover the graphs are attributed, $G(V, E, attr_v, attr_e)$, where $attr_v : V \to \mathbb{R}^+$ and $attr_e : E \to \mathbb{R}^+$, i.e., content of the graph is stored in attributes attached to both vertices and edges. In general a graph pyramid can be generated bottom-up as shown in Algorithm 1.

Algorithm 1. Constructing dual-graph pyramid

Input: Graphs $(G_0, \overline{G_0})$

1: $k \leftarrow 0$.
2: **while** further abstraction is possible **do**
3:   determine contraction kernels, $N_{k,k+1}$.
4:   perform dual-graph contraction and simplification of dual graphs, $(G_{k+1}, \overline{G_{k+1}}) = C[(G_k, \overline{G_k}), N_{k,k+1}]$.
5:   apply reduction functions to compute content $attr : G_{k+1} \to \mathbb{R}^+$ of new reduced level.
6:   $k \leftarrow k + 1$.
7: **end while**

Output: Graph pyramid – $(G_k, \overline{G_k})$, $0 \leq k \leq h$.

Let the building of the dual-graph pyramid be explained by using the image in Fig. 12. For the sake of simplicity of the presentation, in the figures afterward, the dual graphs are not shown explicitly as well as intralevel relations. An example of this intralevel relation is shown in Fig. 8b with the contraction kernel shadowed. In the example from Fig. 13 initially the attributes of the vertices receive the gray values of the pixels. The first step determines what information in the current top level is important and what can be dropped. A contraction kernel is a (small) subtree of the top level, the root of which is chosen to survive (black circles in Fig. 13b). Figure 13a shows the window and the selected contraction kernels with gray. Selection criteria in this case contracts only edges inside connected components having the same gray value. All the edges of the contraction trees are dually contracted during step 3 from Algorithm 1. Dual contraction of an edge e (formally denoted by $G/\{e\}$) consists of

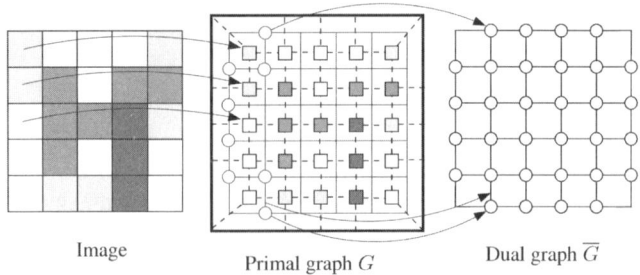

Fig. 12. Image to dual graphs

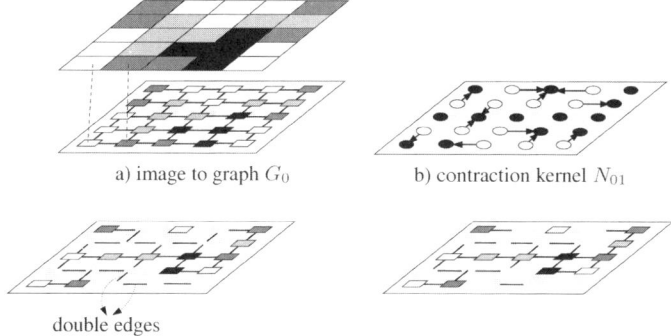

a) image to graph G_0 b) contraction kernel N_{01}

double edges

c) G' after dual-edge contraction of G_0 d) G_1 after the removal of redundant edges in G'

Fig. 13. Dual-graph contraction in G_0 and the creation of the G_1 of the pyramid

contracting e and removing the corresponding dual edge $\overline{e}$ from the dual graph (formally denoted by $\overline{G} \setminus \{\overline{e}\}$). This preserves duality and the dual graph needs not be constructed from the contracted primal graph G' at the next level. Since the contraction of an edge may yield multiedges (an example shown with arrows in Fig. 13c) and self-loops there is a second simplification phase of step 3 which removes all redundant multiedges and self-loops. Note that not all such edges can be removed without destroying the topology of the graph: if the cycle formed by the multiedge or the self-loop surrounds another part of the data its removal would corrupt the connectivity! Fortunately this can be decided locally by the dual graph since *faces of degree two* (having the double-edge as boundary) and *faces of degree one* (boundary = self-loop) cannot contain any connected elements in its interior. Since removal and contraction are dual operations, the removal of a self-loop or of one of the double edges can be done by contracting the corresponding dual edges in the dual graph (which are not depicted in our example for the sake of simplicity). The dual contraction from our example remains a simple graph G_1 without self-loops and multiedges (Fig. 13d). Step 3 generates a reduced pair of dual graphs. Their contents is derived in step 4 from the level later using the reduction function. In our example reduction is very simple: the surviving vertex inherits the color of its sons. The following table summarizes dual-graph contraction in terms of the control parameters used for abstraction and the conditions to preserve topology:

level	representation	contract / remove	conditions
0	$(G_0, \overline{G_0})$		
	↓	contraction kernel $N_{0,1}$	forest, depth 1
	$(G_0/N_{0,1}, \overline{G_0} \setminus \overline{N_{0,1}})$		
	↓	redundant edges $S_{0,1}$	$\deg \overline{v} \leq 2$
1	$(G_1 = G_0/N_{0,1} \setminus S_{0,1},$ $\overline{G_1} = \overline{G_0} \setminus \overline{N_{0,1}}/\overline{S_{0,1}})$		
	↓	contraction kernel $N_{1,2}$	forest, depth 1
	⋮		

6 A Hierarchy of Partitions

The segmentation problem is supposed to find natural groupings of the pixel set given as input. The first question that comes in mind is how these natural groupings are found. In other words what makes pixels in a partition be more like one another than pixels in other segments. This observation pours down into two issues [34] (1) how to measure the similarity between pixels, and (2) how to evaluate a partitioning of the pixels into segments.

It is expected that, these measures of dissimilarity capture the expectation that the distance in a feature space of pixels within a segment is less than the distance between pixels in different segments. The second issue is defining the criterion function to be optimized. The goal is to find the groups or segments that have strong internal similarities, which optimize the criterion function. But before we continue with the presentation of the algorithm for hierarchical image partitioning, let we recall the idea of minimum spanning tree (MST) and Borůvka's algorithm.

Algorithm 2. Borůvka's Algorithm

Input: graph $G(V, E)$
1: $MST \leftarrow$ empty edge list
2: all vertices $v \in V$ make a list of trees L
3: **while** there is more than one tree in L **do**
4:     each tree $T \in L$ finds the edge e with the minimum weight which connects T to $G \setminus T$ and add edge e to MST.
5:     using edge e merge pairs of trees in L
6:     clean the graph from self-loops if necessary
7: **end while**

Output: minimum weight spanning tree - edge induced subgraph on MST.

6.1 Minimum Weight Spanning Tree (MST)

The minimum spanning tree, called afterward MST, is the simplest and best-studied optimization problem in computer science. According to [35] the "Minimum spanning tree is a cornerstone problem of combinatorial optimization and in a sense its cradle." The problem is defined as follows. Let $G = (V, E)$ be a undirected connected plane graph consisting of the finite set of vertices V and the finite set of edges E. Each edge $e \in E$ is identified with a pair of vertices $v_i, v_j \in V$ such that $v_i \neq v_j$. Let each edge $e \in E$ be associated with a *unique* weight $w(e) = w(v_i, v_j)$, from the totally ordered universe (it is assumed that weights are distinct, if not, ties can be broken arbitrarily). Note that parallel edges, for e.g., $e_1 = (v_1, v_2)$ and $e_2 = (v_1, v_2)$ $e_1 \neq e_2$, have different weights. The problem is formulated as construction of a minimum total weight spanning tree of G.

6.2 Borůvka's Algorithm

The idea of Borůvka [9] is to do steps like in Prim's algorithm [36], in parallel over the graph at the same time. This algorithm constructs a spanning tree in iterations composed of the steps shown in Algorithm 2. First create a list L of trees, each a single vertex $v \in V$. For each tree T of L find the edge e with the *smallest weight*, which connects T to $G \setminus T$. The trees T are then connected to $G \setminus T$ with the edges e. In this way the number of trees in L is reduced, until there is only one, the MST.

Observation 0.1 *In the 3rd step of Algorithm 2, each tree $T \in L$ finds the edge with the minimal weight, and as trees become larger, the process of finding these edges takes longer.*

6.3 Minimum Spanning Tree with DGC

Taking the Observation 0.1 into consideration, the contraction of the edge e, which connects T and $G \setminus T$ in the 4th step of Algorithm 2 will speed up the process of searching for minimum weight edges in Borůvka's algorithm. If the graphs are represented as adjacency lists then a vertex with degree d can enumerate its incident edges in its neighborhood in time $O(d)$. Since in the level $k + 1$, after edge contraction, each tree (from level k) will be represented by a vertex, the search for the edge with the minimum weight would be a local search, and the resulting graph is smaller (in the sense of less vertices and less edges), thus the next pass can run faster.

The dual-graph contraction algorithm [10] is used to contract edges and create *super vertices* i.e., it creates father–son relations between vertices in subsequent levels (vertical relation), whereas Borůvka's algorithm is used to create son–son relations between vertices in the same level (horizontal relation). Here we expand Borůvka's algorithm with the steps that contract edges, remove parallel edges and self loops (if the connectivity of the graph is not changed), see Algorithm 3. In the section later we will refine the son–son relation to simulate the pop-out phenomena [37], and to find region borders quickly and effortlessly in a bottom-up "stimulus-driven" way based on local differences in a specific feature (e.g., color).

Algorithm 3. Borůvka's Algorithm with DGC

Input: attributed graph $G_0(V, E)$

1: $k \leftarrow 0$
2: **repeat**
3:     for each vertex $v \in G_k$ find the minimum-weight edge $e \in G_k$ incident to the vertex v and mark the edges e to be contracted
4:     determine CC_i^k as the connected components of the marked edges e
5:     contract connected components CC_i^k in a single vertex and eliminate the parallel edges (except the one with the minimum weight) and self-loops and create the graph $G_{k+1} = C[G_k, CC_i^k]$
6:     $k \leftarrow k + 1$
7: **until** all connected components of G are contracted into one single vertex

Output: a graph pyramid with an apex.

6.4 Building a Hierarchy of Partitions

Hierarchies are a significant tool for image partitioning as they are naturally combined with homogeneity criteria. Horowitz and Pavlidis [38] define a consistent homogeneity criteria over a set V as a boolean predicate P over its parts $\Phi(V)$ that verifies the consistency property: $\forall (x,y) \in \Phi(V) \quad x \subset y \Rightarrow (P(y) \Rightarrow P(x))$. In image analysis this states that the subregions of a homogeneous region are also homogeneous. It follows that if Pyr is a hierarchy and P a consistent homogeneity criteria on V then the set of maximal elements of Pyr that satisfy P defines a unique partition of V. Thus the combined use of a hierarchy and homogeneity criteria allows to define a partition in a natural way.

The goal is to find partitions of connected components $P_k = \{CC(u_1),...,CC(u_n)\}$ such that these elements satisfy certain properties. We use the pairwise comparison of neighboring vertices (partitions) to check for similarities [7, 39, 40]. A pairwise comparison function, $B(CC(u_i), CC(u_j))$ is true, if there is evidence for a boundary between $CC(u_i)$ and $CC(u_j)$, and false when there is no boundary. Note that $B(\cdot,\cdot)$ is a boolean comparison function for pairs of partitions. The definition of $B(\cdot,\cdot)$ depends on the application. The pairwise comparison function $B(\cdot,\cdot)$ that we use measures the difference along the boundary of two components relative to the differences of component's internal differences. This definition tries to encapsulate the intuitive notion of contrast: a contrasted zone is a region containing two components whose inner differences (*internal contrast*) are less then the differences between them (*external contrast*). We define an *external contrast* between two components and an *internal contrast* of each component. These measures are defined analogously to [7, 39, 40].

Every vertex $u \in G_k$ is a representative of a connected component $CC(u)$ of the partition P_k. The equivalent contraction kernel [10] of a vertex $u \in G_k$, $N_{0,k}(u)$ is a set of edges on the base level that are contracted, i.e., applying $N_{0,k}(u)$ on the base level contracts the subgraph $G' \subseteq G$ onto the vertex u. The *internal contrast* of $CC(u) \in P_k$ is the *largest dissimilarity* inside the component $CC(u)$ i.e., the largest edge weight of $N_{0,k}(u)$ of vertex $u \in G_k$, that is

$$Int(CC(u)) = max\{attr_e(e), e \in N_{0,k}(u)\}. \qquad (7)$$

Let $u_i, u_j \in V_k, u_i \neq u_j$ be the end vertices of an edge $e \in E_k$. The *external contrast* between two components $CC(u_i), CC(u_j) \in P_k$ is the *smallest dissimilarity* between component $CC(u_i)$ and $CC(u_j)$ i.e., the smallest edge weight connecting $N_{0,k}(u_i)$ and $N_{0,k}(u_j)$ of vertices $u_i, u_j \in G_k$:

$$\begin{aligned}&Ext(CC(u_i), CC(u_j))\\&= min\{attr_e(e), e = (u_i, u_j) : u_i \in N_{0,k}(u_i) \wedge w \in N_{0,k}(u_j)\}.\end{aligned} \qquad (8)$$

This definition is problematic since it uses only the smallest edge weight between the two components, making the method very sensitive to noise. But in practice this limitation works well as shown in Sect. 6.5. In Fig. 14 an example of $Int(\cdot)$ and $Ext(\cdot,\cdot)$ is given. The $Int(CC(u_i))$ of the component $CC(u_i)$ is

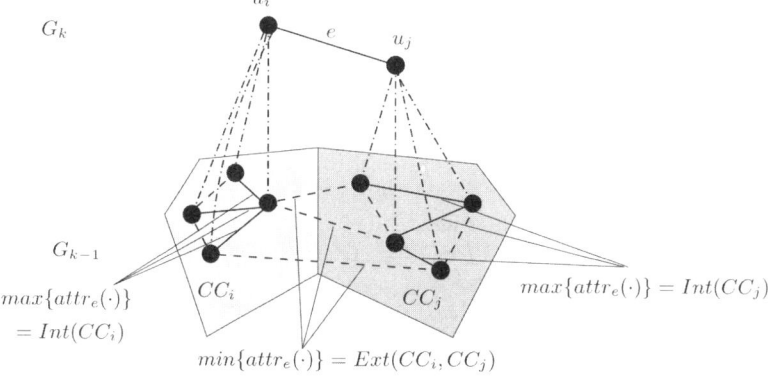

Fig. 14. Internal and external contrast

the $maximum$ of the weights of the solid edges (analogously for $Int(CC(u_j))$), whereas $Ext(CC(u_i), CC(u_j))$ is the $minimum$ of the weights of the dashed edges connecting component $CC(u_i)$ and $CC(u_j)$. Vertices u_i and u_j are the representatives of the components $CC(u_i)$ and $CC(u_j)$, i.e., by contracting the edges $N_{0,k}(u_i)$ one arrives to the vertex u_i. The pairwise comparison function $B(\cdot, \cdot)$ between two connected components $CC(u_i)$ and $CC(u_j)$ can now be defined as:

$$B(CC(u_i), CC(u_j)) = \begin{cases} \text{True} & \text{if } Ext(CC(u_i), CC(u_j)) > PInt(CC(u_i), CC(u_j)), \\ \text{False} & \text{otherwise,} \end{cases} \quad (9)$$

where the minimum internal contrast difference between two components, $PInt(\cdot, \cdot)$, reduces the influence of too small components and is defined as:

$$PInt(CC(u_i), CC(u_j)) \\ = min\{Int(CC(u_i)) + \tau(CC(u_i)), Int(CC(u_j)) + \tau(CC(u_j))\} \quad (10)$$

For the function $B(\cdot, \cdot)$ to be true i.e., for the border to exist, the external contrast difference must be greater than the internal contrast differences. The reason for using a threshold function $\tau(CC(\cdot))$ is that for small components $CC(\cdot)$, $Int(CC(\cdot))$ is not a good estimate of the local characteristics of the data, in the extreme case when $|CC(\cdot)| = 1$, $Int(CC(\cdot)) = 0$. Any nonnegative function of a single component $CC(\cdot)$, can be used for $\tau(CC(\cdot))$.

The algorithm to build the hierarchy of partitions is shown in Algorithm 4. Each vertex $u_i \in G_k$ defines a *connected region* $CC(u_i)$ on the base level of the pyramid, and since the presented algorithm is based on Borůvka's algorithm [9], it builds a MST(u_i) of each region, i.e., $N_{0,k}(u_i) =$ MST(u_i) [41]. The idea is to collect the smallest weighted edges e (4th step) that could be part of the MST, and then to check if the edge weight $attr_e(e)$ is smaller than the internal contrast of both of the components (MST of end vertices of e) (5th step). If these conditions are fulfilled then these two components are merged (7th step). All the edges to be contracted form the contraction kernels $N_{k,k+1}$, which are then used to create the graph

Algorithm 4. Hierarchy of Partitions

Input: Attributed graph G_0.

1: $k \leftarrow 0$
2: **repeat**
3:    **for all** vertices $u \in G_k$ **do**
4:       $E_{min}(u) \leftarrow argmin\{attr_e(e) \,|\, e = (u,v) \in E_k \text{ or } e = (v,u) \in E_k\}$
5:    **end for**
6:    **for all** $e = (u_i, u_j) \in E_{min}$ with
      $Ext(CC(u_i), CC(u_j)) \leq PInt(CC(u_i), CC(u_j))$ **do**
7:       include e in contraction edges $N_{k,k+1}$
8:    **end for**
9:    contract graph G_k with contraction kernels, $N_{k,k+1}$: $G_{k+1} = C[G_k, N_{k,k+1}]$.
10:   **for all** $e_{k+1} \in G_{k+1}$ **do**
11:      set edge attributes $attr_e(e_{k+1}) \leftarrow min\{attr_e(e_k) \,|\, e_{k+1} = C(e_k, N_{k,k+1})\}$
12:   **end for**
13:   $k \leftarrow k+1$
14: **until** $G_k = G_{k-1}$

Output: A region adjacency graph (RAG) pyramid.

$G_{k+1} = C[G_k, N_{k,k+1}]$ [20]. In general $N_{k,k+1}$ is a forest. We update the attributes of those edges $e_{k+1} \in G_{k+1}$ with the minimum attribute of the edges $e_k \in E_k$ that are contracted into e_{k+1} (9th step). The output of the algorithm is a pyramid where each level represents a RAG, i.e., a partition. Each vertex of these RAGs is the representative of a MST of a region in the image. The algorithm is greedy since it collects only the nearest neighbor with the minimum edge weights and merges them if the pairwise comparison (9) evaluates to "false." Some properties of the algorithm are given in [42].

6.5 Experiments on Image Graphs

The base level of our experiments is the trivial partition, where each pixel is a homogeneous region. The attributes of edges can be defined as the difference between features of end vertices, $attr_e(u_i, u_j) = |F(u_i) - F(u_j)|$, where F is some feature. Other attributes could be used as well e.g., [6] $attr_e(u_i, u_j) = \exp\{\frac{-||F(u_i)-F(u_j)||_2^2}{\sigma_I}\}$, where F is some feature, and σ_I is a parameter, which controls the scale of proximity measures of F. F could be defined as $F(u_i) = I(u_i)$, for gray value intensity images, or $F(u_i) = [v_i, v_i \cdot s_i \cdot \sin(h_i), v_i \cdot s_i \cdot \cos(h_i)]$, for color images in HSV color distance [6]. However the choice of the definition of the weights and the features to be used is in general a hard problem, since the grouping cues could conflict with each other [43].

For our experiments we use, as attributes of edges, the difference between pixel intensities $F(u_i) = I(u_i)$, i.e., $attr_e(u_i, u_j) = |I(u_i) - I(u_j)|$. For color images we run the algorithm by computing the distances (weights) in RGB color space. We choose this simple color distances in order to study the properties of the algorithm. To compute the hierarchy of partitions we define $\tau(CC)$ to be a function of the size of CC e.g., $\tau(CC) := \alpha/|CC|$, where $|CC|$ is the size of the component CC and α is a constant. The algorithm has one running parameter α, which is used to compute the function τ. A larger constant α sets the preference for larger components. A more complex definition of $\tau(CC)$, which is large for certain shapes and small otherwise would produce a partitioning which prefers certain shapes. To speed up the computation, vertices are attributed ($attr_v$) with the internal differences, average color and the size of the region they represent. Each of these attributes is computed for each level of the hierarchy. Note that the height of the pyramid depends only on the image content.

We use indoor and outdoor RGB images. We found that $\alpha := 300$ produces the best hierarchy of partitions of the images as shown in Monarch,[16] Object45 and Object11[17] Fig. 15 (I, III, IV) and $\alpha := 1000$ for the woman image in Fig. 15 (II), after the average intensity attribute of vertices is down projected onto the base grid. Figure 15 shows some of the partitions on different levels of the pyramid and the number of components. Note that in all images there are regions of large intensity variability and gradient. This algorithm copes with this kind of gradient and variability.

The algorithm is capable of grouping perceptually important regions despite of large intensity variability and gradient. In contrast to [7] the result is a hierarchy of partitions at multiple resolutions suitable for further goal driven, domain-specific analysis. On lower levels of the pyramid the image is over-segmented whereas in higher levels it is under-segmented. Since the algorithm preserves details in low-variability regions, a noisy pixel would survive through the hierarchy, see Fig. 15 (Id). Image smoothing in low-variability regions would overcome this problem. We do not smooth the images, as this would introduce another parameter into the method. The robustness of topology is discussed in Sect. 6.6. The hierarchy of partitions can also be built from an over-segmented image to overcome the problem of noisy pixels. Note that the influence of τ in the decision criterion is smaller as the region gets bigger for a constant α. The constant α is used to produce a kind of over-segmented image and the influence of τ decays with each new level of the pyramid. For an over-segmented image, where the size of the regions is large, the algorithm becomes parameterless.

6.6 Robustness of Graph Pyramids

There are several places in the construction of a graph pyramid where noise can affect the result (1) the input data; (2) during selection of contraction kernels; and (3) when summarizing the content of a reduction window by the reduction function.

The effects on the topology can be the following: a connected region falls into parts; two regions merge into one; break inclusion, create new inclusions; two adjacent regions become separated; two separated regions become adjacent. All these changes reflect in the Euler characteristic which we will use to judge the topological

---

[16] Waterloo image database.
[17] Coil 100 image database.

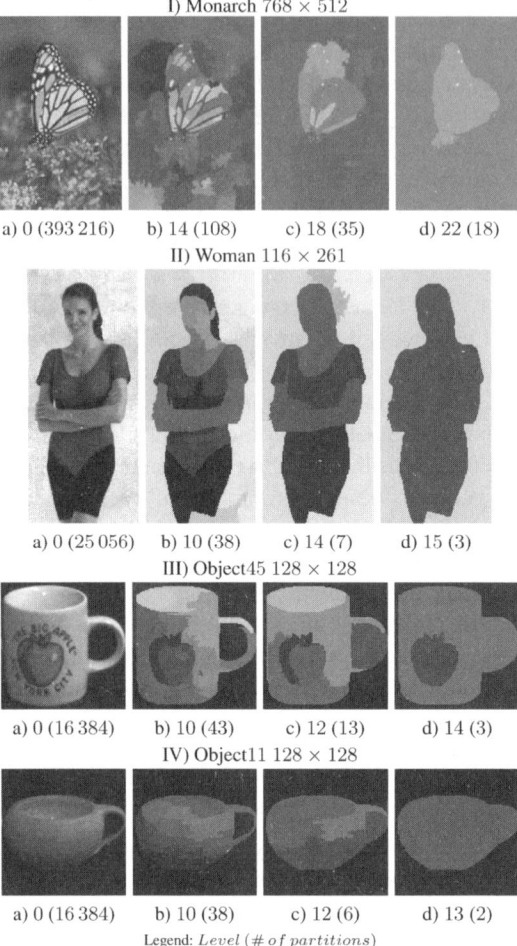

Fig. 15. Partitioning of images

robustness of graph pyramids. Let us start with the influence of a wrong pixel on the connectivity structure. A wrong pixel adjacent to a region can corrupt its connectivity (and the property of inclusion in $2D$) if it falls on a one pixel wide branch of the Figure. The consequence can be that the region breaks into two parts which increases the Euler characteristic by 1. A noisy pixel inside a region creates a new connected component which is a topological change (e.g., a new inclusion) but it can be easily recognized and eliminated by its size. However the change is again not very drastic since one noisy pixel can change the Euler characteristic only by 1. If all regions of the picture both foreground and background are at least 2 pixels wide a single wrong pixel changes their size but not their connectivity.

For a branch of two pixels in width, two noisy pixels in a particular spatial position relative to each other are needed to modify the topology. More generally to break the connectivity across an n-pixel wide branch of a region noisy pixels are needed, forming a connected path from one side of the branch to the other. This can be considered as the consequence of the sampling theorem (see [44]). All these topological modifications happen in the base of our pyramid. As long as we use topology-preserving constructions and/or consider identified noise pixels as nonsurvivors the topology is not changed in higher levels.

Different criteria and functions can be used for selecting contraction and reduction kernels. In contrast to data, noise errors are introduced by the specific operations and may be the consequence of numerical instabilities or quantizations errors. There is no general property allowing to derive an overall property like robustness of all possible selection or reduction functions. Hence operational robustness needs to be checked for any particular choice.

7 Evaluation of Segmentations

The segmentation process results in "homogeneous" regions with respect to the low-level cues using some similarity measures. Problems emerge because the homogeneity of low-level cues does not always lead to semantics and the difficulty of defining the degree of homogeneity of a region. Also some of the cues can contradict each other. Thus, low-level cue image segmentation cannot produce a complete final "good" segmentation [45], leading researchers to look at the segmentation only in the context of a task, as well as the evaluation of the segmentation methods. However in [46] the segmentation is evaluated purely[18] as segmentation by comparing the segmentation done by humans with those done by a particular method. As can be seen in 2, 3, 4 of Fig. 16 there is a consistency in segmentations done by humans (already demonstrated empirically in [46]), even thought humans segment images at different granularity (refinement or coarsening). This refinement or coarsening could be thought as hierarchical structure of the image, i.e., the pyramid.

Evaluation of the segmentation algorithms is difficult because it depends on many factors [47] among them: the segmentation algorithm; the parameters of the algorithm; the type(s) of images used in the evaluation; the method used for evaluation of the segmentation algorithms, etc. Our evaluation copes with these facts:

1. Real world images should be used, because it is difficult to extrapolate conclusion based on synthetic images to real images [48].
2. The human should be the final evaluator [49].

There are two general methods to evaluate segmentations:

- Qualitative
- Quantitative methods

Qualitative methods involve humans for doing the evaluation, meaning that different observers would give different opinions about the segmentations (e.g., already encountered in edge detection evaluation [47], or in image segmentation [46]). On the

---
[18] The context of the image is not taken into consideration during segmentation.

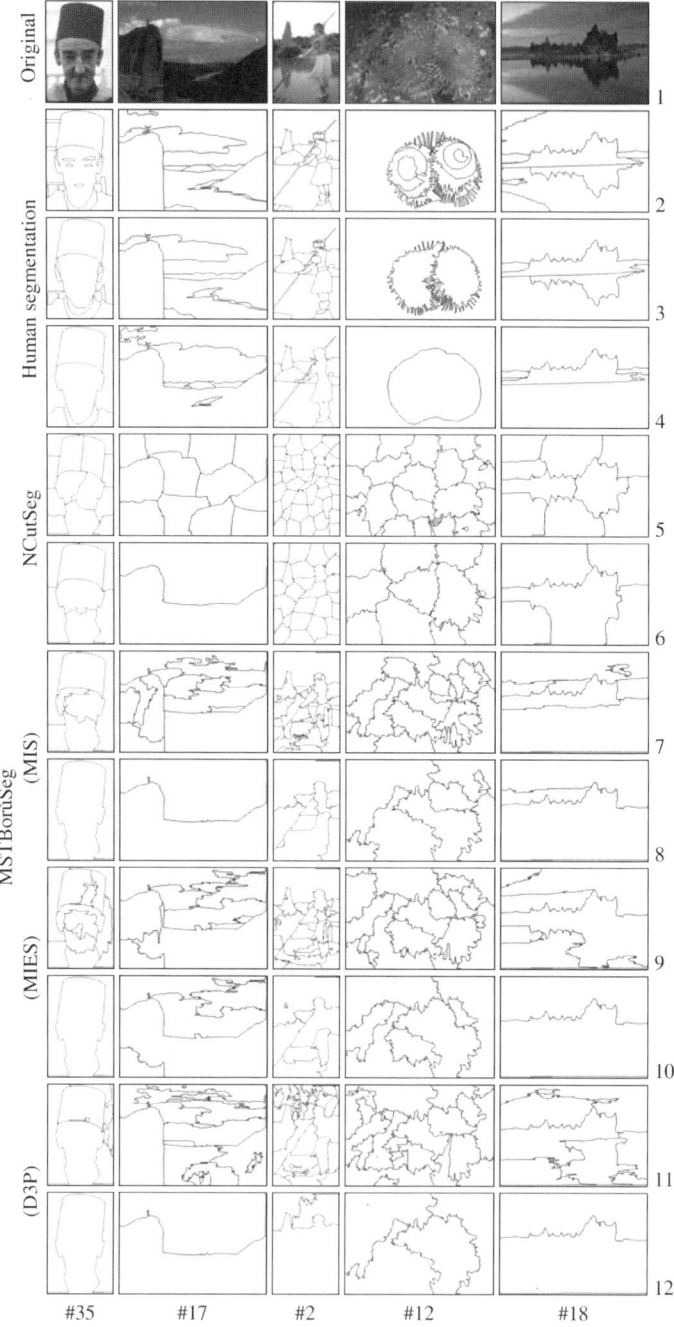

Fig. 16. Segmentation of Humans, NCutSeg, and MSTBorůSeg (MIS, MIES, D3P)

other hand, quantitative methods are classified into analytical and empirical methods [50]. Analytical methods study the principles and properties of the algorithm, like processing complexity, efficiency, and so on. Empirical methods study properties of the segmentations by measuring how "good" a segmentation is close to an "ideal" one, by measuring this "goodness" with some function of parameters. Qualitative and empirical methods depend on the subjects, the first one in coming up with the reference (perfect) segmentation[19] and the second one defining the function. The difference between the segmented image and the reference (ideal) one can be used to asses the performance of the algorithm [50]. The reference image could be a synthetic image or manually segmented by humans. Higher value of the discrepancy means bigger error, signaling poor performance of the segmentation method. In [50], it is concluded that evaluation methods based on *mis-segmented pixels should be more powerful than other methods using other measures*. In [46] the error measures used for segmentation evaluation "count" the mis-segmented pixels.

Note that the segmented image #35/2 in Fig. 16 can be coarsened to obtain the image in #35/4, this is called *simple refinement*; whereas to obtain image in #35/3 from #35/2 (or vice versa) we must coarsen in one part of the image and refine in the other (notice the chin of the man in #35/3, this is called *mutual refinement*. Therefore in [46] a segmentation consistency measure that does not penalize this granularity difference is defined (Sect. 7.1).

The segmentation results of NCutSeg [6] on gray value images are shown in Fig. 16 in 5 and 6 of BorůSeg with MIS [24] decimation strategy in 7 and 8; with MIES [8] in 9, and 10; and with D3P [52] in 11 and 12. Note that the NCutSeg and BorůSeg methods are capable of producing a hierarchy of images. These methods use only local contrast based on pixel intensity values. As it is expected, and can be seen from the Fig. 16, segmentation methods which are based only on low-level local cues cannot create segmentation results as good as humans. Even thought it looks like the NCutSeg method produces more regions, actually the overall number of regions 6, 8, 10, and 12 of Fig. 16 is almost the same, but BorůSeg produces a bigger number of small regions. The methods (see Fig. 16) were capable of segmenting the face of a man satisfactory (image #35). The BorůSeg method did not merge the statue on the top of the mountain with the sky (image #17), but it merged it with the mountain, compared to humans which do segment this statue as a single region. All methods have problems segmenting the see creatures (image #12). Note that the segmentation done by humans on the image of rocks (image #18), contains the axis of symmetry, even thought there is no "big" change in the local contrast, therefore the NCutSeg and BorůSeg methods fail in this respect. It must be mentioned that none of the methods is "looking" for this axis of symmetry.

In the rest of this section, we evaluate two graph-based segmentation methods, the normalized cut [6] (NCutSeg) and the method based on the Borůvka's minimum spanning tree (MST) [41] (BorůSeg). In fact we evaluate three flavors of the BorůSeg depending on the decimation strategy used: MIS, MIES, or D3P, denoted by BorůSeg (MIS), BorůSeg (MIES), and BorůSeg (D3P). See [25] for details on these

---

[19] Also called a gold standard [51].

decimation strategies. We compare these methods following the framework of [46] i.e., comparing the segmentation result of the two graph-based methods with the human segmentations. The results of the evaluation are reported in Sect. 7.1 later. Also the variation of regions sizes is shown in this section.

Some examples of applying BorůSeg on color images are shown in Sect. 6, where for visualization purposes each region has the mean color value. In this section we use the region borders to highlight the regions. Note that, two pixel wide borders are used only for better visualization purposes, and are not produced by these segmentation methods nor are part of the evaluation process.

7.1 Segmentation Benchmarking

In [46] segmentations made by humans are used as a reference and basis for benchmarking segmentations produced by different methods. The concept behind this is the observation that even though different people produce different segmentations for the same image, the obtained segmentations differ, mostly, only in the local refinement of certain regions. This concept has been studied on the human segmentation database (see 2, 3, 4 of Fig. 16) by [46] and used as a basis for defining two error measures, which do not penalize a segmentation if it is coarser or more refined than another. In this sense, a *pixel error measure* $E(S_1, S_2, p)$, called the local refinement error, is defined as:

$$E(S_1, S_2, p) = \frac{|R(S_1, p) \setminus R(S_2, p)|}{|R(S_1, p)|}, \qquad (11)$$

where $\setminus$ denotes set difference, $|x|$ the cardinality of a set x, and $R(S, p)$ is the set of pixels corresponding to the region in segmentation S that contains pixel p. Using the local refinement error $E(S_1, S_2, p)$ the following error measures are defined [46]: the global consistency error (GCE), which forces all local refinements to be in the same direction, and is defined as:

$$GCE(S_1, S_2) = \frac{1}{|I|} \min \left\{ \sum_{p \in I} E(S_1, S_2, p), \sum_{p \in I} E(S_2, S_1, p) \right\}, \qquad (12)$$

and the local consistency error (LCE), which allows refinement in different directions in different parts of the image, and is defined as:

$$LCE(S_1, S_2) = \frac{1}{|I|} \sum_{p \in I} \min \left\{ E(S_1, S_2, p), E(S_2, S_1, p) \right\}, \qquad (13)$$

where $|I|$ is the number of pixels in the image I. Notice that LCE $\leq$ GCE for any two segmentations. GCE is a tougher measure than LCE, because GCE tolerates only simple refinements, while LCE tolerates mutual refinement as well.

We have used the GCE and LCE measures presented earlier to do an evaluation of the BorůSeg method using the human segmented images from the Berkley humans segmented images database [46]. The results of comparison of the NCutSeg method vs. humans and humans vs. humans are confirmed [46].

7.2 Evaluation of Segmentations on the Berkley Image Database

As mentioned in [46] a segmentation consisting of a single region and a segmentation where each pixel is a region, is the coarsest and finest possible of any segmentation. In this sense, the LCE and GCE measures should not be used when the number of regions in the two segmentations differs a lot. Taking into consideration that both methods can produce segmentations with different number of regions, we have taken for each image as a region count reference number the average number of regions from the human segmentations available for that image. We instructed the NCutSeg to produce the same number of regions and for the BorůSeg we have taken the level of the pyramid that has the region number closest to the same region count reference number.

As data for the experiments, we take 100 gray level images from the Berkley Image Database.[20] For segmentation, we have used the normalized cuts implementation available on the Internet[21] and for the BorůSeg we have implementations based on combinatorial pyramids [53].[22]

For each of the images in the test, we have calculated the GCE and LCE using the results produced by the two methods and all the human segmentations available for that image. Having more then one pair of GCE and LCE for the methods NCutSeg and BorůSeg (all its versions) and each image, we have calculated the mean and the standard deviation.

In Fig. 17, the histogram of error values LCE (a) and GCE (b) ($[0\ldots 1]$, where zero means no error) of Humans vs. Humans, NCutSeg vs. Human, BorůSeg (all versions) vs. Human are shown. $\widehat{\mu}$ represents the mean value of the error. Notice that the humans are consistent in segmenting the images and the Human vs. Human histogram shows a peak very close to 0. i.e., a small $\widehat{\mu} = 0.0592$ for LCE and $\widehat{\mu} = 0.0832$ for GCE. For the NCutSeg and BorůSeg there is not a significant difference between the values of LCE and GCE (see the mean values of the respective histograms). One can conclude that the quality of segmentation of these methods seen over the whole database is not different.

We wanted to also tested how produced region sizes vary from one method to the other and how this variation depends on the content of the segmented images. For this, we have normalized the size of each region by dividing it to the size of the segmented image it belonged to (number of pixels), and for each segmentation, we have calculated the standard deviation (σ_S) of the normalized region sizes. For the case of human segmented images, we have done separately the calculation for each segmentation and taken the mean of the results for the segmentations of the same image. Figure 18a shows the resulting σ_S for 70 images (a clear majority for which the σ_S order Humans>MSTBorůSeg>NCutSeg existed). Results are shown sorted by the sum of the 3 σ_S for each image. The average region size variation for the whole dataset is: 0.1537 for Humans, 0.0392 for NCutSeg, and 0.0872 for MSTBorůSeg (MIES). Note, that the size variation is smallest and almost content

---
[20] http://www.cs.berkeley.edu/projects/vision/grouping/segbench/.
[21] http://www.cis.upenn.edu/~jshi/software/.
[22] http://www.prip.tuwien.ac.at/Research/FSPCogVis/Software/.

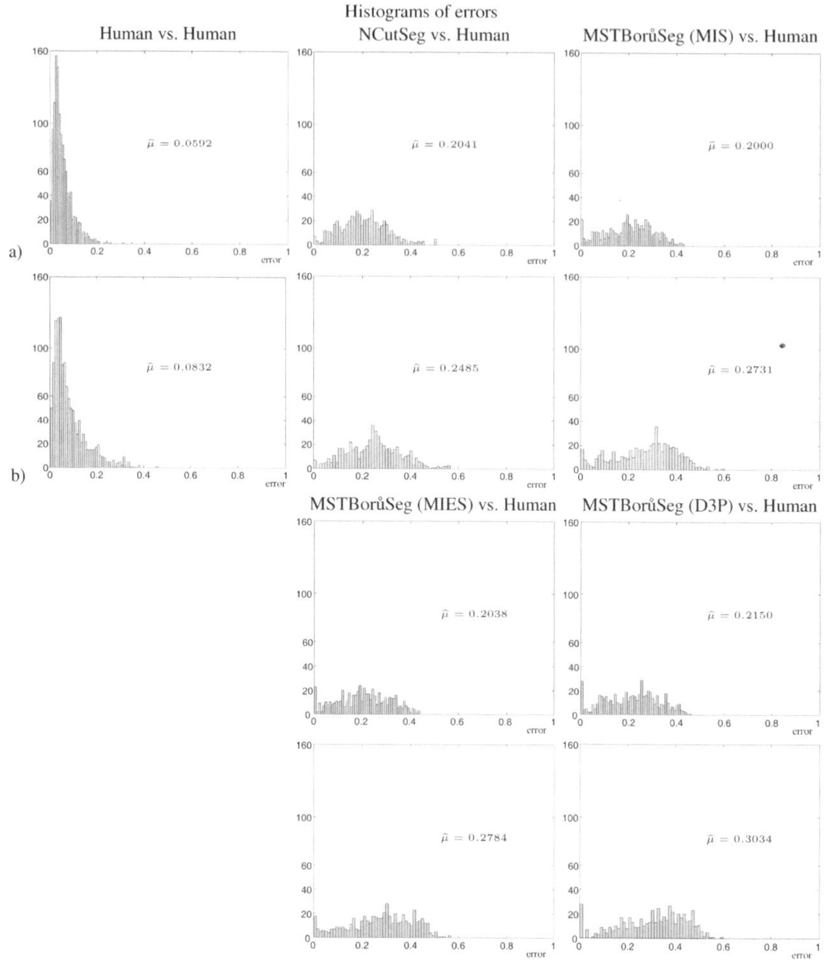

Fig. 17. Histograms of discrepancy measure: LCE (**a**) and GCE (**b**)

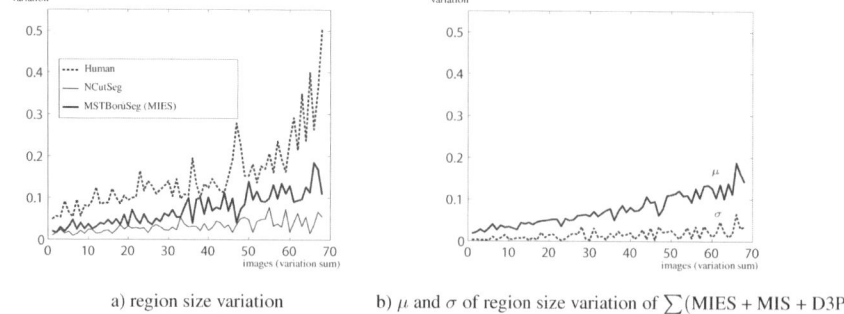

a) region size variation   b) μ and σ of region size variation of $\sum$(MIES + MIS + D3P)

Fig. 18. Variation of region sizes σ_S

independent for the NCutSeg and largest for Humans. We calculated the variation of regions sizes for the different decimation strategies MIS, MIES, and D3P. The average region size variation for the whole data set is 0.0893 for MSTBorůSeg (MIS) and 0.1037 for MSTBorůSeg (D3P). In Fig. 18b a solid line represents the mean region size variation of the three decimation strategies MIES, MIS, and D3P, and the doted line the standard deviation. Note that the standard deviation stays small for the whole spectrum which shows the region size variation consistency between the three decimation methods.

8 Conclusion

Image segmentation aggregates sets of pixels into connected regions that satisfy a certain homogeneity criteria. All such regions partition a given image into homogeneous areas. Real objects are composed of such homogeneous regions but there are no globally unified criteria to aggregate the smaller homogeneous regions into the larger regions corresponding to objects. We therefore need a representation able to aggregate smaller regions into larger regions using different criteria on different levels of abstraction. Starting with the dual graphs created for the input image, the irregular graph pyramid is constructed bottom-up by repeatedly applying dual-graph contraction. This progressively simplifies the graphs, level by level, obtaining a topmost level usually made out of one single vertex, called the apex. Dual-graph contraction involves concepts from graph theory like edge contraction and it is dual, edge removal to simplify a pair of dual graphs while preserving planarity and duality. The edges to be removed/contracted build up contraction kernels which form a spanning forest of the input graph. Repeated contraction steps can be combined in a single contraction using large equivalent contraction kernels. The receptive field of a high-level vertex is spanned by the tree of the equivalent contraction kernel. The corresponding regions are connected and form an inclusion hierarchy well suited to hold the intended segmentations. In this chapter, we presented a hierarchical image partitioning method using a pairwise similarity function. The function encapsulates the intuitive notion of contrast by measuring the difference along the boundary of two components, relative to a measure of differences of the components' internal variation. Two components are merged if there is an edge with low-cost connection between them. Borůvka's MST algorithm together with the dual-graph contraction algorithm is used for building an MST, and at the same time, preserving the connectivity of the input graph. For vision tasks, in natural systems, the topological relations seem to play a role even more important than precise geometrical position. Even though the MST algorithm makes local greedy decisions, it produces perceptually important partitions by finding region borders quickly and effortlessly in a bottom-up "stimulus-driven" way based only on local differences in a specific feature. The framework is general and can handle large variation and gradient intensity in images. Experimental results prove the validity of the theoretical concept. We evaluated quantitatively the segmentation result produced by different methods. The evaluation is done by using discrepancy measures, that do not penalize segmentations that are coarser or more refined in certain regions. We used only gray images to evaluate the quality of results on one feature. It is shown that the graph-based method presented produce qualitatively similar results.

Acknowledgment

Supported by the Austrian Science Found (FWF) under grants P18716-N13 and S9103-N04.

References

1. A. Shokoufandeh, Y. Keselman, F. Demirci, D. Macrini, and S. Dickinson. Many-to-Many Feature Matching in Object Recognition. In H. Christensen and H.-H. Nagel, editors, *Cognitive Vision Systems: Sampling Spectrum of Approaches*, Lecture Notes in Computer Science. Springer, Berlin Heidelberg New York, 2004
2. W.G. Kropatsch. Abstraction pyramids on discrete representations. In A.J.-P. Braquelaire, J.-O. Lachaud, and A. Vialard, editors, *Proceedings of Discrete Geometry for Computer Imagery*, pages 1–21, Bordeaux, France. Springer, Berlin Heidelberg New York, April 3–5, 2002
3. M. Wertheimer. Über gestaltheorie. *Philosophische Zeitschrift für Forschung und Aussprache*, 1: 30–60, 1925. Reprint in Gestalt Theory vol. 7, 1985
4. Y. Keselman and S. Dickinson. Generic model abstraction from examples. In *Proceedings of IEEE Conference on Computer Vision and Pattern Recognition*, volume 1, pages 856–863, Kauai, Hawaii, December 2001. IEEE Computer Society
5. C.-S. Fu, S.-W. Cho, and K. Essig. Hierarchical color image region segmentation for content-based image retrieval system. *IEEE Transaction on Image Processing*, 9(1): 156–62, 2000
6. J. Shi and J. Malik. Normalized cuts and image segmentation. *IEEE Transactions on Pattern Analysis and Machine Intelligence*, 22(8): 888–905, 2000
7. P.F. Felzenszwalb and D.P. Huttenlocher. Efficient graph-based image segmentation. *International Journal of Computer Vision*, 59(2): 167–181, 2004
8. Y. Haxhimusa, R. Glantz, M. Saib, G. Langs, and W.G. Kropatsch. Logarithmic tapering graph pyramid. In L. van Gool, editor, *Proceedings of German Pattern Recognition Symposium*, volume 2449 of *Lecture Notes in Computer Science*, pages 117–124, Switzerland, Springer, Berlin Heidelberg New York, 2002
9. O. Borůvka. Příspěvek k řešení otázky ekonomické stavby elektrovodných sítí (contribution to the solution of a problem of economical construction of electrical networks). *Elektrotechnický Obzor*, 15: 153–154, 1926
10. W.G. Kropatsch. Building irregular pyramids by dual-graph contraction. *IEEE Proceedings Vision, Image and Signal Processing*, 142(6): 366–374, December 1995
11. Y. Haxhimusa. *The Structurally Optimal Dual-Graph Pyramid and its Application in Image Partitioning*. PhD thesis, Vienna University of Technology, Faculty of Informatics, Institute of Computer Aided Automation, Pattern Recognition and Image Processing Group, Vienna, 2006
12. R. Diestel. *Graph Theory*. Springer, Berlin Heidelberg New York, 1997
13. F. Harary. *Graph Theory*. Addison-Wesley, Reading, MA, 1969

14. K. Thulasiraman and M.N.S. Swamy. *Graphs: Theory and Algorithms.* Wiley, New York, 1992
15. J.-M. Jolion and A. Rosenfeld. *A Pyramid Framework for Early Vision.* Kluwer, Dordecht, 1994
16. P.J. Burt and E.H. Adelson. The laplacian pyramid as a compact image code. *IEEE Transactions on Communications,* 31(4): 532–540, 1983
17. S.G. Mallat. A theory for multiresolution signal decomposition: The wavelet representation. *IEEE Transactions on Pattern Analysis and Machine Intelligence,* 11(7): 674–693, 1989
18. W.G. Kropatsch. Image pyramids and curves – an overview. Technical Report PRIP-TR-2, Vienna University of Technology, Faculty of Informatics, Institute of Computer Aided Automation, Pattern Recognition and Image Processing Group, http://www.prip.tuwien.ac.at/ftp/pub/publications/trs/, 1991
19. M. Bister, J. Cornelis, and A. Rosenfeld. A critical view of pyramid segmentation algorithms. *Pattern Recognition Letters,* 11(9): 605–617, 1990
20. W.G. Kropatsch, A. Leonardis, and H. Bischof. Hierarchical, adaptive and robust methods for image understanding. *Surveys on Mathematics for Industry,* 9: 1–47, 1999
21. B.B. Bederson. *A Miniature Space-variant Active Vision System.* PhD thesis, New York University, Courant Insitute, New York, 1992
22. H. Bischof. *Pyramidal Neural Networks.* Lawrence Erlbaum Associates, 1995
23. A. Rosenfeld. Arc colorings, partial path groups, and parallel graph contractions. Technical Report TR-1524, University of Maryland, Computer Science Center, July 1985
24. P. Meer. Stochastic image pyramids. *Computer Vision, Graphics, and Image Processing,* 45(3): 269–294, 1989
25. W.G. Kropatsch, Y. Haxhimusa, Z. Pizlo, and G. Langs. Vision pyramids that do not grow too high. *Pattern Recognition Letters,* 26(3): 319–337, 2005
26. W.G. Kropatsch and A. Montanvert. Irregular versus regular pyramid structures. In U. Eckhardt, A. Hbler, W. Nagel, and G. Werner, editors, *Geometrical Problems of Image Processing,* pages 11–22. Springer, Berlin Heidelberg New York, 1991
27. G.H. Granlund. The complexity of vision. *Signal Processing,* 74(1): 101–126, 1999
28. R.S. Wallace, P.-W. Ong, B.B. Bederson, and E.L. Schwatz. Space variant image processing. *International Journal of Computer Vision,* 13(1): 71–90, 1994
29. A. Rojer and E.L. Schwartz. Design considerations for a space variant visual sensor complex-logarithmic geometry. In *Proceeding of Internation Conference in Pattern Recognition,* volume 2, pages 278–285, Los Alamitos, California, 1990
30. W.G. Kropatsch. Building irregular pyramids by dual-graph contraction. Technical Report PRIP-TR-35, Vienna University of Technology, Faculty of Informatics, Institute of Computer Aided Automation, Pattern Recognition and Image Processing Group, http://www.prip.tuwien.ac.at/ftp/pub/publications/trs/, 1994
31. W.G. Kropatsch. Equivalent contraction kernels and the domain of dual irregular pyramids. Technical Report PRIP-TR-42, Vienna University of Technology, Faculty of Informatics, Institute of Computer Aided Automation, Pattern Recognition and Image Processing Group, http://www.prip.tuwien.ac.at/ftp/pub/publications/trs/, 1995
32. T.H. Cormen, C.E. Leiserson, R.L. Rivest, and C. Stein. *Introduction to Algorithms.* MIT, USA, 2001
33. R. Glantz and W.G. Kropatsch. Plane embedding of dually contracted graphs. In G. Borgefors, I. Nyström, and G. Sanniti di Baja, editors, *Proceedings of Discrete Geometry for Computer Imagery,* volume 1953 of *Lecture Notes in Computer Science,* Uppsala, Sweden, Springer, Berlin Heidelberg New York, pages 348–357, December 2000

34. R.O. Duda, P.E. Hart, and D.G. Stork. *Pattern Classification*. Wiley, New York, 2001
35. J. Nešťřil. A few remarks on the history of MST-problem. *Archivum Mathematicum Brno*, 33: 15–22, 1997
36. R.C. Prim. Shortest connection networks and some generalizations. *The Bell System Technical Journal*, 36: 1389–1401, 1957
37. B. Julesz. Textons, the elements of texture perception and their interactions. *Nature*, 290: 91–97, 1981
38. S. Horowitz and T. Pavlidis. Picture segmentation by a tree traversal algorithm. *Journal of the Association for Computer and Machinery*, 2(23): 368–388, 1976
39. B. Fischer and J.M. Buhmann. Data resampling for path based clustering. In L. van Gool, editor, *Proceedings of German Pattern Recognition Symposium*, volume 2449 of *Lecture Notes in Computer Science*, Switzerland, Springer Berlin Heidelberg New York, pages 206–214, 2002
40. L. Guigues, L.M. Herve, and J.-P. Cocquerez. The hierarchy of the cocoons of a graph and its application to image segmentation. *Pattern Recognition Letters*, 24(8): 1059–1066, 2003
41. Y. Haxhimusa and W.G. Kropatsch. Hierarchy of partitions with dual-graph contraction. In B. Milaelis and G. Krell, editors, *Proceedings of German Pattern Recognition Symposium*, volume 2781 of *Lecture Notes in Computer Science*, Germany, Springer Berlin Heidelberg New York, pages 338–345, 2003
42. Y. Haxhimusa and W.G. Kropatsch. Hierarchical image partitioning with dual-graph contraction. Technical Report PRIP-TR-81, Vienna University of Technology, Faculty of Informatics, Institute of Computer Aided Automation, Pattern Recognition and Image Processing Group, http://www.prip.tuwien.ac.at/ftp/pub/publications/trs/, July 2003
43. J. Malik, S. Belongie, T. Leung, and J. Shi. Contour and texture analysis for image segmentation. *International Journal of Computer Vision*, 43(1): 7–27, 2001
44. U. Koethe and P. Stelldinger. Shape preserving digitization of ideal and blurred binary images. In I. Nyström, G.S. di Baja, and S. Svensson, editors, *International Conference on Discrete Geometry for Computer Imagery (DGCI)*, volume 2886 of *Lecture Notes in Computer Science*, Springer, Berlin Heidelberg New York, pages 82–91, 2003
45. B. Sudhir and S. Sarkar. A framework for performance characterization of intermediate-level grouping modules. *IEEE Transactions on Pattern Analysis and Machine Intelligence*, 19(11): 1306–1312, 1997
46. D. Martin, C. Fowlkes, D. Tal, and J. Malik. A database of human segmented natural images and its application to evaluating segmentation algorithms and measuring ecological statistics. In *Proceedings of International Conference on Computer Vision*, volume 2, pages 416–423, July 2001
47. M.D. Heath, S. Sarkar, and K.W. Sanocki, T. Bowyer. A robust visual method for assessing the relative performance of edge-detection algorithms. *IEEE Transactions on Pattern Analysis and Machine Intelligence*, 19(12): 1338–1359, 1997
48. Y. Zhou, V. Venkateswar, and R. Chellappa. Edge detection and linear feature extraction using a 2-d random field model. *IEEE Transactions on Pattern Analysis and Machine Intelligence*, 11(1): 84–95, 1989
49. L. Cinque, C. Guerra, and L. Levialdi. Reply: On the paper by R. Haralick. *CVGIP: Image Understanding*, 60(2): 250–252, 1994
50. Y. Zhang. A survey on evaluation methods for image segmentation. *Pattern Recognition*, 29(8): 1335–1346, 1996
51. C.N. Graaf, A.S.E. Koster, K.L. Vincken, and M.A. Viergever. Validation of the interleaved pyramid for the segmentation of $3d$ vector images. *Pattern Recognition Letters*, 15(5): 469–475, 1994

52. J.-M. Jolion. Stochastic pyramid revisited. *Pattern Recognition Letters*, 24(8): 1035–1042, 2003
53. Y. Haxhimusa, A. Ion, W.G. Kropatsch, and L. Brun. Hierarchical image partitioning using combinatorial maps. In *Proceeding of the Joint Hungarian-Austrian Conference on Image Processing and Pattern Recognition*, pages 179–186, 2005

A Graphical Model Framework for Image Segmentation

Rui Huang, Vladimir Pavlovic and Dimitris N. Metaxas

Summary. Graphical models are probabilistic models defined in terms of graphs. The intuitive and compact graph representation and its ability to model complex probabilistic systems make graphical models a powerful modeling tool in various research areas. In this paper we introduce a graphical model framework for image segmentation based on the integration of Markov random fields (MRFs) and deformable models. A graphical model is constructed to represent the relationship of the observed image pixels, the true region labels and the underlying object contour. We then formulate the problem of image segmentation as the one of joint region-contour inference and learning in the graphical model. The graphical model representation allows us to use an approximate structured variational inference technique to solve this otherwise intractable joint inference problem. Using this technique, the MAP solution to the original model is obtained by finding the MAP solutions of two simpler models, an extended MRF model and a probabilistic deformable model, iteratively and incrementally. In the extended MRF model, the true region labels are estimated using the BP algorithm in a band area around the estimated contour from the probabilistic deformable model, and the result in turn guides the probabilistic deformable model to an improved estimation of the contour. Finally, we generalize our method from 2D to 3D. Experimental results on both synthetic and real images, in both 2D and 3D, show that our new hybrid method outperforms both the MRF-based and the deformable model-based methods using only homogeneous constraints.

1 Introduction

Graphical models are a marriage between probability theory and graph theory [1]. A graphical model is a probabilistic model defined in terms of a graph in which the nodes represent random variables and the edges describe the probabilistic relationships among these variables. In particular, these probabilistic relationships are usually defined by conditional probabilities among the related variables or potential functions on the cliques of the graph, depend on whether the graph is directed or undirected. The joint probability distribution of a set of variables or the whole system can then be computed by taking products over the functions defined on relevant nodes. The graph theoretic side of graphical models provides an intuitive and compact representation for the complex probabilistic system, as well as well-defined data structures and efficient general-purpose algorithms. Probability theory, on the other hand, ensures the consistency of the whole system, and provides various statistical inference and learning methods to analyze the data.

Graphical models have recently received extensive attention from many different research communities, including artificial intelligence, machine learning, computer vision, etc. In this paper, we apply graphical models to the image segmentation problem, one of the most important and difficult tasks in computer vision area. We are able to integrate two fundamentally different traditional segmentation methods and take advantage of both using graphical models. Furthermore, the graphical model theory allows us to employ an approximate, computationally efficient solution to the otherwise intractable inference problem. We will focus on the graphical model representation and inference (mainly approximate inference) techniques for image segmentation. See [2] for a more comprehensive introduction to graphical models and [1] for more advanced topics.

The rest of this paper is organized as follows: Section 2 defines the segmentation problem and reviews the previous work; Sect. 3 introduces a new integrated model and its decoupled approximation using the variational inference method; detailed inferences on the decoupled models are described in Sect. 4; the 2D model is then generalized to 3D in Sect. 5; Sect. 6 shows the experimental results on both synthetic and real 2D images and 3D volumes; and finally Sect. 7 summarizes the paper.

2 Previous Work

Image segmentation is one of the most important and difficult preliminary processes for high-level computer vision and pattern recognition problems. The main goal of image segmentation is to divide an image into its constituent parts that have a strong correlation with objects or areas of the real world depicted by the image.

Region-based and edge-based segmentations are the two major classes of segmentation methods. Though one can label regions according to edges or detect edges from regions, these two kinds of methods are naturally different and have respective advantages and disadvantages.

Region-based methods assign image pixels to a region according to some image property (e.g., region homogeneity). These methods work well in noisy images, where edges are usually difficult to detect while the region homogeneity is preserved. The disadvantages of region-based methods are that they may generate rough edges and holes inside the objects, and they do not take account of object shape.

On the other hand, edge-based methods generate boundaries of the segmented objects. A prior knowledge of object shape and topology can be easily incorporated to constrain the segmentation result. While this often leads to sufficiently smooth boundaries, the oversmoothing may be excessive. Because edge-based methods rely on edge detecting operators, they are sensitive to image noise and need to be initialized close to the actual region boundaries.

Most segmentation methods are either region-based or edge-based. Among region-based methods, besides the classical region growing method [3], the Markov random field (MRF) model has been extensively used. Because the exact MAP inference in MRF models is computationally infeasible, various techniques for approximating the MAP estimation have been proposed, such as Markov Chain Monte Carlo (MCMC) [4], iterated conditional modes (ICM) [5], maximizer of posterior marginals (MPM) [6], etc. [7] presents a comparative analysis of some of these methods. Two of the more recent algorithms, Belief Propagation (BP) [8,9] and Graph Cuts [10] are compared in [11]. The estimation of the MRF model parameters is another related problem, often solved using the EM algorithm [12].

In edge-based methods, since Kass et al. introduced Snakes [13], deformable models have attracted much attention. Variants of deformable models have been proposed to address different problems. For instance, Balloons [14] and Gradient Vector Flow (GVF) Snakes [15] introduces different external forces, and Topologically Adaptable Snakes [16] allow changes in the model's topology. See [17] for a review of deformable models and [3] for other edge-based methods and some basic edge detecting operators.

Hybrid approaches [18–20] attempt to combine region-based and edge-based segmentations to alleviate deficiencies of the individual methods and improve the segmentation results. There are different choices of the combination. For instance, [20] proposes a way of integrating MRFs and deformable models. MRFs are used to initially estimate the boundary of objects in noisy images. Balloons are then fitted to the estimated boundary. The result of the fitting is in turn used to update the MRF parameters. Final segmentation is achieved by iteratively integrating these processes. While this hybrid method attempted to take advantage of both MRFs and deformable models, the model coupling was loose. This may cause failure of deformable models if the initial estimation of the boundary by MRF is not closed, and it may also yield oversmoothed boundaries.

We propose a new framework to combine the MRF-based and the deformable model-based segmentation methods. To tightly couple the two models, we construct a graphical model to represent the relationship of the observed image pixels, the true region labels and the underlying object contour. Exact inference in the graphical model is intractable because of the large state spaces and the couplings of model variables. To tackle this problem we use a variational inference method to seemingly decouple the graphical model into two simpler models: one extended MRF model and one probabilistic deformable model. Then we obtain the MAP solution in the original model by solving the MAP problems of the two simpler models iteratively and incrementally. In the extended MRF model, the true region labels are estimated using the BP algorithm in a band area around the estimated contour from the probabilistic deformable model, and the result in turn guides the probabilistic deformable model to an improved estimation of the contour.

3 Our Method

The goal of our segmentation method is to find one specific region with a smooth and closed boundary. A seed point is manually specified and the region containing it is then segmented automatically. Thus, without significant loss of modeling generality, we simplify the MRF model and avoid possible problems caused by segmenting multiple regions simultaneously.

In this section, we first briefly review MRFs and deformable models, define the notation, and then introduce our hybrid framework.

3.1 MRF-Based Segmentation

MRF models are a special case of undirected graphical models. They are often used for image analysis, because of their ability to capture the context of an image (i.e., dependencies among neighboring image pixels) and deal with the noise.

A typical MRF model for image segmentation, as shown in Fig. 1, is a graph with two types of nodes: observable nodes (shaded nodes in Fig. 1, representing image pixel values) and hidden nodes (clear nodes in Fig. 1, representing region labels). The edges in the graph depict the relationships among the nodes.

Let n be the number of the hidden/observable states (i.e., the number of pixels in the image). A configuration of the hidden layer is:

$$\mathbf{x} = (x_1, ..., x_n), x_i \in L, i = 1, ..., n$$

where L is a set of region labels, such as $L = \{inside, outside\}$.

Similarly, a configuration of the observable layer is:

$$\mathbf{y} = (y_1, ..., y_n), y_i \in D, i = 1, ..., n$$

where D is a set of pixel values, e.g., gray values 0–255.

The relationship between the hidden states and the observable states (also known as local evidence) can be described by the potential (or compatibility) function: $\phi(x_i, y_i)$, which is often a conditional Gaussian to handle the image noise; the relationship between the neighboring hidden states is described by the second potential function: $\psi(x_i, x_j)$, which usually penalizes differences between the states to keep region smoothness. The detailed definitions will be discussed later.

Now the segmentation problem can be viewed as a problem of estimating the MAP solution of the MRF model:

$$\mathbf{x}_{MAP} = \arg\max_{\mathbf{x}} P(\mathbf{x}|\mathbf{y}) \tag{1}$$

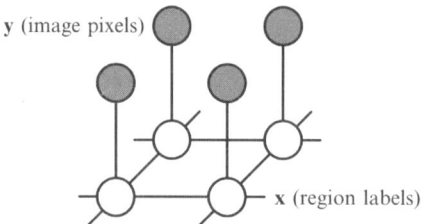

Fig. 1. MRF model

where
$$P(\mathbf{x}|\mathbf{y}) \propto P(\mathbf{y}|\mathbf{x})P(\mathbf{x}) \propto \prod_i \phi(x_i, y_i) \prod_{(i,j)} \psi(x_i, x_j)$$

As mentioned previously, the exact MAP inference in MRFs is computationally infeasible, and various techniques have been used for approximating the MAP estimation. In our method, we use the BP algorithm. The MRF model parameters (i.e., the parameters in the potential functions) are learned using the EM algorithm. However, in the presence of multiple regions in the image, the automatic determination of the number of regions and the initial guess of the parameters could be difficult. More importantly, like other region-based methods, MRFs do not take account of object shape and may generate rough edges and even holes inside the objects.

3.2 Deformable Model-Based Segmentation

Many deformable model-based methods have also been used in image segmentation. A deformable model is usually a parameterized geometric primitive, whose deformation is determined by geometry, kinematics, dynamics, and other constraints (e.g., material properties, etc.) [21]. Snakes [13], a special case of deformable models, are a parametric contour:

$$\Omega = [0, 1] \to \Re^2$$
$$s \to \mathbf{c}(s) = (x(s), y(s))$$

where s is the parametric domain and x and y are the coordinate functions. The energy of the contour:

$$E(\mathbf{c}) = E_{\text{int}}(\mathbf{c}) + E_{\text{ext}}(\mathbf{c}) = \int_\Omega \omega_1(s) \left|\frac{\partial \mathbf{c}}{\partial s}\right|^2 + \omega_2(s) \left|\frac{\partial^2 \mathbf{c}}{\partial s^2}\right|^2 + F(\mathbf{c}(s)) ds$$

where $\omega_1(s)$ and $\omega_2(s)$ control the "elasticity" and "rigidity" of the contour, and F is the potential associated to the external forces. The final shape of the contour corresponds to the minimum of this energy.

To minimize the above energy term, one can use the discretized first-order Lagrangian dynamics equation:

$$\dot{\mathbf{d}} + \mathbf{K}\mathbf{d} = \mathbf{f}$$

where $\mathbf{d}$ is discretized version of $\mathbf{c}$, $\mathbf{K}$ is the stiffness matrix calculated from $\omega_1(s)$ and $\omega_2(s)$, and $\mathbf{f}$ is the generalized force vector.

Image gradient forces are usually used to attract a deformable model to edges. However, when far from the true boundary, the model often gets attracted to spurious image edges. Balloon forces have been introduced to solve this problem [14]. Namely, the deformable model is considered a balloon, which is inflated by an additional force and stopped by strong edges. The initial contour need no longer be close to the true boundary. Mathematically, a force along the normal direction to the curve at point $\mathbf{c}(s)$ with some appropriate amplitude k is added to the original forces.

$$\mathbf{f}' = \mathbf{f} + k\vec{\mathbf{n}}(s)$$

Deformable models can also be viewed in a probabilistic framework [17]. The internal energy $E_{\text{int}}(\mathbf{c})$ leads to a Gibbs prior distribution of the form:

$$P(\mathbf{c}) = \frac{1}{Z_i} \exp(-E_{\text{int}}(\mathbf{c})) \qquad (2)$$

while the external energy $E_{\text{ext}}(\mathbf{c})$ can be converted to a sensor model with conditional probability:

$$P(\mathbf{I}|\mathbf{c}) = \frac{1}{Z_e} \exp(-E_{\text{ext}}(\mathbf{c})) \qquad (3)$$

where $\mathbf{I}$ denotes the image, and $E_{\text{ext}}(\mathbf{c})$ is a function of the image $\mathbf{I}$.

The deformable models can now be fitted by solving the MAP problem:

$$\mathbf{c}_{\text{MAP}} = \arg\max_{\mathbf{c}} P(\mathbf{c}|\mathbf{I}) \qquad (4)$$

where

$$P(\mathbf{c}|\mathbf{I}) \propto P(\mathbf{c})P(\mathbf{I}|\mathbf{c})$$

One limitation of the deformable model-based method is its sensitivity to image noise, a common drawback of edge-based methods. This may result in the deformable model being "stuck" in a local energy minimum of a noisy image.

3.3 Integrated Model

As shown in (1) and (4), both the MRF-based and the deformable model-based segmentations can be viewed as the MAP estimation problems. In previous work [20], these two models were loosely coupled. Our new framework uses the graphical model theory to tightly couple the two models. This is achieved, as depicted in Fig. 2, by adding a new hidden state to the traditional MRF model to represent the underlying contour.

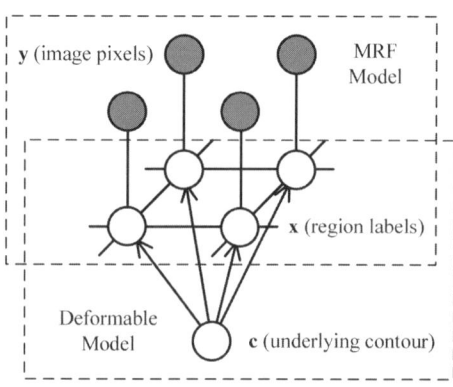

Fig. 2. Integrated model

In the new model, the segmentation problem can also be viewed as a *joint* MAP estimation problem:
$$(\mathbf{c}, \mathbf{x})_{MAP} = \arg\max_{\mathbf{c},\mathbf{x}} P(\mathbf{c}, \mathbf{x}|\mathbf{y})$$
where
$$P(\mathbf{c}, \mathbf{x}|\mathbf{y}) \propto P(\mathbf{y}|\mathbf{x})P(\mathbf{x}|\mathbf{c})P(\mathbf{c})$$

To define the joint distribution of the integrated model, we model the image likelihood term $P(\mathbf{y}|\mathbf{x})$ as:
$$P(\mathbf{y}|\mathbf{x}) = \prod_i \phi(x_i, y_i)$$
identical to the traditional MRF model. The second term $P(\mathbf{x}|\mathbf{c})$, modeling the distribution of the region labels conditioned on the contour, is defined as:
$$P(\mathbf{x}|\mathbf{c}) = \prod_{(i,j)} \psi(x_i, x_j) \prod_i P(x_i|\mathbf{c})$$
where we incorporated a shape prior $\mathbf{c}$ to constrain the region labels $\mathbf{x}$, in addition to the original Gibbs distribution.

Since we only segment one specific region at one time, we need only consider the pixels near the contour, and label them either *inside* or *outside* the contour.

We model the dependency between the contour $\mathbf{c}$ and the region labels $\mathbf{x}$ using the softmax function:
$$P(x_i = inside|\mathbf{c}) = \frac{1}{1 + \exp(-dist(i, \mathbf{c}))} \quad (5)$$

$$P(x_i = outside|\mathbf{c}) = 1 - P(x_i = inside|\mathbf{c}) \quad (6)$$

induced by the signed distance of pixel i from the contour $\mathbf{c}$:
$$dist(i, \mathbf{c}) = sign(i) \min_{s \in \Omega} \|loc(i) - \mathbf{c}(s)\| \quad (7)$$

where $sign(i) = 1$ if pixel i is inside contour $\mathbf{c}$, $sign(i) = -1$ when it is outside, and $loc(i)$ denotes the spatial coordinates of pixel i.

Lastly, the prior term $P(\mathbf{c})$, as in (2), can be represented as a Gibbs distribution when the shape prior is given by a parametric contour $\mathbf{c}$.

Despite the compact graphical representation of the integrated model, the exact inference in the model is computationally intractable. One reason for this is the large state space size and the complex dependency structure introduced by the Gibbs distribution of the prior $P(\mathbf{c})$. The second reason is the existence of loops in the graphical model, which preclude polynomial-time inference. To deal with these problems we propose an approximate, yet tractable, solution based on structured variational inference.

3.4 Approximate Inference Using Structured Variational Inference

Structured variational inference techniques [22, 23] consider parameterized distribution which is in some sense close to the desired posterior distribution, but is easier to compute. Namely, for a given image $\mathbf{y}$, a distribution $Q(\mathbf{c}, \mathbf{x}|\mathbf{y}, \theta)$ with an additional set of *variational parameters* θ is defined such that the Kullback–Leibler (KL) divergence between $Q(\mathbf{c}, \mathbf{x}|\mathbf{y}, \theta)$ and $P(\mathbf{c}, \mathbf{x}|\mathbf{y})$ is minimized with respect to θ:

$$\theta^* = \arg\min_\theta \sum_{\mathbf{c},\mathbf{x}} Q(\mathbf{c}, \mathbf{x}|\mathbf{y}, \theta) \log \frac{P(\mathbf{c}, \mathbf{x}|\mathbf{y})}{Q(\mathbf{c}, \mathbf{x}|\mathbf{y}, \theta)}$$

The dependency structure of Q is chosen such that it closely resembles the dependency structure of the original distribution P. However, unlike P the dependency structure of Q *must* allow a computationally efficient inference.

In our case we define Q by decoupling the MRF model and the deformable model components of the original integrated model in Fig. 2. The original distribution is factorized into two independent distributions: an extended MRF model Q_M with variational parameter $\mathbf{a}$ and a probabilistic deformable model Q_D with variational parameter $\mathbf{b}$ (Fig. 3). The *extended* MRF model means we have an additional layer to the traditional MRF model to deal with the shape prior, and the *probabilistic* deformable model means the contour is not fitted to the image directly, but to the probabilistic label image.

Because Q_M and Q_D are independent,

$$Q(\mathbf{c}, \mathbf{x}|\mathbf{y}, \mathbf{a}, \mathbf{b}) = Q_M(\mathbf{x}|\mathbf{y}, \mathbf{a}) Q_D(\mathbf{c}|\mathbf{b})$$

According to the extended MRF model, we have:

$$Q_M(\mathbf{x}|\mathbf{y}, \mathbf{a}) \propto Q_M(\mathbf{y}|\mathbf{x}) Q_M(\mathbf{x}|\mathbf{a})$$

$$Q_M(\mathbf{y}|\mathbf{x}) = \prod_i \phi(x_i, y_i)$$

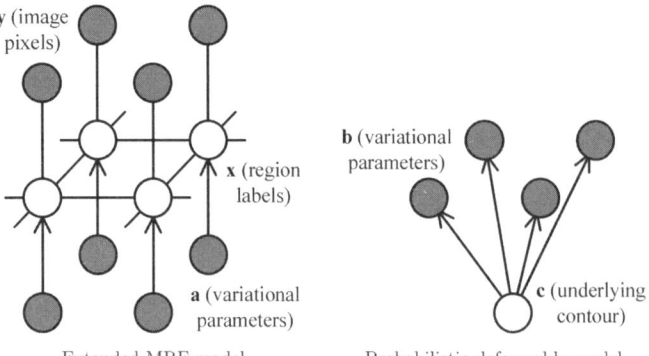

Fig. 3. Decoupled models

$$Q_M(\mathbf{x}|\mathbf{a}) = \prod_{(i,j)} \psi(x_i, x_j) \prod_i P(x_i|a_i)$$

Hence,

$$Q_M(\mathbf{x}|\mathbf{y}, \mathbf{a}) \propto \prod_i \phi(x_i, y_i) \prod_{(i,j)} \psi(x_i, x_j) \prod_i P(x_i|a_i) \quad (8)$$

On the other hand, the probabilistic deformable model yields:

$$Q_D(\mathbf{c}|\mathbf{b}) \propto Q_D(\mathbf{b}|\mathbf{c}) Q_D(\mathbf{c})$$

$$Q_D(\mathbf{b}|\mathbf{c}) = \prod_i P(b_i|\mathbf{c})$$

leading to

$$Q_D(\mathbf{c}|\mathbf{b}) \propto \prod_i P(b_i|\mathbf{c}) Q_D(\mathbf{c}) \quad (9)$$

The optimal values of the variational parameters $\theta = (\mathbf{a}, \mathbf{b})$ are obtained by minimizing the KL-divergence. It can be shown, using e.g., [24], that the optimal parameters $\theta^* = (\mathbf{a}^*, \mathbf{b}^*)$ should satisfy the following equations:

$$\log P(x_i|a_i^*) = \sum_{\mathbf{c}} Q_D(\mathbf{c}|\mathbf{b}^*) \log P(x_i|\mathbf{c}) \quad (10)$$

$$\log P(b_i^*|\mathbf{c}) = \sum_{x_i \in L} Q_M(x_i|\mathbf{y}, \mathbf{a}^*) \log P(x_i|\mathbf{c}) \quad (11)$$

Notice that the inference solutions, (8) and (9), together with the parameter optimizations, (10) and (11), form a set of *fixed-point equations*. Solution of this fixed-point set yields a tractable approximation to the intractable original posterior.

Since the state space of $\mathbf{c}$ (all possible contour configurations in the image plane) is too large, (10) is still intractable. We simply use the winner-take-all strategy and approximate $Q_D(\mathbf{c}|\mathbf{b})$ as a delta function:

$$Q'_D(\mathbf{c}|\mathbf{b}) = \begin{cases} 1 \text{ if } \mathbf{c} = \arg\max_{\mathbf{c}} Q_D(\mathbf{c}|\mathbf{b}) \\ 0 \text{ else} \end{cases}$$

and (10) can be simplified to:

$$P(x_i|a_i) = P(x_i|\mathbf{c}) \quad (12)$$

where $\mathbf{c} = \arg\max_{\mathbf{c}} Q_D(\mathbf{c}|\mathbf{b})$.

3.5 Algorithm Description

The variational inference algorithm for the hybrid MRF-DM model can now be summarized as:

> Initialize contour **c**;
> **while** ($error > maxError$) {
>   1. Calculate a band area B around **c**. Perform remaining steps inside B;
>   2. Calculate $P(x_i|a_i)$ based on (12) using **c**;
>   3. Estimate the MRF-MAP solution $Q_M(x_i|\mathbf{y},\mathbf{a})$ based on (8) using $P(x_i|a_i)$;
>   4. Calculate $\log P(b_i|\mathbf{c})$ based on (11) using $Q_M(x_i|\mathbf{y},\mathbf{a})$;
>   5. Estimate the DM-MAP solution $Q_D(\mathbf{c}|\mathbf{b})$ based on (9) using $\log P(b_i|\mathbf{c})$;
> }

Steps 2 and 4 follow directly from (12) and (11). The details of steps 1, 3, and 5 are discussed in Sect. 4.

4 Implementation Issues

4.1 Solve MRF-MAP with EM and BP

Step 3 of our algorithm solves the MAP problem in the extended MRF model. The EM algorithm is used to estimate both the MAP solution of region labels **x** and the parameters of the model (i.e., the parameters in the potential functions).

Particularly, in E step, the MAP solution of region labels **x** is estimated based on current parameters. Unlike most of the previous work mentioned in Sect. 2, we solve this MRF-MAP estimation problem using the BP algorithm. BP is an inference method proposed by Pearl [8] to efficiently estimate Bayesian beliefs in the network by the way of iteratively passing messages between neighbors. It is an exact inference method in the network without loops. Even in the network with loops, the method often leads to good approximate and tractable solutions [25].

There are two variants of the BP algorithm: sum–product and max–product. The sum–product message passing rule can be written as:

$$m_{ij}(x_j) = \sum_{x_i} \Psi_{ij}(x_i, x_j)\Phi_i(x_i) \prod_{k \in \aleph(i)\setminus j} m_{ki}(x_i)$$

The max–product has analogous formula, with the marginalization replaced by the maximum operator. At convergence:

$$x_{iMAP} = \arg\max_{x_i} \Phi_i(x_i) \prod_{j \in \aleph(i)} m_{ji}(x_i)$$

According to our extended MRF model the compatibility functions are:

$$\Phi_i(x_i) = \phi(x_i, y_i)P(x_i|a_i)$$

$$\Psi_{ij}(x_i, x_j) = \psi(x_i, x_j)$$

We again note the difference from a traditional MRF model, due to the incorporated shape prior. $P(x_i|a_i)$ is calculated in step 2 of the algorithm. $\phi(x_i, y_i)$ and $\psi(x_i, x_j)$ can be calculated using current MRF parameters.

In this model we assume the image pixels are corrupted by white Gaussian noise:

$$\phi(x_i, y_i) = \frac{1}{\sqrt{2\pi\sigma_{x_i}^2}} \exp\left(-\frac{(y_i - \mu_{x_i})^2}{2\sigma_{x_i}^2}\right)$$

On the other hand, to penalize differences between the neighboring labels (i.e., to keep local region smoothness),

$$\psi(x_i, x_j) = \frac{1}{Z} \exp\left(\frac{\delta(x_i - x_j)}{\sigma^2}\right)$$

where $\delta(x) = 1$ if $x = 0$; $\delta(x) = 0$ if $x \neq 0$, σ controls the similarity of neighboring hidden states, and Z is a normalization constant.

As shown in step 1 in our algorithm, belief propagation is restricted to a single band of model variables around the current contour estimates. A reason for this is that, in practice, we only need to care about the statistics of pixels near the boundary. More importantly, the banded inference significantly speeds up the whole algorithm. Although convergence of the banded algorithm is not guaranteed, in our experiments, the BP algorithm does converge, usually in only one or two iterations.

In M step, the MRF parameters are updated based on the MAP solution of the region labels **x** using the following equations:

$$\mu_l = \frac{\sum_i Q_M(x_i = l | y_i, a_i) y_i}{\sum_i Q_M(x_i = l | y_i, a_i)}$$

$$\sigma_l^2 = \frac{\sum_i Q_M(x_i = l | y_i, a_i)(y_i - \mu_l)^2}{\sum_i Q_M(x_i = l | y_i, a_i)}$$

where $l \in \{inside, outside\}$.

4.2 Probabilistic Deformable Model

In step 5, according to (3), we use the negative log term, $-\log P(\mathbf{b}|\mathbf{c})$, as the external energy in the deformable model. Given this "label image" energy landscape, the image force is simply $\nabla(\log P(\mathbf{b}|\mathbf{c}))$. With the additional balloon forces, this leads to the discretized first-order Lagrangian dynamics equation:

$$\dot{\mathbf{d}} + \mathbf{K}\mathbf{d} = \nabla(\log P(\mathbf{b}|\mathbf{c})) + k\vec{\mathbf{n}}(s)$$

We note that this formulation is different from that of [20] where the deformable model is fitted to a *binary* label image obtained from the MAP configuration of **x**. In our method, we use a *probabilistic* measurement of the label of each pixel as specified in (11).

Finally, following the definition in (5)–(7), we note that the gradient of the coupling energy at pixel i, $\nabla(\log P(\mathbf{b}|\mathbf{c}))$, can be shown to be:

$$\frac{\partial \log P(\mathbf{b}|\mathbf{c})}{\partial \mathbf{c}} = -\frac{\partial \log P(\mathbf{b}|\mathbf{c})}{\partial loc(i)}$$

5 3D Image Segmentation

Increasing availability of high-resolution 3D image data using modalities such as magnetic resonance (MR) and computed tomography (CT) has prompted the need for 3D segmentation approaches. However, 3D image segmentation remains an extremely difficult problem, due to the complex topology of 3D objects, the massive data, and demanding computational algorithms. Many 3D approaches are often 2D in nature (i.e., applying the 2D algorithm slice by slice to the 3D volume data [26]). The lack of interaction among individual slice solutions, however, leads to results that are inferior to true 3D-based solutions [27].

In this section, we generalize our framework to 3D image segmentation based on the integration of 3D MRFs and deformable surface models. The proposed method is a true 3D method that fully exploits the structure of the 3D data, resulting in improved object segmentation. The generalization is straightforward using the graphical model representation, and the variational inference in the graphical model also leads to computationally more efficient solutions, which, in the 3D case, is still of main concern.

A 3D MRF model is shown in Fig. 4. The hidden nodes are positioned at the vertices of a regular 3D grid of the same size as the volume data (Fig. 4 left). Each hidden node x_i is connected to six neighboring hidden nodes (more neighbors can be connected by adding diagonals in the grid) and one observable node y_i (Fig. 4 right). Again, the observable nodes represent the voxel values of the 3D volume data and the hidden nodes represent the region labels of corresponding voxels.

As to the deformable models, Finite-Element Method (FEM)-based balloon models [27] and Polygonal Geometrically Deformed Model (GDM) [28] are commonly used for representation of 3D surfaces and segmentation of volume data.

Similar to the 2D case, a new hidden node representing the underlying boundary surface is added to the 3D MRF model (Fig. 5 left, only one pair of voxel/label nodes is drawn for simplicity). We again use the structured variational inference technique to seemingly decouple the integrated model into two simpler models (Fig. 5 right): one extended 3D MRF model with shape prior constraints and one probabilistic deformable surface model.

The 3D algorithm is similar to the 2D one. However, the expansion process of the 3D balloon model far away from the true boundary can be time-consuming and needs frequent reparametrization, and often suffers from local energy minima in

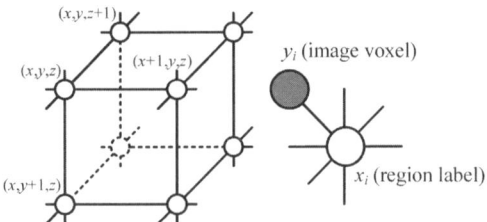

Fig. 4. 3D MRF model

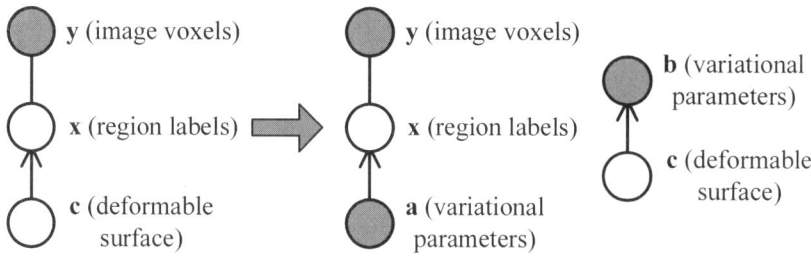

Fig. 5. Integrated and decoupled models

noisy images. An interactive initialization procedure or a learned shape prior would be helpful. When there is no shape prior, one can use the 3D MRF segmentation algorithm alone to generate an initial region segmentation and apply the Marching Cubes algorithm [29] to the 3D belief image to generate an initial surface. Marching Cubes is an algorithm for constructing triangle models of constant density surfaces from discrete volume data. The resulting surface representation is suitable for the FEM-based balloon model. The rest of the 3D algorithm is a straightforward generalization of the 2D one.

6 Experiments

6.1 2D Synthetic Images

The initial study of properties and utility of our method was conducted on a set of synthetic images. The images were synthesized in a way similar to [7]. In [7], the 64×64 perfect images contain only two gray levels representing the *object* (gray level is 160) and the *background* (gray level is 100) respectively. In our experiments, we made the background more complicated by introducing a gray level gradient. The gray levels of the background are increasing from 100 to 160, along the normal direction of the object contour (Fig. 6a). Figure 6b shows the result of a traditional MRF-based method. The object is segmented correctly, however some regions in the background are misclassified. On the other hand, the deformable model fails because of the leaking from the high-curvature part of the object contour, where the gradient in the normal direction is too weak (Fig. 6c). Our hybrid method, shown in Fig. 6d, results in a significantly improved segmentation.

We next generated a test image (Fig. 6e) by adding Gaussian noise with mean 0 and standard deviation 60 to Fig. 6a. The result of the MRF-based method on the noisy image (Fig. 6f) is somewhat similar to that in Fig. 6b, which shows the MRF can deal with image noise to some extent. But significant misclassification occurred because of the complicated background and noise levels. The deformable model either sticks to spurious edges caused by image noise or leaks (Fig. 6g) because of the weakness of the true edges. Unlike the two independent methods, our hybrid algorithm, depicted in Fig. 6h, correctly identifies the object boundaries despite the

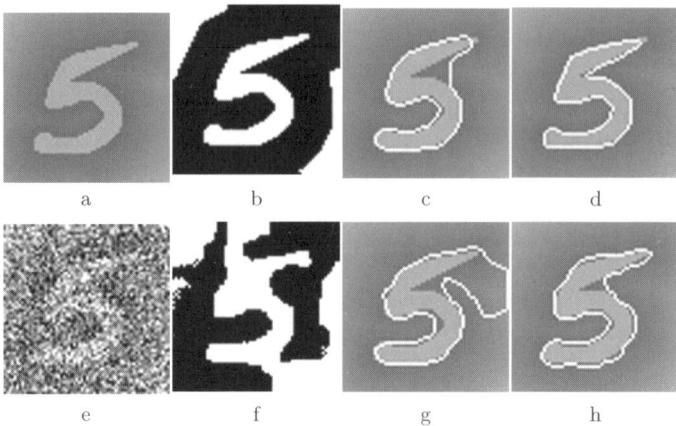

Fig. 6. Experiments on 2D synthetic images

excessive image noise. For visualization purposes we superimpose the contour on the original image (Fig. 6a) to show the quality of the result in Fig. 6g and h.

6.2 2D Medical Images

Experiments with synthetic images in Sect. 6.1 outlined some of the benefits of our hybrid method. The real world images usually have significant, often nonwhite noise and contain multiple regions and objects, rendering the segmentation task a great deal more difficult. In this section we show results of applying our method to real medical images on which we can hardly get satisfying results with either the MRF-based or the deformable model-based methods alone.

In the following comparisons, we manually specified the inside/outside regions to get an initial guess of the parameters for the MRF-only method. For the deformable model method, we started the balloon model at several different initial positions and use the best results for the comparison. On the other hand, our hybrid method is significantly less sensitive to the initialization of the parameters and the initial seed position.

Figure 7a shows a 2D MR image of the left ventricle of the human heart. Figure 7b is the result of the MRF-based method. While it is promising, the result still exhibits rough edges and holes. Figure 7c depicts the result of the deformable model-based method. Although we carefully chose the magnitude of the balloon forces, parts of the contour begin to leak others stick to spurious edges. Our hybrid method, started from the initial contour shown in Fig. 7e, generated better result (Fig. 7d). One of the intermediate iterations is shown in Fig. 7f. The corresponding external energy in the band area is depicted in Fig. 7g (image intensities are proportional to the magnitude of the energy), showing a more useful profile than the traditional edge energy $-|\nabla(G_\sigma * I)|^2$ shown in Fig. 7h.

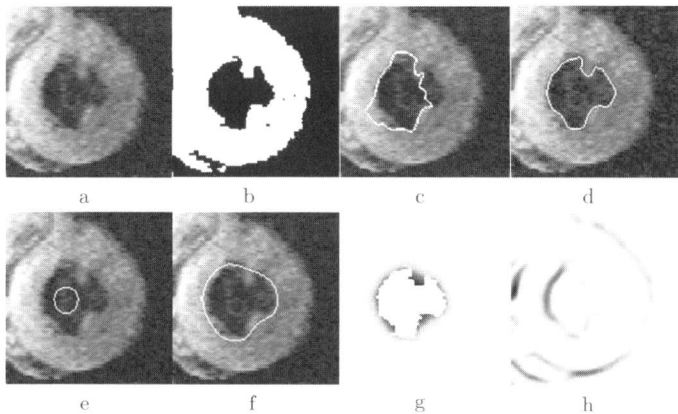

Fig. 7. Experiments on 2D medical images (1)

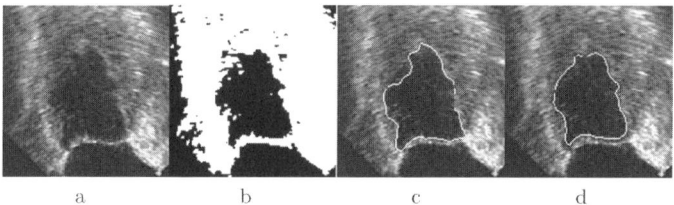

Fig. 8. Experiments on 2D medical images (2)

Figure 8a is an ultrasound image. The MRF gets rough edges and holes in the objects (Fig. 8b) while the deformable model cannot escape a local minimum (Fig. 8c). Our hybrid method eliminates the rough edges and holes caused by the MRF while outlining the region more accurately than the deformable model (Fig. 8d).

Finally, Figs. 9a and 10a are both examples of difficult images with complicated global properties, requiring the MRF-based method to automatically determine the number of regions and the initial values of the parameters. Figure 9b is obtained by manually initializing the MRF model. Our method avoids this problem by creating and updating an MRF model locally and incrementally. The images are also difficult for deformable models because the boundaries of the objects to be segmented are either high-curvature (Fig. 9a) or low-gradient (Fig. 10a). Figure 9c exemplifies the oversmoothed deformable models. Our method's results, shown in Figs. 9d and 10b, do not suffer from either of the problems.

6.3 3D Synthetic Images

Our 3D method was also first experimented on a set of synthetic images. The perfect image contains two gray levels representing the object (gray level is 160) and the background (gray level is 100), respectively. Gaussian noise with mean 0 and standard deviation 60 is added to the whole image to generate the test image.

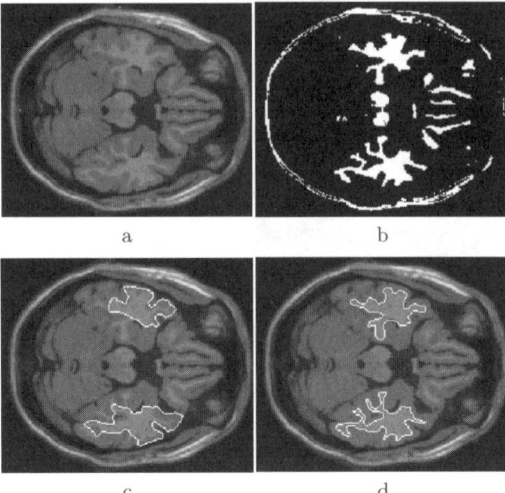

Fig. 9. Experiments on 2D medical images (3)

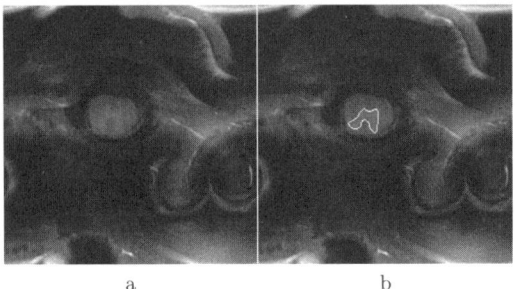

Fig. 10. Experiments on 2D medical images (4)

The first experiment intended to show the advantages of the true 3D method over the 2D slice-based method. In this experiment, we generated a $100 \times 100 \times 100$ 3D image containing a ball-like object. Figure 11a shows several slices of the perfect image. Our test image is generated by cutting out a quarter of the pie-like object from the 50th frame and adding the Gaussian noise (Fig. 11b). The segmentation results by 2D MRFs and 3D MRFs are shown in Fig. 11c, d. Both models handled noise successfully. The 3D MRF model obviously recovered the pie-like object in the 50th frame by retaining region smoothness in the direction perpendicular to the frame. The 2D MRF model cannot achieve this due to the lack of interaction between neighboring frames. The boundary of the results from 3D MRFs also look smoother.

The second experiment was performed on a $64 \times 64 \times 64$ volume containing a "5"-like object similar to Fig. 6a. The thickness of the object is 8 (i.e., frames 29 to 36 contain the object). Besides the zero mean Gaussian noise, extra noise with mean 160 is also added to a part of the two successive frames 32 and 33. The test

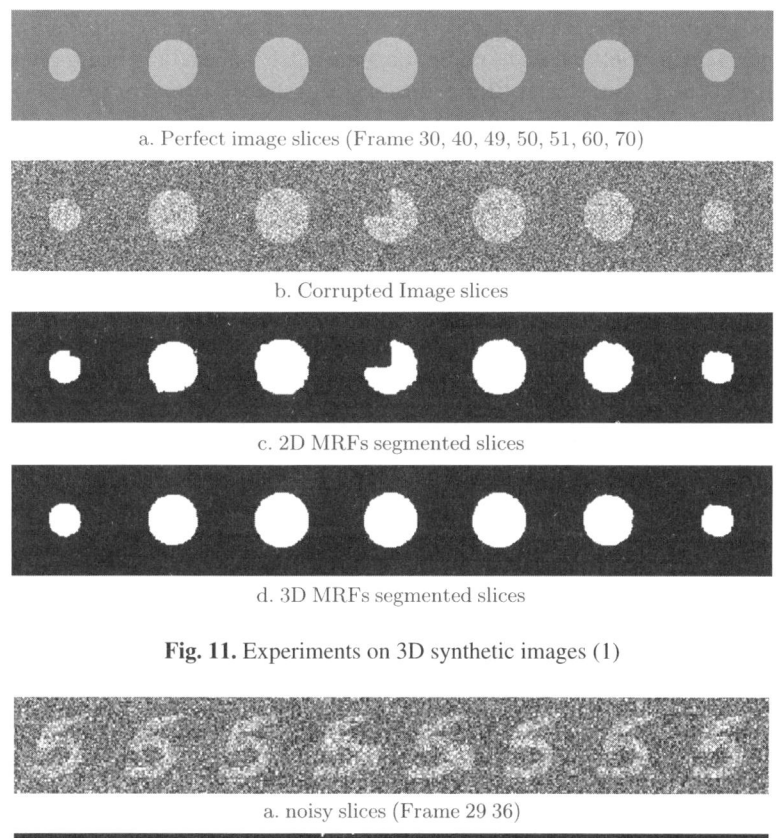

a. Perfect image slices (Frame 30, 40, 49, 50, 51, 60, 70)

b. Corrupted Image slices

c. 2D MRFs segmented slices

d. 3D MRFs segmented slices

Fig. 11. Experiments on 3D synthetic images (1)

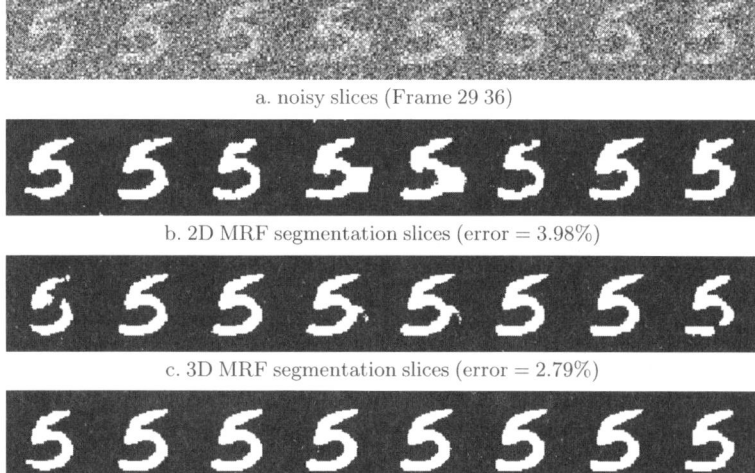

a. noisy slices (Frame 29 36)

b. 2D MRF segmentation slices (error = 3.98%)

c. 3D MRF segmentation slices (error = 2.79%)

d. 3D MRF + DM segmentation slices (error = 1.62%)

Fig. 12. Experiments on synthetic 3D images (2)

image slices are shown in Fig. 12a. The results of 2D MRFs are shown in Fig. 12b. Each slice looks different from others, especially for the two frames with extra noise. The slices in Fig. 12c (results of 3D MRFs), however, are smoother and similar to

60   R. Huang et al.

their neighbors, except for the first and last frames, which suffered more interference from the background. These two outermost frames are improved by coupling the DM with the 3D MRF model, and other frames are also slightly smoother (Fig. 12d). The average error rates of the three methods are 3.98%, 2.79%, and 1.62%.

6.4 3D Medical Images

Experiments with synthetic images in Sect. 6.3 outlined the advantages of both the 3D method over the 2D method and the hybrid method over the MRF-only method. In this section, we show experimental results of applying our methods to 3D medical images. We do not show the results of the slice-based method with 2D MRFs as in previous experiments mainly because this method is sensitive to initialization and we cannot get satisfying results on these medical images. While our 3D method also needs manual initialization when the shape prior is not given, the slice-based method requires manual initialization for almost each single slice.

We first test our algorithms on simulated brain MRI data from BrainWeb [30]. The database contains simulated brain MRI data based on two anatomical models: normal and multiple sclerosis. For both of these, full 3D data volumes have been simulated using three sequences (T1-, T2-, and proton-density (PD)-weighted) and a variety of slice thicknesses, noise levels, and levels of intensity nonuniformity. We segmented the white matter from three different normal brain data volumes using the hybrid method. Figure 13a shows a slice from the ground truth data of the white matter. Figure 13d is the result from our hybrid method. The second column of Fig. 13 shows the segmentation results on T1 image without noise and intensity nonuniformity (RF inhomogeneity). The segmented white matter is slightly thicker than the results from the ground truth, because some of the grey matter is misclassified due to its similar grey value to the white matter. Same misclassification can be observed in

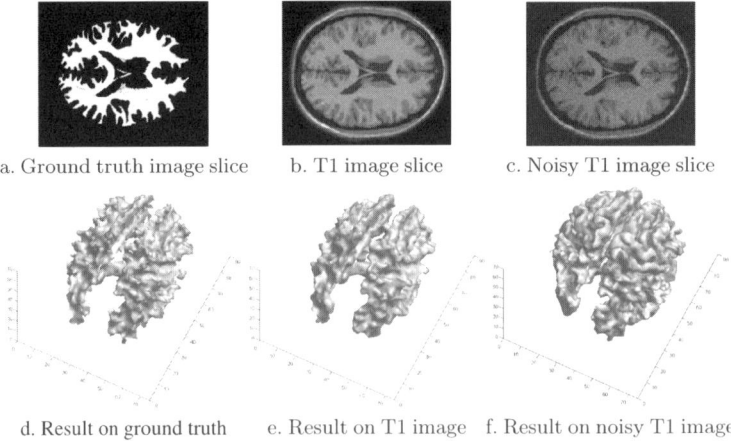

a. Ground truth image slice   b. T1 image slice   c. Noisy T1 image slice

d. Result on ground truth   e. Result on T1 image   f. Result on noisy T1 image

Fig. 13. Experiments on 3D medical images (1)

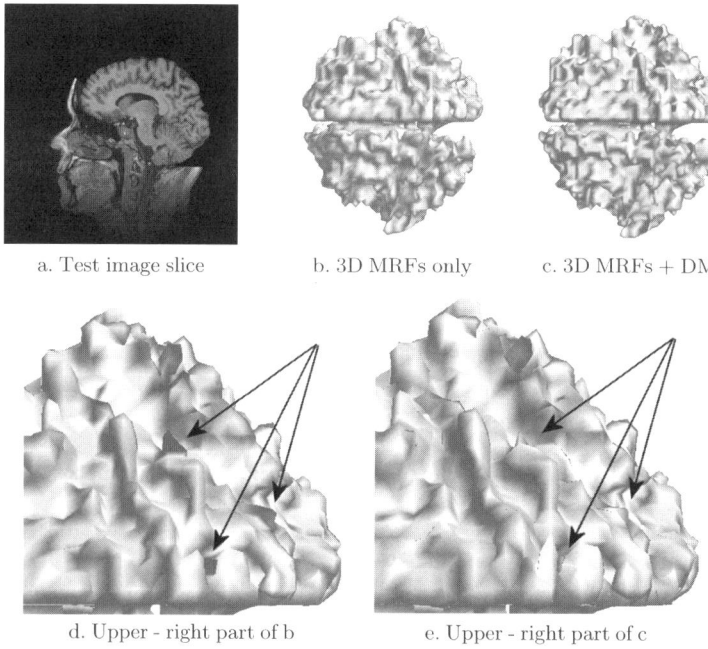

Fig. 14. Experiments on 3D medical images (1)

the third column, which is the segmentation result on T1 image with 9% noise and 40% intensity nonuniformity. One possible solution to the misclassification problem is using the 3D MRF-only algorithm to do a multiregion segmentation first.

Finally, we show some results on a real medical image [31], which is an MR image of a head with the skull partially removed to reveal the brain. Figure 14a is one of the slices from the volume. The results of our methods are shown in Fig. 14b, c. To show the difference between the two algorithms (i.e., the effect of adding deformable models), the upper-right parts of Fig. 14b, c are magnified in Fig. 14d, e. The arrows show that some incorrect patches are eliminated by the deformable fitting process. Surface smoothness can be easily controlled by tuning the parameters in the stiffness matrix. Because the white matter itself is a complicated object with high curvature, the parameters are usually chosen according to experts' opinion.

7 Conclusions

We proposed a new framework to combine the MRF-based and the deformable model-based segmentation methods. The framework was developed under the auspices of the graphical model theory allowing us to employ a well-founded set of statistical inference and learning techniques. In particular, we employed the variational inference method, an approximate, computationally efficient solution, to otherwise intractable inference of region boundaries. Experimental results on both synthetic and real 2D images and 3D volumes show that the hybrid methods outperforms both the MRF-based and the deformable model-based methods using only homogeneous constraints.

References

1. Jordan, M.I. ed.: Learning in Graphical Models. The MIT, Cambridge, MA (1998)
2. Jordan, M.I.: Graphical models. Statistical Science (Special Issue on Bayesian Statistics) **19**(1) (2004) 140–155
3. Sonka, M., Hlavac, V., Boyle, R.: Image Processing, Analysis and Machine Vision, Second Edition. Thomson Learning (1998)
4. Geman, S., Geman, D.: Stochastic relaxation, Gibbs distributions and the Bayesian restoration of images. IEEE Transaction on Pattern Analysis and Machine Intelligence **6**(6) (1984) 721–741
5. Besag, J.E.: On the statistical analysis of dirty pictures. Journal of the Royal Statistical Society Series B **48**(3) (1986) 259–302
6. Marroquin, J., Mitter, S., Poggio, T.: Probabilistic solution of ill-posed problems in computational vision. Journal of the American Statistical Association **82**(397) (1987) 76–89
7. Dubes, R., Jain, A., Nadabar, S., Chen, C.: MRF model-based algorithm for image segmentation. In: Proceedings of International Conference on Pattern Recognition. Volume 1. (1990) 808–814
8. Pearl, J.: Probabilistic Reasoning in Intelligent Systems: Networks of Plausible Inference. Morgan Kaufmann, Stanford, CA (1988)
9. Yedidia, J., Freeman, W.T., Weiss, Y.: Understanding belief propagation and its generalizations. In: International Joint Conference on Artificial Intelligence, Distinguished Presentations Track. (2001)
10. Boykov, Y., Veksler, O., Zabih, R.: Fast approximate energy minimization via graph cuts. IEEE Transactions on Pattern Analysis and Machine Intelligence **23**(11) (2001) 1222–1239
11. Tappen, M., Freeman, W.: Comparison of graph cuts with belief propagation for stereo, using identical MRF parameters. In: Proceedings of IEEE International Conference on Computer Vision. Volume 2. (2003) 900–907
12. Zhang, Y., Brady, M., Smith, S.: Segmentation of brain MR images through a hidden Markov random field model and the expectation–maximization algorithm. IEEE Transaction on Medical Imaging **20**(1) (2001) 45–57
13. Kass, M., Witkin, A., Terzopoulos, D.: Snakes: Active contour models. International Journal of Computer Vision **1**(4) (1988) 321–331
14. Cohen, L.: On active contour models and balloons. Computer Vision, Graphics, and Image Processing: Image Understanding **53**(2) (1991) 211–218
15. Xu, C., Prince, J.: Gradient vector flow: A new external force for snakes. In: Proceedings of IEEE Conference on Computer Vision and Pattern Recognition. (1997) 66–71
16. McInerney, T., Terzopoulos, D.: Topologically adaptable snakes. In: Proceedings of IEEE International Conference on Computer Vision. (1995) 840–845
17. McInerney, T., Terzopoulos, D.: Deformable models in medical image analysis: A survey. Medical Image Analysis **1**(2) (1996) 91–108
18. Ronfard, R.: Region-based strategies for active contour models. International Journal of Computer Vision **13**(2) (1994) 229–251

19. Jones, T., Metaxas, D.: Image segmentation based on the integration of pixel affinity and deformable models. In: Proceedings of IEEE Conference on Computer Vision and Pattern Recognition. (1998) 330–337
20. Chen, T., Metaxas, D.: Image segmentation based on the integration of Markov random fields and deformable models. In: Proceedings of International Conference on Medical Image Computing and Computer-Assisted Intervention. (2000) 256–265
21. Metaxas, D.: Physics-based Deformable Models: Applications to Computer Vision, Graphics and Medical Imaging. Kluwer, Dordecht (1997)
22. Jordan, M., Ghahramani, Z., Jaakkola, T., Saul, L.: An introduction to variational methods for graphical models. Machine Learning **37**(2) (1999) 183–233
23. Pavlovic, V., Frey, B., Huang, T.: Variational learning in mixed-state dynamic graphical models. In: Proceedings of Conference on Uncertainty in Artificial Intelligence. (1999) 522–530
24. Ghahramani, Z.: On structured variational approximations. Technical Report CRG-TR-97-1 (1997)
25. Weiss, Y.: Belief propagation and revision in networks with loops. Technical Report MIT A.I. Memo 1616 (1998)
26. Choi, S., Lee, J., Kim, J., Kim, M.: Volumetric object reconstruction using the 3D-MRF model-based segmentation. IEEE Transaction on Medical Imaging **16**(6) (1997) 887–892
27. Cohen, L., Cohen, I.: Finite-element methods for active contour models and balloons for 2-D and 3-D images. IEEE Transactions on Pattern Analysis and Machine Intelligence **15**(11) (1993) 1131–1147
28. Miller, J., Breen, D., Lorensen, W., O'Bara, R., Wozny, M.: Geometrically deformed models: A method for extracting closed geometric models from volume data. Computer Graphics (Proceedings of SIGGRAPH) **25**(4) (1991) 217–226
29. Lorensen, W., Cline, H.: Marching cubes: A high resolution 3D surface construction algorithm. Computer Graphics (Proceedings of SIGGRAPH) **21**(4) (1987) 163–169
30. BrainWeb: Simulated Brain Database. (http://www.bic.mni.mcgill.ca/brainweb/)
31. Stanford: Volume Data Archive. (http://graphics.stanford.edu/data/voldata/)

Digital Topologies on Graphs

Alain Bretto

Summary. In this chapter we focus on the relationship between graph theory and topology. Topologies on vertices of graphs became much more essential in topology, with the development of computer science, especially with the development of computer graphic and image analysis. Digital topology is the study of the topological properties of digital images. In most of the literature a digital image has been endowed with a graph model; the vertices being the points of the image, and the edges giving the connectivity between the points. This has led to the investigation of topology on graph [7–12]. We study compatible topologies on graphs, (here compatibility is to be understood as connectivity). We describe some properties of these particular topological spaces. We discuss the relation between T_0-spaces (which play an important role) and other compatible topologies on graph. We develop some applications to digital geometry. Other results related to compatible topologies on graphs is developed.

Key words: Graph theory, Topology, Computer sciences, Image processing

1 Introduction

Because data structures in computer sciences are enumerable, the only set which can be used in this case are *discrete* or *digital*. Roughly speaking discrete or digital is used in this chapter as opposed to *continuous*. For instance the space $\mathbb{R}^n$, $n \geq 1$ is where we do both continuous geometry and continuous topology, while the space $\mathbb{Z}^n$, $n \geq 1$ is an example of a space where we do both digital geometry and digital topology. Graphs have particular significance in computational sciences because of their presence in applications such as solid modeling, molecular biology, computer graphic, image analysis, etc. The most popular approach to define a discrete analog of the topologies of the Euclidean space is the *graph-theoretic* approach. Actually a *digital d-space* is the set of d-tuple of the real Euclidean d-space having integer coordinates. Such a point is called a *digital* point. Moreover a digital space is equipped with a graph structure based on the local adjacency relations. So the graph theoretic approach gives directly the connectedness but it is difficult to handle some topological concepts such as continuity, compacity and so on. Hence some important problems arise:

- What are the topological or geometrical properties of a discrete set?
- What does connected component mean for a digital set?
- What does continuous mapping mean between two discrete spaces?
- Which other properties has a digital set?

Discrete topological spaces, can be defined as a topological space such that any point has a smallest neighborhood. These types of topological spaces were first study by Alexanfroff [1]. Some applications of these topologies have been developed, where topological spaces are used to model discrete situations:

- In "Graphs, topologies and simple games" J. M. Bilbao [2] uses discrete topological spaces on a finite set to study the existence of connected coalitions in a simple game. Thanks to the digital topology he gives some sufficient conditions for the existence of winning coalitions.
- Baik and Miller [3] give a topological approach for testing equivalence in heterogeneous relational databases.
- By introducing "pretopology" M. Brissaud [4] models preference structures in economy.

We can find other applications in data structures, logics, complexity theory....
Any digital image can be interpreted as a graph, (it is "embedded" in a discrete space) whose vertices are the pixels (geometric points and grey level intensity) and whose edges define nearness and connectedness. Because a digital image can be viewed as a discrete set, hence a graph, it is interesting to study the problems enumerate below. For this we have to define a notion of compatibility between graphs and topologies. The natural way to define the compatible topologies on graphs is the connectedness: Let $G = (V, E)$ be a graph and let $\mathcal{T}$ be a topology on V, $\mathcal{T}$ is called a *compatible topology* on V if it satisfies the following conditions:

(a) For every connected induced subgraph $G(V')$, V' is connected for $\mathcal{T}$.
(b) For each $V' \subseteq V$ connected set for $\mathcal{T}$, the induced subgraph $G(V')$ is a connected graph.

Another important problem in image processing is the *digitalization*: let a subset A of $\mathbb{R}^n$, $n \geq 1$ and let $f : \mathbb{R}^n \longrightarrow \mathbb{Z}^n$ be a map which associates to A a discrete set $f(A)$. This map is called a *digitizer*. From this definition some questions arise about the continuity of f or f^{-1}, the structural properties of f, etc. Consequently compatible topologies have been intensively studied and there are a lot of contributions by many authors [5–15].

In the first part of this chapter we investigate compatible topologies on graphs. We study the relation between connectedness and Alexandroff space. We show that any locally finite Alexandroff topology can be "embedded" into a T_0-locally finite Alexandroff topology such that both have the same connected set. We use these results to characterize the graphs which have compatible topologies on the set of vertices. We introduce an example of compatible topology on a bipartite graph and we study exhaustively this one. In a second part, to illustrate the first part, we give some applications to digital plane and digital spaces; others applications of compatible topologies will be given.

2 Definitions

All graphs are finite or infinite, undirected without isolated vertices. We consider that these graphs are simple (graphs without no loop or multiple edge). We denote them $G = (V; E)$. Given a graph G, we denote the *neighborhood* of a vertex x by $\Gamma(x)$, i.e. the set formed by all the vertices adjacent to x:

$$\Gamma(x) = \{y \in V, \{x, y\} \in E\}$$

The number of neighbors of x is the *degree* of x (denoted by dx). For all $x \in V$, if dx is finite one will say that G is a *locally finite graph*. A *chain* (or *path*) from x_0 to x_k is a sequence of distinct vertices: $V = \{x_0, x_1, \ldots, x_k\}$ such that $\{x_0 x_1, x_1 x_2, \ldots, x_{k-1} x_k\} \subseteq E$, where $x_i x_{i+1}$, stand for the edge $\{x_i, x_{i+1}\}$. The number of edges is the *length* of this chain. A graph is *connected* if for all $x, y \in V$ there exists a chain from x to y. A *cycle* is a chain such that the first vertex and the last vertex are the same. We denote a cycle with a length equal to n by C_n. Let C_n be a cycle, a *chord* of C_n is an edge linking two nonconsecutive vertices of C_n. A *circle* is a cycle without chord. A graph $G' = (V'; E')$ is a *subgraph* of G when it is a graph satisfying $V' \subseteq V$ and $E' \subseteq E$. If $V' = V$ then G' is a *spanning subgraph*. An *induced subgraph* (generated by A) $G(A) = (A; U)$, with $A \subseteq V$ and $U \subseteq E$ is a subgraph such that for $x, y \in A$: $\{x, y\} \in E$ implies $\{x, y\} \in U$. An *orientation* of $G = (V; E)$ is a preorder, (reflexive and transitive relation) on its vertices such that $\{x; y\} \in E$ if and only if $x < y$ or $y < x$. A simple undirected graph $G = (V; E)$ is a *comparability graph* if there exists an orientation of G. A graph $G = (V, E)$ is *bipartite* if $V = V_1 \cup V_2$ with $V_1 \cap V_2 = \emptyset$ and every edge joins a vertex of V_1 to a vertex V_2. We denote a bipartite graph by $G = (V_1, V_2; E)$.

A *topology* on a set X is a nonempty collection $\mathcal{T}$ of subsets of X, called *open*, such that any union of open sets is open, any finite intersection of open sets is open, and both X and the empty set are open. A set together with a topology on it is called a *topological space*. We denote a topological space by $(X, \mathcal{T})$.

Let $(X, \mathcal{T}_1)$ and $(X, \mathcal{T}_2)$ be two topological spaces. If any open set of $\mathcal{T}_1$ is a an open set of $\mathcal{T}_2$ one will say that $\mathcal{T}_1$ is more thin than $\mathcal{T}_2$ and one will denote that by

$T_1 \leq T_2$. A *neighborhood* of $x \in X$ is a subset V of X containing an open subset which contains the point x. The set of neighborhoods of a point x will be denoted by $\mathcal{V}(x)$, or $\mathcal{V}_T(x)$.

Suppose we have a topology on a set X, and a collection of neighborhoods $\mathcal{V}' = (V_i')_{i \in I}$ of a point $x \in X$ such that any neighborhood of x contains an element of $\mathcal{V}'$. Then $\mathcal{V}'$ is called a *fundamental system of neighborhood* of x. Suppose we have a topology on a set X, and a collection $\mathcal{O}$ of open sets such that every open set is a union of members of $\mathcal{O}$: then $\mathcal{O}$ is called a *base* for the topology and elements of $\mathcal{O}$ are called *basic open sets*.

A family $\mathcal{U} = \{U_i\}$ of (open) subsets of X is an (open) *covering* if each element in X belongs to at least one $U_i \in \mathcal{U}$. The *closure* of $A \subseteq X$ is $\phi(A) = \{x \in X; \forall V \in \mathcal{V}(x) \; V \cap A \neq \emptyset\}$.

$(X, \mathcal{P}(X))$ is a topological space called *discrete topological space*, we will denote it by $(X, \mathcal{D})$. The set $\{X, \emptyset\} = \mathcal{G}$ defines a topology on X, we call it *trivial topology*.

Let V' be a subset of X and let $\mathcal{T}$ be a topology on X; the collection of sets $\mathcal{T}' = \{V' \cap U; U \in \mathcal{T}\}$ defines a topology on X' called *induced topology*. We will call *subspace* of X this topological space.

A topological space $(X, \mathcal{T})$ is *connected* if X cannot be expressed as the union of two disjoint nonempty open sets. A *connected subset* of X is a connected subspace of X. The *connected component* $C(x)$ of x is the biggest connected subset of X containing x. $\{C(x), x \in X\}$ is a closed partition of X.

X is *totally disconnected* if $C(x) = \{x\}$ for every $x \in X$. A topological space X is *path connected*, if given any two points a and b in X, there exists a continuous path between them, that is a continuous map $\gamma : [0,1] \longrightarrow X$ such that $\gamma(0) = a$ and $\gamma(1) = b$, where $[0,1]$ is equipped with the usual topology.

A topological space X is *locally connected* if there exists a fundamental system of connected neighborhood for every $x \in X$. An *Alexandroff topology* is one in which every intersection of open sets is open. So in an Alexandroff space any point x has a smallest neighborhood denoted by $\mathcal{N}(x)$, or $\mathcal{N}_T(x)$: it is a open set.

A topological space $(X, \mathcal{T})$ is a T_0-*space* if, for any distinct points $x, y \in X$ if $x \in \phi(\{y\})$, then $y \notin \phi(\{x\})$.

Let $G = (V; E)$ be a graph and let $\mathcal{T}$ be a topology on V, $\mathcal{T}$ is called a *compatible topology* on G if it satisfies the following conditions:

(a) For every connected induced subgraph $G'(V')$, V' is connected for $\mathcal{T}$.
(b) For each $V' \subseteq V$ connected set for $\mathcal{T}$, $G(V')$ is a connected graph.

3 Connectivity and Alexandroff Spaces

We give first a preliminary lemma.

Lemma 1 *Let $(X, \mathcal{T})$ be a topological space,*

(1) $\{x, y\}$ is connected if and only if $y \in \phi(\{x\})$ or $x \in \phi(\{y\})$.

(2) If $(X,\mathcal{T})$ is connected then for any open covering $\mathcal{U}$, one has the following property: for any $x,y \in X$ either there exists $V_1 \in \mathcal{U}$ such that $x,y \in V_1$, or there exists $n \geq 2$ and $V_1, V_2, \ldots, V_n \in \mathcal{U}$ such that $x \in V_1 \setminus V_2, y \in V_n \setminus V_{n-1}$ and satisfying: $V_i \cap V_j \neq \emptyset$ if and only if $|i-j| \leq 1$.

Moreover if $(X,\mathcal{T})$ is an Alexandroff space then:

(a) $\{x,y\}$ is connected if and only if $x \in \mathcal{N}(y)$ or $y \in \mathcal{N}(x)$.
(b) For any x, if $x \in A \subseteq \mathcal{N}(x)$, then A is connected; so $(X,\mathcal{T})$ is locally connected. In particular for any x, $C(x)$ is a set both open and closed.
(c) If X is connected, then X is path connected.

Proof. (1) Let us suppose that $\{x,y\}$ is nonconnected. There exists $O_1, O_2 \in \mathcal{T}$ with $O_1' = O_1 \cap \{x,y\}$ and $O_2' = O_2 \cap \{x,y\}$ such that $O_1' \cap O_2' = \emptyset$ and $\{x,y\} = O_1' \cup O_2'$. So $y \notin O_1$ (for example) and O_1 is an open neighborhood of x in X. Consequently $x \notin \phi(\{y\})$. In the same way $x \notin O_2$, so O_2 is an open neighborhood of y, that leads to $y \notin \phi(\{x\})$.

Suppose that $x \notin \phi(\{y\})$ and $y \notin \phi(\{x\})$. There exists an open neighborhood V of x (respectively, open neighborhood W of y) such that $y \notin V$ (respectively, $x \notin W$). So $\{x,y\} = (V \cap \{x,y\}) \cup (W \cap \{x,y\})$ and $(V \cap \{x,y\}) \cap (W \cap \{x,y\}) = \emptyset$.

(2) See [16], ex 11, p 188.

Suppose now that $(X,\mathcal{T})$ is an Alexandroff space.

(a) Because $y \in \mathcal{N}(x) \iff x \notin \phi(y)$ and (1).
(b) $A = \bigcup_{y \in A} \{x,y\}$ and any $\{x,y\}$ is connected by (a).
(c) It is well known that if $(X,\mathcal{T})$ is locally connected then $C(x)$ is open.

Suppose that X is connected. Take the open covering $\mathcal{U} = \{\mathcal{N}(x), x \in X\}$. Let us suppose $u, v \in X$; X being connected we have $u \in \mathcal{N}(x_0), \ldots, \mathcal{N}(x_i), \ldots, \mathcal{N}(x_n) \ni v$, with $\mathcal{N}(x_i) \cap \mathcal{N}(x_j) \neq \emptyset$ if and only if $|i-j| \leq 1$. So it is sufficient to prove that if $x \in \mathcal{N}(y)$ then there exists a path from x to y: $\gamma(t) = x$ if $0 \leq t < 1$ and $\gamma(1) = y$ is suitable because $[0,1] \subseteq \gamma^{-1}(\mathcal{N}(x))$ and $[0,1] \subseteq \gamma^{-1}(\mathcal{N}(y))$. □

3.1 Generation of Alexandroff Spaces

If $\mathcal{T}$ is an Alexandroff topology then it is easy to see that the map $x \longmapsto \mathcal{N}(x)$ satisfies

(a) $x \in \mathcal{N}(x)$.
(b) $y \in \mathcal{N}(x)$ involves $\mathcal{N}(y) \subseteq \mathcal{N}(x)$.

Conversely if we have a map $\mathcal{N}$ verifying the conditions (a), (b), then there exists an Alexandroff topology $\mathcal{T}$ where the open sets containing x are defined in the following way: $\{Y, \mathcal{N}(x) \subseteq Y\}$. $\mathcal{T}$ is the Alexandroff topology associated to $\mathcal{N}$. The following assertion characterizes generated Alexandroff space.

Theorem 1 *Let $(X, \mathcal{T})$ be a topological space, and $\mathcal{N}(x) = \bigcap_{V \in \mathcal{V}(x)} V$, $x \in X$:*

(i) *$\mathcal{N}$ produces an Alexandroff topology on X, denoted by $\mathcal{AT}$.*
(ii) *$\mathcal{AT} = \inf\{\mathcal{T}', \mathcal{T}' \text{ Alexandroff topology and } \mathcal{T} \leq \mathcal{T}'\}$ ($\mathcal{AT}$ is the Alexandroff topology generated by $\mathcal{T}$).*
(iii) *$\mathcal{T}$ is $T_0 \iff \mathcal{AT}$ is $T_0 \iff$ the map $\mathcal{N}$ is injective.*
(iv) *The connected sets with two elements are the same for $\mathcal{T}$ and $\mathcal{AT}$.*
(v) *$\phi_\mathcal{T}(\{x\}) = \phi_{\mathcal{AT}}(\{x\})$ for all $x \in X$.*
(vi) *$\mathcal{AT}$ can have fewer connected sets than $\mathcal{T}$.*

Proof. (i) It is obvious that $x \in \mathcal{N}(x)$. Let us show that if $y \in \mathcal{N}(x)$ then $\mathcal{N}(y) \subseteq \mathcal{N}(x)$: let $t \in \mathcal{N}(y)$: for every $W \in \mathcal{V}(x)$, W open set, we have $y \in W$ (because $y \in \mathcal{N}(x)$); W being open set is a neighborhood of y, hence $t \in W$ so $t \in \mathcal{N}(x)$.

(ii) One has $\mathcal{T} \leq \mathcal{AT}$ because if $V \in \mathcal{V}_\mathcal{T}(x)$ then $V \supseteq \mathcal{N}(x)$, so $V \in \mathcal{V}_{\mathcal{AT}}(x)$. Moreover if $\mathcal{T} \leq \mathcal{T}'$, $\mathcal{T}'$ an Alexandroff topology, for all $V \in \mathcal{V}_\mathcal{T}(x)$ one has $V \supseteq \mathcal{N}_{\mathcal{T}'}(x)$; consequently $\mathcal{N}(x) \supseteq \mathcal{N}_{\mathcal{T}'}(x)$, so $\mathcal{AT} \leq \mathcal{T}'$.

(iii) If $y \in \mathcal{N}(x)$ then for all $V \in \mathcal{V}_\mathcal{T}(x)$, $y \in V$. $\mathcal{T}$ being T_0 there exists $W \in \mathcal{V}(y)$ such that $x \notin W$, so $x \notin \mathcal{N}(y)$.

(iv) If $\{x, y\}$ is a connected set for $\mathcal{AT}$, it is a connected set for $\mathcal{T}$ because $\mathcal{T} \leq \mathcal{AT}$.

If $\{x, y\}$ is a nonconnected set for $\mathcal{AT}$, we have $x \notin \phi_{\mathcal{AT}}(\{y\})$ and $y \notin \phi_{\mathcal{AT}}(\{x\})$, so $y \notin \mathcal{N}(x)$ and $x \notin \mathcal{N}(y)$. Consequently there exists $V \in \mathcal{V}_\mathcal{T}(x)$ such that $y \notin V$ and there exists $W \in \mathcal{V}_\mathcal{T}(y)$ such that $x \notin W$, that leads to $x \notin \phi_\mathcal{T}(\{y\})$ and $y \notin \phi_\mathcal{T}(\{x\})$, and $\{x, y\}$ is a nonconnected set for $\mathcal{T}$.

(v) Easy from the fact that $y \in \phi_\mathcal{T}(\{x\})$ if and only if $x \in \bigcap_{V \in \mathcal{V}_\mathcal{T}(y)} V$.

(vi) For instance if $\mathcal{T}$ is the usual topology on $\mathbb{R}$ then $\mathcal{AT}$ is the discrete topology. □

Theorem 2 gives more precisions about the generation of Alexandroff spaces.

Theorem 2 *Let $(X, \mathcal{T})$ be a Alexandroff space, there exists $\mathcal{T}'$ such that:*

(i) *$\mathcal{T} \leq \mathcal{T}'$ and $\mathcal{T}'$ is an T_0 Alexandroff space.*
(ii) *$\mathcal{T}$ and $\mathcal{T}'$ have the same connected sets.*
(iii) *$\mathcal{T}'$ is a minimal element of the set:*
  $\mathcal{M} = \{\mathcal{S}, \mathcal{S} \text{ is an } T_0 \text{ Alexandroff topology on } X, \mathcal{T} \leq \mathcal{S}, \text{ and } \mathcal{S}, \mathcal{T}$
  have the same connected subsets$\}$.

Proof. (i) Let us define the map $\mathcal{N} : X \longrightarrow \mathcal{P}(X)$, with $\mathcal{N}(x)$ the smallest neighborhood of x. For $V \in \mathcal{P}(X)$ one denotes $\mathcal{N}^{-1}(V) = \{x, \mathcal{N}(x) = V\}$. So $X = \bigcup_{V \in \mathcal{P}(X)} \mathcal{N}^{-1}(V)$ is a partition of X. For every nonempty element of this partition:

– If $\mathcal{N}^{-1}(V) = \{x\}$ one has $\mathcal{N}(x) = V$, and one choose $\mathcal{N}'(x) = V$.
– If $\#(\mathcal{N}^{-1}(V)) \geq 2$, by the well-ordered axiom, V can be well ordered, so one takes the smallest element ω of V and one sets $\mathcal{N}'(\omega) = V$. Now suppose

that $\mathcal{N}'(x)$ is defined for any $x < \alpha$, $(x, \alpha \in V)$, one takes $\mathcal{N}'(\alpha) = V \setminus \{t, t < \alpha\}$.

It is obvious that $x \in \mathcal{N}'(x) \subseteq \mathcal{N}(x)$, for all $x \in X$.

Let us suppose $y \in \mathcal{N}'(x)$, either $\mathcal{N}'(x) = \mathcal{N}(x)$ and $\mathcal{N}'(y) \subseteq \mathcal{N}'(x)$, or $\mathcal{N}'(x) \subsetneq \mathcal{N}(x) = V$, $\mathcal{N}'(x) = V \setminus \{t, t < x\}$, V being well ordered, we have $y \geq x$. Consequently $\mathcal{N}'(y) = V \setminus \{t, t < y\} \subseteq \mathcal{N}'(x) = V \setminus \{t, t < x\}$. That is to say $\mathcal{N}'(y) \subseteq \mathcal{N}'(x)$. One can conclude that $\mathcal{N}'$ generates an Alexandroff topology $\mathcal{T}'$, and $\mathcal{T} \leq \mathcal{T}'$. Now suppose $x \neq y$:

- If $\mathcal{N}(x) \neq \mathcal{N}(y)$ then $x \notin \mathcal{N}(y)$ or $y \notin \mathcal{N}(x)$; "a fortiori" $y \notin \mathcal{N}'(x)$ or $x \notin \mathcal{N}'(y)$.
- If $\mathcal{N}(x) = \mathcal{N}(y) = V$ then $\mathcal{N}'(x) = V \setminus \{t, t < x\}$ and $\mathcal{N}'(y) = V \setminus \{t, t < y\}$. Moreover $x < y$ or $y < x$: Consequently $y \notin \mathcal{N}'(x)$ or $x \notin \mathcal{N}'(y)$.

So $(X, \mathcal{T}')$ is a T_0 Alexandroff space and $\mathcal{T} \leq \mathcal{T}'$.

(ii) Let us show that $\mathcal{T}$ and $\mathcal{T}'$ have the same connected sets. It is obvious that any connected set for $\mathcal{T}'$ is a connected set of $\mathcal{T}$.

If $\{x, y\}$ is connected for $\mathcal{T}$, it is equivalent to say that $y \in \mathcal{N}(x)$ or $x \in \mathcal{N}(y)$ (from Lemma 1). Without losing generality, suppose $y \in \mathcal{N}(x)$. We have two cases:

- If $\mathcal{N}'(x) = \mathcal{N}(x)$ then $\{x, y\}$ is a connected set for $\mathcal{T}'$.
- If $\mathcal{N}'(x) \subsetneq \mathcal{N}(x) = V$ then $\mathcal{N}'(x) = V \setminus \{t, t < x\}$. If $y \in \mathcal{N}'(x)$ then $\{x, y\}$ will be a connected set for $\mathcal{T}'$. If $y \notin \mathcal{N}'(x)$ then $y < x$, consequently $x \in \mathcal{N}'(y)$ and $\{x, y\}$ is a connected set for $\mathcal{T}'$.

Let us now suppose that C is a connected set for $\mathcal{T}$, from the proof of Lemma 1, it is path connected. Let us suppose $x, y \in C$ with $y \in \mathcal{N}(x)$. From above $\{x, y\}$ is a connected set for both topologies, so $\{x, y\}$ is a connected set for $\mathcal{T}'$. From lemma 1 one can conclude that C is a connected set for $\mathcal{T}'$.

(iii) If $\mathcal{S} \in \mathcal{M}$ and $\mathcal{S} \leq \mathcal{T}'$ one has $\mathcal{N}_{\mathcal{T}'}(x) \subseteq \mathcal{N}_{\mathcal{S}}(x) \subseteq \mathcal{N}_{\mathcal{T}}(x)$; and if $y \in \mathcal{N}_{\mathcal{S}}(x)$ necessarily $x \notin \mathcal{N}_{\mathcal{S}}(y)$, a fortiori $x \notin \mathcal{N}_{\mathcal{T}'}(y)$; but $y \in \mathcal{N}_{\mathcal{S}}(x)$ implies $\{x, y\}$ is connected for $\mathcal{S}$, hence $\{x, y\}$ is connected for $\mathcal{T}'$, so since $x \notin \mathcal{N}_{\mathcal{T}'}(y)$, necessarily $y \in \mathcal{N}_{\mathcal{T}'}(x)$: consequently $\mathcal{N}_{\mathcal{S}}(x) \subseteq \mathcal{N}_{\mathcal{T}'}(x)$. □

Proposition 1 *Let $(X, \mathcal{T})$ be a T_0-Alexandroff space, $\mathcal{T}$ is maximal in the set of topologies on X for which the connected sets are the same as for $\mathcal{T}$.*

Proof. Let us suppose that $\mathcal{T} < \mathcal{T}'$, there is an open set U for $\mathcal{T}'$ which is not an open set for $\mathcal{T}$. So there exists $x \in X$ such that $U \in \mathcal{V}_{\mathcal{T}'}(x)$ and $U \notin \mathcal{V}_{\mathcal{T}}(x)$. Consequently $U \not\supseteq \mathcal{N}(x)$ and there exists $y \in \mathcal{N}(x)$ such that $y \notin U$: $\{x, y\}$ is a connected set for $\mathcal{T}$.

$\mathcal{T}$ is a T_0 space, so $x \notin \mathcal{N}(y)$ which means that $y \notin \phi_{\mathcal{T}}(x)$, consequently $y \notin \phi_{\mathcal{T}'}(x)$. Moreover $y \notin U$ implies $x \notin \phi_{\mathcal{T}'}(y)$ therefore $\{x, y\}$ is not a connected set for $\mathcal{T}'$. □

4 Dual Alexandroff Topologies

Let $(X, \mathcal{T})$ be an Alexandroff space. Because any intersection of open sets is an open set, the closed sets of $\mathcal{T}$ are the open sets of a topology on X. We denote this topology by $\tilde{\mathcal{T}}$, and we will call it the *dual topology* of $\mathcal{T}$. If $\mathcal{T} = \tilde{\mathcal{T}}$ one say that $\mathcal{T}$ is *self-dual*.

Theorem 3 provides us with a way of building dual Alexandroff topologies.

Theorem 3 *Let $\sigma : X \longrightarrow \mathcal{P}(X)$ be a map verifying:*

- $\forall\ x \in X,\ x \in \sigma(x)$.
- $y \in \sigma(x)$ *implies* $\sigma(y) \subseteq \sigma(x)$.

There exists two Alexandroff topologies associated with σ:
$\mathcal{T}$ *for which* $\mathcal{N}_{\mathcal{T}}(x) = \sigma(x)$, *and* $\mathcal{T}'$ *for which* $\phi_{\mathcal{T}'}(\{x\}) = \sigma(x)$.

- $\mathcal{T}$ *and* $\mathcal{T}'$ *are dual:* $\mathcal{T}' = \tilde{\mathcal{T}}$.
- $\mathcal{T}$ *is* T_0 *if and only if* $\tilde{\mathcal{T}}$ *is* T_0 *if and only if* σ *is injective.*
- $\mathcal{T}$ *and* $\tilde{\mathcal{T}}$ *have the same connected sets, hence these topologies have the same connected components.*

Proof. Let $\mathcal{T}$ the Alexandroff topology associated to σ. The map ψ, (see Sect. 3.1): $\mathcal{P}(X) \longrightarrow \mathcal{P}(X)$ defined by $\psi(A) = \bigcup_{x \in A} \sigma(x)$ is a closure operator. Hence this closure operator generates a topology $\mathcal{T}'$.

Let us suppose that U an open set for $\mathcal{T}$. For all $x \in U$ $\sigma(x) \subseteq U$, so $U = \bigcup_{x \in U} \sigma(x) = \psi(U)$ and U is a closed set for $\mathcal{T}'$. If F is a closed set for $\mathcal{T}'$ we have $\psi(F) = F = \bigcup_{x \in F} \sigma(x)$. Consequently $x \in F$ implies that $\sigma(x) \subseteq F$. Hence F is a open set for $\mathcal{T}$. So $\mathcal{T}' = \tilde{\mathcal{T}}$.

Suppose that $\mathcal{T}$ is T_0, so $y \in \mathcal{N}(x)$ implies $x \notin \mathcal{N}(y)$, but $\mathcal{N}(x) = \sigma(x)$, so $y \in \sigma(x)$ means that $x \notin \sigma(y)$ and $\tilde{\mathcal{T}}$ is T_0. The converse can be obtained in the same way. $\mathcal{T}$ and $\tilde{\mathcal{T}}$ have the same connected sets because the open sets for $\mathcal{T}$ are the closed sets for $\mathcal{T}'$. $\square$

Proposition 2 characterizes self-dual topologies.

Proposition 2 *If $\mathcal{T} = \tilde{\mathcal{T}}$ then $B = \{\mathcal{N}(x)\}_{x \in X}$ is a partition of X.*

If $(A_i)_{i \in I}$ is a partition of X, $B = \{A_i, i \in I\}$ is a base of a self-dual Alexandroff topology $\mathcal{T} = \tilde{\mathcal{T}}$.

Moreover X is connected if and only if $\mathcal{T} = \mathcal{G}$.

Proof. Let us suppose $\mathcal{T} = \tilde{\mathcal{T}}$, $\mathcal{N}(x)$ is the smallest open set containing x and we have $\mathcal{N}(x) = \phi(\{x\})$. If $t \in \mathcal{N}(x) \cap \mathcal{N}(y)$ then $\mathcal{N}(t) \subseteq \mathcal{N}(x)$; $t \in \phi(\{x\})$ implies $x \in \mathcal{N}(t)$ therefore $\mathcal{N}(x) \subseteq \mathcal{N}(t)$. Likewise $\mathcal{N}(y) \subseteq \mathcal{N}(t)$: B is a partition of X.

Let $B = \{A_i, i \in I\}$ be a partition: obviously B is a base of one topology denoted by $\mathcal{T}$. Moreover if $x \in A_i$, A_i is the smallest open set containing x. So $\mathcal{T}$ is an Alexandroff topology and $\mathcal{N}(x) = A_i$.

Let us now show that $\phi(\{x\}) = \mathcal{N}(x)$.

A_i is a closed set, consequently $\phi(\{x\}) \subseteq A_i = \mathcal{N}(x)$. We have for all $y \in \mathcal{N}(x), \mathcal{N}(y) = \mathcal{N}(x), x \in \mathcal{N}(y)$ and $y \in \phi(\{x\})$. Consequently $\phi(\{x\}) = \mathcal{N}(x)$ and $\mathcal{T} = \tilde{\mathcal{T}}$.

Any $\mathcal{N}(x)$ is both a closed and open set, so X is a connected set if and only if for every $x \in X, \mathcal{N}(x) = X$ if and only if $\mathcal{T} = \mathcal{G}$. □

Examples

The trivial topology is self-dual, it is connected. The discrete topology is also self-dual, it is totally disconnected if $\operatorname{card} X \geq 2$

We give now some results on the topological lattice of dual topologies on the set X.

Proposition 3 *Let $(X, \mathcal{T})$ be an Alexandroff space.*

- $\sup(\mathcal{T}, \tilde{\mathcal{T}})$ *is the self-dual Alexandroff topology associated with the partition $\{A(x)\}_{x \in X}$ with $A(x) = \{y, \mathcal{N}(y) = \mathcal{N}(x)\}$.*
  Moreover $\mathcal{T}$ is T_0, if and only if $\sup(\mathcal{T}, \tilde{\mathcal{T}}) = \mathcal{D}$, ($\mathcal{D}$ being the discrete topology).
- $\inf(\mathcal{T}, \tilde{\mathcal{T}})$ *is the self-dual Alexandroff topology associated with the partition $\{C(x), x \in X\}$ where $C(x)$ is the connected component of x for $\mathcal{T}$; so $\mathcal{T}$ is connected if and only if $\inf(\mathcal{T}, \tilde{\mathcal{T}}) = \mathcal{G}$, ($\mathcal{G}$ being the trivial topology).*

Proof. Let us $\mathcal{S} = \sup(\mathcal{T}, \tilde{\mathcal{T}})$; for any $x \in X, \mathcal{N}(x) \cap \phi(\{x\}) = A(x)$ is an open neighborhood of x for $\mathcal{S}$. Moreover $\mathcal{N}(x) \cap \phi(\{x\}) = A(x)$ implies that $A(x) = \{y, \mathcal{N}(y) = \mathcal{N}(x)\}$, (because any neighborhood of x is a neighborhood of y and conversely). So $\{A(x)\}_{x \in X}$ is a partition of X and from the last proposition we have a self-dual Alexandroff topology $\mathcal{U}$ associated with it. We have $\mathcal{T} \leq \mathcal{U}$ because $A(x) \subseteq \mathcal{N}(x)$ for all $x \in X$. $\mathcal{N}(x)$ is a neighborhood of x for $\mathcal{U}$, therefore $\mathcal{T} \leq \mathcal{U}$, hence $\tilde{\mathcal{T}} \leq \tilde{\mathcal{U}} = \mathcal{U}$; which means that $\mathcal{S} \leq \mathcal{U}$. $A(x)$ is an open set of $\mathcal{S}$ and since $\mathcal{U} \leq \mathcal{S}$, so $\mathcal{S} = \mathcal{U}$.

$\mathcal{T}$ is T_0 if and only if $\mathcal{N}$ is injective if and only if $A(x) = \{x\}$ if and only if $\mathcal{S} = \mathcal{D}$.

Let $\mathcal{C}$ be the self-dual topology associated with the partition $\{C(x), x \in X\}$; each $C(x)$ is both open and closed for $\mathcal{T}$ (from Theorem 3 or Lemma 1), so $C(x)$ is open set for $\tilde{\mathcal{T}}$, hence $\mathcal{C} \leq \mathcal{T}$ and $\mathcal{C} \leq \tilde{\mathcal{T}}$.

Assume now $\mathcal{E} \leq \mathcal{T}$ and $\mathcal{E} \leq \tilde{\mathcal{T}}$; every open set U for $\mathcal{E}$ is an open and closed set for $\mathcal{T}$. So for all $x \in U$ we have $C(x) \subseteq U$, (because $C(x)$ is a subset of any open–closed set containing x). As a consequence $U = \bigcup_{x \in U} C(x)$ is an open set for $\mathcal{C}$. So $\mathcal{E} \leq \mathcal{C}$. □

In Sect. 5 we study compatible topologies on a graph $G = (V; E)$.

5 Compatible Topologies on Graph

Recall, [15], that a topology $\mathcal{T}$ on the set of vertices V of a graph $G = (V; E)$ is *compatible* if the connected subspaces of $(V, \mathcal{T})$ are the same as the connected induced subgraphs of $G = (V; E)$.

Some preliminary results are given by:

Lemma 2 *Let T be a compatible topology on G, we have the following properties:*
(a) $\phi(\{x\}) \cup \bigcap_{V \in \mathcal{V}(x)} V = \{x\} \cup \Gamma(x)$.
(b) T is T_0 if and only if $\phi(\{x\}) \cap \bigcap_{V \in \mathcal{V}(x)} V = \{x\}$.
(c) If T is an Alexandroff topology then $\phi(\{x\}) \cup \mathcal{N}(x) = \{x\} \cup \Gamma(x)$ and $\{x\} \cup \Gamma(x)$ is a connected neighborhood of x, and T is locally connected.

Proof. (a) Let us suppose that $y \in \phi(\{x\})$, $x \neq y$, so $\{x, y\}$ is a connected set for T. Hence $\{x, y\} \in E$ and $y \in \{x\} \cup \Gamma(x)$.

Let us suppose that $y \in \bigcap_{V \in \mathcal{V}(x)} V$, one has for all $V \in \mathcal{V}(x)$, $y \in V$, so $x \in \phi(\{y\})$ and $y \in \{x\} \cup \Gamma(x)$.

Now suppose that $y \in \Gamma(x)$, $\{x, y\} \in E$ and $\{x, y\}$ is a connected set for T, from Lemma 1: $y \in \phi(\{x\})$, or $x \in \phi(\{y\})$: consequently for all $V \in \mathcal{V}(x)$, $y \in V$ and $y \in \bigcap_{V \in \mathcal{V}(x)} V$.

(b) Suppose that T is T_0. Let us suppose $y \in \phi(\{x\}) \cap \bigcap_{V \in \mathcal{V}(x)} V$, so $x \in \phi(\{y\})$ and $y \in \phi(\{x\})$, consequently $x = y$ and $\phi(\{x\}) \cap \bigcap_{V \in \mathcal{V}(x)} V = \{x\}$.

(c) Indeed $\mathcal{N}(x) = \bigcap_{V \in \mathcal{V}(x)} V$. □

An important result can be inferred from Theorem 4. The assertion (b) is to be found in [15].

Theorem 4 *Let T be a compatible topology on G, we have the following properties:*
(a) If T is an Alexandroff topology then $\{x\} \cup \Gamma(x)$ is a neighborhood of x.
(b) If G is locally finite then $\{x\} \cup \Gamma(x)$ is a neighborhood of x.

Proof. For $x \in X$ the connected component of $X \setminus \Gamma(x)$ are $\{x\}$ and $(C_i)_{i \in I}$. $\{x\} \cup C_i$ is not a connected set (by construction), so $x \notin \phi(C_i)$. From this, there exists $V_i \in \mathcal{V}(x)$ such that $V_i \cap C_i = \emptyset$. Let us set $W := \bigcap_{i \in I} V_i$, one has for all $i \in I$, $W \cap C_i = \emptyset$ so $W \subseteq \{x\} \cap \Gamma(x)$.

(a) If T is an Alexandroff topology then $W \in \mathcal{V}(x)$ and $\{x\} \cap \Gamma(x)$ is a neighborhood of x.
(b) If G is locally finite then $\#(I) \leq \#(\Gamma(\Gamma(x))) < \aleph_0$, so I is a finite set. Consequently $W \in \mathcal{V}(x)$ and $\{x\} \cup \Gamma(x)$ is a neighborhood of x. □

Corollary 1 *If G is locally finite then T is an Alexandroff topology.*

Proof. Because for all x, $(V \cap (\{x\} \cup \Gamma(x)))_{V \in \mathcal{V}(x)}$ is a fundamental neighborhood system of x with a finite number of elements, (the number of subsets of $\{x\} \cup \Gamma(x)$ at most), consequently the intersection of these elements is the smallest neighborhood of x. □

Problem 1. Which are the compatible topologies such that $\{x\} \cup \Gamma(x)$ is neighborhood of x?

The following result can be found in [14, 15]. The proof given here is shorter.

Theorem 5 *Let $G = (V, E)$ be a graph, the following properties are equivalent:*
(i) G has a compatible topology $\mathcal{T}$.
(ii) G is a comparability graph.

Proof. Under hypothesis (i), and from Theorem 1 there exists a compatible T_0-Alexandroff topology, $\mathcal{A}_0$, such that $\mathcal{T} \leq \mathcal{A}_0$. Let $\leq$ the binary relation defined by $x \leq y$ if and only if $y \in \phi(\{x\})$. This relation is a partial order relation, so

- If $\{x, y\} \in E$, then $\{x, y\}$ is a connected set for $\mathcal{A}_0$, (by (iv) from Theorem 1) consequently $y \in \phi(\{x\})$ or $x \in \phi(\{y\})$, and $x \leq y$ or $y \leq x$.
- If $x \leq y$, $y \in \phi(\{x\})$, so $\{x, y\}$ is a connected set for $\mathcal{T}$, so for $\mathcal{A}_0$, and $\{x, y\} \in E$.

Under hypothesis (ii), let $\leq$ be a preorder on V verifying: $[x \leq y$ or $y \leq x]$ if and only if $\{x, y\} \in E$. To this preorder one can associate an Alexandroff topology $\mathcal{A}$ defined by: $\phi(\{x\}) = \{y \in V, x \leq y\}$. So $\{x, y\}$ is a connected set for $\mathcal{A}$ if and only if $y \in \phi(\{x\})$ or $x \in \phi(\{y\})$, if and only if $y \leq x$ or $x \leq y$. □

Proposition 4 links compatible topologies and generated Alexandroff topologies.

Proposition 4 *Let $\mathcal{T}$ be a compatible topology, we denote by $\mathcal{A}$ the Alexandroff topology generated by $\mathcal{T}$ and the T_0 Alexandroff topology by $\mathcal{A}_0$. We have the following properties:*

- *$\mathcal{A}, \tilde{\mathcal{A}}$ and $\mathcal{A}_0, \tilde{\mathcal{A}}_0$ are compatible.*
- *$\sup\{\mathcal{A}_0, \tilde{\mathcal{A}}_0\} = \mathcal{D}$ (not compatible if $\#(E) > 1$).*
- *$\inf\{\mathcal{A}, \tilde{\mathcal{A}}\} = \inf\{\mathcal{A}_0, \tilde{\mathcal{A}}_0\} = \mathcal{G}$. (not compatible if $\#(E) > 2$).*

Proof. Obvious from Theorems 1–3. □

Proposition 5 *If $\mathcal{T}$ is a T_0 compatible Alexandroff topology, it is maximal in the compatible topologies.*

Proof. Obvious from Proposition 1. □

Proposition 6 *Let $G = (V, W; A)$ be a locally finite connected bipartite graph, one has:*

(a) If $\#(A) = 1$ there are three compatible topologies on V: $\mathcal{G}$, $\mathcal{A} = \{\emptyset, V, V \cup W\}$ and $\tilde{\mathcal{T}} = \{\emptyset, W, V \cup W\}$.
(b) If $\#(A) \geq 2$ there are two compatible topologies:
  (1) the Alexandroff topology $\mathcal{A}$ associated with $\mathcal{N}_\mathcal{A}(x) = \{x\}$ for $x \in V$ and $\mathcal{N}_\mathcal{A}(x) = \{x\} \cup \Gamma(x)$ for $x \in W$.
  (2) the dual Alexandrov topology $\tilde{\mathcal{A}}$ with $\mathcal{N}_{\tilde{\mathcal{A}}}(x) = \{x\}$ for $x \in W$ and $\mathcal{N}_{\tilde{\mathcal{A}}}(x) = \{x\} \cup \Gamma(x)$ for $x \in V$.

These two topologies are T_0.

Proof. (a) Obvious.
(b) By Corollary 1 any topology $\mathcal{U}$ on the set of vertices of G is an Alexandroff topology. From Theorem 4, $\{x\} \cup \Gamma(x)$ is a neighborhood of x. Hence $\{x\} \cup \Gamma(x)$

contains an open set containing x; the graph being a bipartite graph and $\mathcal{U}$ being a compatible topology this open set is either $\{x\} \cup \Gamma(x)$ or $\{x\}$. By connectivity, if $x \in V$ and $\mathcal{N}(x) = \{x\}$ then $\mathcal{N}(y) = \{y\}$ for all $y \in V$. This topologies are T_0. Let us $x \neq y$ and $\{z, y\} \in E$: if $x \in \phi(y)$ then $y \notin \phi(x)$, otherwise x, y, z should be an odd cycle. □

To illustrate this, we give some examples.

6 Examples

Let $G = (V; E)$ be the bipartite graph defined by: $V = \mathbb{Q} \cup \{\infty\}$ and $\{q, y\} \in E$ if and only if $q \in \mathbb{Q}$ and $y = \infty$ ($\mathbb{Q}$ being the set of rational numbers).

6.1 Construction of Alexandroff Topologies

Let $\mathcal{A}_\infty$ be the Alexandroff topology defined by: $\mathcal{N}(\infty) = \{\infty\}$ and $\mathcal{N}(q) = \{q, \infty\}$. So $\phi(\{\infty\}) = V$ (because for all $q \in \mathbb{Q}$, $V \in \mathcal{V}(q)$ implies $\infty \in V$, it is equivalent to $\mathbb{Q} \subset \phi(\{\infty\})$); $\phi(\{q\}) = \{q\}$. The open set are $\emptyset$ and the subsets $A \subseteq V$ such that $\infty \in A$.

It goes without saying that $\mathcal{A}_\infty$ is a T_0 topology.

V is a connected set for $\mathcal{A}_\infty$ because $\{\infty\}$ is connected and $\phi(\{\infty\}) = V$.

6.2 Construction of $\tilde{\mathcal{A}}_\infty$

$\phi(\{q\}) = \{q\}$ and $\phi(\{\infty\}) = V$, so $\tilde{\mathcal{N}}(q) = \{q\}$ and $\tilde{\mathcal{N}}(\infty) = V$. Consequently the open sets of $\tilde{\mathcal{A}}_\infty$ are: $\emptyset$, V and the subsets $A \subseteq \mathbb{Q}$.

So we have: $\sup\{\mathcal{A}_\infty, \tilde{\mathcal{A}}_\infty\} = \mathcal{D}$, because $\mathcal{A}_\infty$ is T_0. $\inf\{\mathcal{A}_\infty, \tilde{\mathcal{A}}_\infty\} = \mathcal{G}$, because V is a connected set for $\mathcal{A}_\infty$.

6.3 Connected Sets of $\mathcal{A}_\infty$ and $\tilde{\mathcal{A}}_\infty$

By hypothesis $\mathcal{N}(q) = \{q, \infty\}$ is a connected set. Moreover if $\{x, y\}$ is a connected set then $x \in \mathcal{N}(y)$ or $y \in \mathcal{N}(x)$. Consequently, if $x \neq y$ there exists just one possibility $\{x, y\} = \{x, \infty\}$. Any connected set A with $\#(A) > 1$ contains ∞. Indeed, let A be a connected set, A is path connected; for $x, y \in A$ so there exists a path from x to y, this one contains ∞.

6.4 Construction of Other Topologies Having the same Connected Sets with two Elements

Let $\mathcal{T}$ be a compatible topology, (not necessarily an Alexandroff topology) on V. For this topology the subspace $\mathbb{Q}$ is totally disconnected: indeed, if $A \subseteq \mathbb{Q}$ is a connected set for the induced topology on $\mathbb{Q}$, A is a connected set for the topology $\mathcal{T}$ on V. But $\infty \notin A$, so the cardinality of A is less or equal to one.

Conversely if $\mathcal{T}$ is a totally disconnected topology on $\mathbb{Q}$, one can associate two topologies with $\mathcal{T}$:

For the first one topology $\mathcal{T}_\infty$, the set of open sets from this topology is: $\{\emptyset\} \cup \{U \cup \{\infty\},\ U$ open set of $\mathcal{T}\}$.

For the second $\mathcal{T}_\infty^*$, the set of open sets from this topology is: $\{V\} \cup \{U,\ U$ openset of $\mathcal{T}\}$. (These topologies are not necessarily Alexandroff topologies).

We have a lot of choices for $\mathcal{T}$, for example one can choose the discrete topology $\mathcal{D}$ (which gives $\mathcal{T}_\infty = \mathcal{A}_\infty$ from Sect. 6.1), or the topology $\mathcal{T}_+$ whose an open base is given by $[x,y[,\ x,y \in \mathbb{Q}$, or the topology $\mathcal{T}_-$ whose an open base is given by $]x,y]$, $x,y \in \mathbb{Q}$, or the usual topology $\mathcal{T}_0$ on $\mathbb{Q}$ (associated with the usual distance on $\mathbb{Q}$), or the p-adic topology (associated with the p-adic distance on $\mathbb{Q}$), etc.

7 Applications to Digital Topology

The main focus of digital topology and digital geometry is to determine geometrical and topological properties between the discrete nature of a computational objects and their theoretical representation in terms of continuous space $\mathbb{R}^n$. To use the geometrical and topological notions in digital topology and digital geometry it is necessary to define an analog to the continuous space $\mathbb{R}^n$, $(n > 1)$.

7.1 Digital Spaces

Usually there is two ways to define a digital space: Let $x = (x_1, x_2, x_3, \ldots, x_p)$ be a point of $\mathbb{Z}^p$. The $(3^p - 1)$-neighbors of x are all points $y = (y_1, y_2, y_3, \ldots, y_p) \in \mathbb{Z}^p$ such that:

$$\max |x_i - y_i| = 1$$

The discrete space with a dimension equal to n, $(n \geq 2)$ defined thanks to the equation below will be denote by (d_∞, n)-space.

In the case where $p = 2$ we obtain the "8-connected" digital plane, see Fig. 1.

The $2p$-neighbors of x are all points $y = (y_1, y_2, y_3, \ldots, y_p) \in \mathbb{Z}^p$ such that:

$$\sum_{i=1}^{p} |x_i - y_i| = 1$$

The discrete space with a dimension equal to n, $(n \geq 2)$ defined thanks to the equation below will be denote by (d_1, n)-space.

In the case where $p = 2$ we obtain the "4-connected" digital plane, see Fig. 2.

It is well known that an induced graph of a comparability graph is a comparability graph. So if a graph Γ contains an induced subgraph which is not a comparability graph then Γ is not a comparability graph. In [5] we showed that:

Proposition 7 *Let $\Gamma = (V; E)$ be a graph with a compatible topology on V. For every induced subgraph H the topology restricted to the vertices of H is a compatible topology on H.*

Proof. From remark above and Theorem 5. □

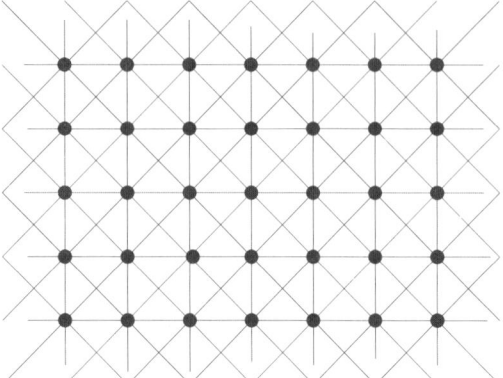

Fig. 1. A piece of 8-connected discrete plane

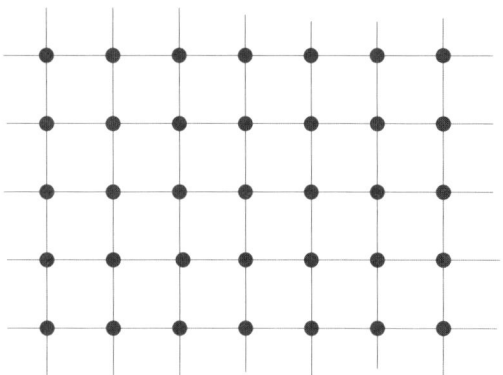

Fig. 2. A piece of 4-connected discrete plane

In Fig. 3 we see that the 8-connected digital plane is not a comparability graph because we are able to display an induced subgraph which is not a comparability graph. From Theorem 5 we find a generalization of the well known CHASSERY' theorem [5, 7, 17, 18]:

Theorem 6 *The (d_∞, n)-space has no compatible topology.*

The next theorem can be found in [5, 17]

Theorem 7 *The (d_1, n)-discrete space has exactly two compatible topologies.*

Proof. It is easy to see that the (d_1, n)-discrete space is a bipartite graph, (is the Cartesian product of n chains [5]), hence from Proposition 6 the result follow. □

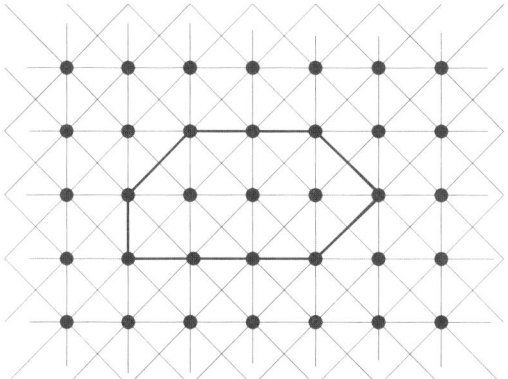

Fig. 3. Circle with an odd length

Fig. 4. An example of a part of an abstract complex. The points have a dimension equal to 0, the lines are cells with a dimension equal to 1 and the square are cells with a dimension equal to 2

7.2 Cell Complexes

An *abstract cell complex*, is a triplet $C = (E, B, Dim)$ where E is a set of abstract elements called *cells*, $B \subseteq E \times E$ is an antisymmetric, irreflexive and transitive binary relation and with a dimension function $Dim E \longrightarrow I \subseteq \mathbb{N}$ such that $Dim(e) < Dim(e')$ for all pairs $(e, e') \in B$. The plane $\mathbb{R}^2$ can be see as abstract cell complex, see Fig. 4.

From the abstract cell complex which stands for the plane $\mathbb{R}^n$, $n \geq 1$, we can construct a graph in the following way:

- The set of vertices V is the set of cells.
- Two vertices, $x, y \in V$ form an edge if and only if either $(x, y) \in B$ or $(y, x) \in B$.

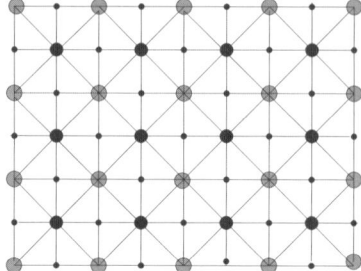

Fig. 5. Induced subgraph associated with a part of the cell complex of the plane

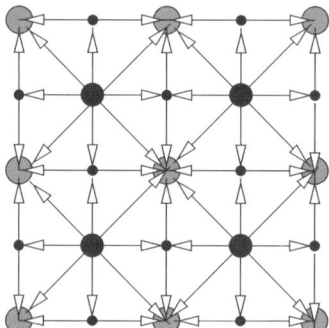

Fig. 6. Orientation of graph Fig. 5

By putting a black point on the cells with a dimension equal 2, a small black point on the cells with a dimension equal 1, and a grey point on the cells with a dimension equal 0, we obtain the graph of Fig. 5.

This graph is denoted by $G(C)$; the associate graph with a cell complex. The binary relation B being transitive, graph Fig. 5 is a comparability graph. An orientation is given Fig. 6. We will denote by $H(x)$ the horizontal neighbors of a small black point x, and by $V(x)$ the vertical neighbors of a small black point x.

Lemma 3 *Let $\mathcal{T}$ be a T_0-Alexandroff compatible topology with $G(C)$, C being the cell complex associated with the plane. For all x black point or grey point we have: either $\mathcal{N}_\mathcal{T}(x) = \{x\} \cup \Gamma(x)$ or $\mathcal{N}_\mathcal{T}(x) = \{x\}$.*

Sketch of Proof. If $\mathcal{N}_\mathcal{T}(x) \neq \{x\}$ one can show thank to T_0 hypothesis than 6 of the 8 neighbors of x are in $\mathcal{N}_\mathcal{T}(x)$ and one can conclude that the two others neighbors are equally in $\mathcal{N}_\mathcal{T}(x)$.

Let $\mathcal{T}$ be a T_0-Alexandroff compatible topology and let x be a grey point, we have two cases.

(a) We suppose that $\mathcal{N}_\mathcal{T}(x) = \{x\}$. One can show that for all grey point y we have $\mathcal{N}_\mathcal{T}(y) = \{y\}$, and for all y small black points we have $\mathcal{N}_\mathcal{T}(x) = H(x)$. Consequently for all black points y we have $\mathcal{N}_\mathcal{T}(x) = \{x\} \cup \Gamma(x)$.

(b) If $\mathcal{N}_{\mathcal{T}}(x) = \{x\} \cup \Gamma(x)$ in the same way we show a similar result.

So from this remark and by applying Theorems 2–5 we have:

Theorem 8 *There are exactly two T_0 Alexandroff compatible topologies, $\mathcal{T}_1$ and $\mathcal{T}_2$ on the graph associated with the cell complex of the plane.*

These two topologies are described in the following way:

- Let x be a black point, $\mathcal{N}_{\mathcal{T}_1}(x) = \{x\} \cup \Gamma(x)$ and $\mathcal{N}_{\mathcal{T}_2}(x) = \{x\}$.
- Let x be a small black points, $\mathcal{N}_{\mathcal{T}_1}(x) = H(x)$ $\mathcal{N}_{\mathcal{T}_2}(x) = V(x)$.
- Let x be a grey points $\mathcal{N}_{\mathcal{T}_1}(x) = \{x\}$ and $\mathcal{N}_{\mathcal{T}_2}(x) = \{x\} \cup \Gamma(x)$.

We are now showing that this two topologies are homeomorphic, (but not equal).

Theorem 9 *The two T_0 Alexandroff compatible topologies, $\mathcal{T}_1$ and $\mathcal{T}_2$ are homeomorphic.*

Proof. Settle:

- The set of grey points $W = \{(2n, 2m+1), n, m \in \mathbb{Z}\}$.
- The set of small black points $SB = \{(n, m), m + n = 2k, k \in \mathbb{N}^*, n, m \in \mathbb{Z}\}$.
- The set of black points $B = \{(2n+1, 2m), n, m \in \mathbb{Z}\}$.

Let us define the following mapping:

$$\phi : (\mathbb{Z}^2; \mathcal{T}_1) \longrightarrow (\mathbb{Z}^2; \mathcal{T}_2)$$

$$(x, y) \longmapsto (y, x)$$

It is a symmetry given Fig. 7. We have to show that

$$\phi(\mathcal{N}_{\mathcal{T}_1}((x, y))) = \mathcal{N}_{\mathcal{T}_2}((y, x)).$$

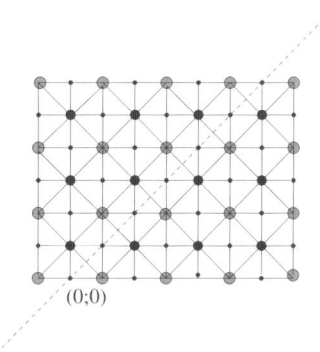

Fig. 7. Symmetry of the plane

(1) Let (x,y) be a point of SB. Because $x+y = 2k = y+x$ it is easy to see that if $(x,y) \in SB$ then $(y,x) \in SB$. Consequently

$$\phi(\mathcal{N}_{T_1}((x,y))) = \phi(\{(x+\epsilon,y), \epsilon = 0, \pm 1\}) = \{(y,x+\epsilon = 0, \pm 1)\} = \mathcal{N}_{T_2}((y,x)).$$

(2) Let $(x,y) = (2n; 2m+1) \in W$, hence $(y,x) = (2n+1; 2m) \in B$, so:

$$\phi(\mathcal{N}_{T_1}((x,y))) = \phi(\{(x,y)\}) = \{(y,x)\} = \mathcal{N}_{T_2}((y,x)).$$

Moreover the identity is not an homeomorphism because $\mathcal{N}_{T_1}((x,y)) \neq \mathcal{N}_{T_2}((y,x))$.
□

References

1. P. Alexandroff, "Diskrete Räume," Math. Sbornik. (N. S.), **2**, 1937, 501–518
2. M. J. Bilbao, Graphs Topologies and Simple Games. Questiio, 24, 2000, 317–331
3. K. H. Baik, L. L. Miller, Topological Approach for Testing Equivalence in Heterogeneous Relational Databases. Computer Journal **33(1)**, 1990, 2–10
4. M. Brissaud, Structures Topologiques Des Espaces Preferencies. Comptes Rendus hebdomadaires des Seances de l'Academie des Sciences, **280**, 1975, Serie A:96–964
5. A. Bretto, "Comparability Graphs and Digital Topology." Computer Vision Graphic and Image Processing (Image Understanding), **82**, 2001, 33–41
6. B. A. Bretto, A. Faisant, and T. Vallée, Compatible topologies on graphs : An application to graph isomorphism problem complexity. Theoretical Computer Science, **362**, 2006, 255–272.
7. J. M. Chassery, "Connectivity and Consecutivity in Digital Pictures." Computer Vision Graphic and Image Processing, **9**, 1979, 294–300
8. A. V. Evako, R. Kopperman, and Y. V. Mukhin, "Dimensional Properties of Graphs and Digital Spaces." Journal of Mathematical Imaging and Vision, **6**, 1996, 109–111
9. E. D. Khalimsky, R. Kopperman, and P. R. Meyer, Computer Graphics and Connected Topologies on finite ordered sets. Topology and its Applications, **36**, 1990 1–17
10. V. A. Kovalevsky, Finite Topology as Applied to Image Processing. Computer Vision Graphics and Image Processing, **46**, 1989, 141–161
11. T. Y. Kong, R. Kopperman, and P. R. Meyer, A Topological Approach to Digital Topology. American Mathematical Monthly, **98**, 1991, 901–917
12. A. Bretto, J. Azema, H. Cherifi, and B. Laget. Combinatoric and image processing. Computer Vision, Graphic and Image Processing (Graphical Model and Image Processing), **59(5)**, 1997, 128–132
13. R. Klette, Topologies on the Planar Orthogonal Grid 2002. Proceedings of the 16th International Conference on Pattern Recognition, **2**, 2002, 354–357
14. V. Neumann-Lara and R. G. Wilson, "Compatible Connectedness in Graphs and Topological Spaces." Order 12, 1995, 77–90
15. P. Prea, "Graphs and Topologies on Discrete Sets." Discrete Mathematics, **103**, 1992, 189–197
16. S. Lipschutz, Theory and Problems of general Topology. McGraw-Hill, New York 1965
17. U. Eckhardt and L. Lateki, Topologies for Digital Spaces $\mathbb{Z}^2$ and $\mathbb{Z}^3$, Computer Vision and Image understanding, **95**, 2003, 261–262
18. L. Lateki, "Topological Connectedness and 8-connectness in Digital Pictures." CVGIP Image Understanding, **57(2)**, 1993, 261–262

Part II

Graph Similarity, Matching, and Learning for High Level Computer Vision and Pattern Recognition

How and Why Pattern Recognition and Computer Vision Applications Use Graphs

Donatello Conte, Pasquale Foggia, Carlo Sansone and Mario Vento

Summary. In this chapter, we present a review of graph-based methodologies for pattern recognition and computer vision, by considering three different points of view: the algorithms, the applications, and the performance evaluation. Preliminarily, a survey of graph-matching approaches, including a synthetic description of a plenty of algorithms and their inspiring rationale, is discussed. Afterward, a detailed taxonomy of pattern recognition applications using graphs is organized, motivating, for each of them, why graph-based approaches can be profitably used and how a specific technique can be exploited. Finally, a section reporting the state-of-the-art of benchmarking activities is present, together with a discussion of the performance issues of well-known graph-based algorithms.

1 Introduction

Starting from the late 1970s, graph-based techniques have been proposed as a powerful tool for pattern representation and classification. After the initial enthusiasm, graphs have been practically left unused for a long period of time and only recently are obtaining a growing attention from the scientific community of pattern recognition (PR) and computer vision.

Due to their expressive power, graphs are conquering a primary role as a smart data structure for representing complex visual patterns, especially in structural methods, whose rationale is a vision of the objects as made of parts suitably connected to each other. Under this assumption, nodes of the graphs, enriched with properly defined attributes, can be thought as descriptors of the component parts of the objects, while the edges of the graphs represent the relationships between the parts. The reason why the literature on graph-based approaches is so wide depends on the fact that description schemes generally lead to a variety of graph representations differing from each other for the graph topology, the nature of the nodes and edges (deterministic or stochastic), the type of the attributes (numeric values, symbols, probabilities), etc. Of course, for each representation scheme, some methods for comparing the obtained graph representations must be defined, so obtaining also a variety of algorithms able to calculate exact or somewhat inexact correspondence between graphs, or a sort of distance between them.

The huge material available on graphs often discourages a researcher who intends to use these smart and promising approaches; in other cases, the complexity of this material is the main cause of unsuccessful attempts. The under-evaluation of the complexity of the literature may suggest to the researcher a quick and superficial choice of an algorithm, because of the wrong convincement that almost all the algorithms differ from each other for minor performance issues.

In the recent past some surveys of graph-based techniques have been published (see for example [1]). Since they are mainly focused to presenting almost all the existing algorithms (even if sometimes they are organized in a taxonomy), they often result really useful only to experienced researchers of the field. The idea of the present paper is the attempt of filling the "knowledge gap" existing between the graph-based techniques and their use in PR applications. This is done by considering both the above-mentioned issues in successive sections of the paper, carefully bridging techniques and applications, retracing the history of almost all the PR applications using graphs in the last decades.

In Sect. 2, a survey of graph-matching approaches, including a synthetic description of a plenty of algorithms with their inspiring rationale, is discussed. The survey groups similar approaches and algorithms in a few categories, each described in terms of the underlying technique, purposely neglecting inessential algorithmic details. This is done for both the classes of exact and inexact graph-matching methods. In a following subsection a commented bibliography is given, presenting for each group of algorithms the most important papers, with the aim of explaining the main differences among them, and reconstructing their publication sequence.

In Sect. 3 a detailed taxonomy of PR applications using graphs is organized; this section highlights, for each application, why graph-based approaches can be profitably used and how a specific technique can be exploited. The taxonomy is organized so as to render more understandable by a practitioner of the field the relationship among the techniques and the applications, so as to help him to choose the more suitable structural method using graphs. To complete the review, a final section reports the state of the art of benchmarking activities, together with a discussion of the performance issues of well-known graph-based algorithms. It is mainly organized in order to give general criteria for selecting the most effective algorithm for dealing with the problem at hand, with respect to some common classes of graphs.

2 Graph-Matching Taxonomy

In this section we will present a review of the algorithms that have been proposed and used in the PR field for the graph-matching problem. We have divided the matching methods into two broad categories: the first contains exact matching methods that require a strict correspondence between the two object being matched or at least between subparts of them. The second category defines inexact matching methods, where a matching can occur even if the two graphs being compared are structurally different to some extent.

2.1 An Introduction to Exact Graph-Matching Problems

Exact graph matching is characterized by the fact that the mapping between the nodes of the two graphs must be *edge-preserving* in the sense that if two nodes in the first graph are linked by an edge, they are mapped to two nodes in the second graph that are linked by an edge as well.

Conceptually, the simplest form of graph matching is *graph isomorphism* (see Fig. 1), where an exact structural correspondence is sought: there must be a bijective mapping between the nodes of the two graphs that preserves the edges of both graphs.

A slightly weaker form of matching is *subgraph isomorphism* (see Fig. 2), that requires the existence of an isomorphism between one of the graphs and a subgraph of the other. In other words, one of the graphs may have extra nodes and extra edges linking these new nodes to the rest. Subgraph isomorphism is often confused with *monomorphism*, which is a little more relaxed matching: in monomorphism extra edges in the larger graph are allowed also between nodes that do have a correspondent in the smaller graph. In subgraph isomorphism, instead, one of the ends of the extra edges must be an extra node. In other words, while isomorphism and subgraph isomorphism impose a two-way constraint on the edges of the graphs, monomorphism imposes a one-way constraint.

A more robust form of graph matching is based on the computation of the maximum common subgraph (MCS - see Fig. 3), that is the largest subgraph of one of the two graphs that is isomorphic to a subgraph of the other. This kind of matching allows both graphs to have extra nodes and edges, but is also significantly more expensive from a computational viewpoint.

	Definition	Example
Graph	A *graph* is a 4-tuple $g = (V, E, \alpha, \beta)$ - V is the finite set of vertices - $E \subseteq V \times V$ is the set of edges - $\alpha : V \to L$ is a function assigning labels to the vertices - $\beta : E \to L$ is a function assigning labels to the edges	$V \equiv \{V_1, V_2, V_3\}$ $E \equiv \{(V_1,V_2),(V_2,V_3)\}$ $\alpha: V_1 \to 1$ $V_2 \to 3$ $V_3 \to 7$ $\beta: (V_1,V_2) \to b$ $(V_2,V_3) \to a$ 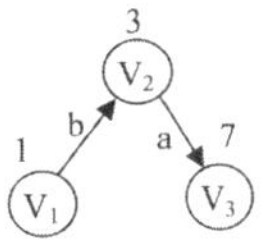
Graph Isomorphism	Let g and g' be graphs. A *graph isomorphism* between g and g' is a bijective mapping $f: V \to V'$ such that - $\alpha(v) = \alpha'(f(v)) \ \forall v \in V$ - for any edge $e=(u,v) \in E$ there is an edge $e'=(f(u), f(v)) \in E'$ such that $\beta(e)=\beta'(e')$, and for any edge $e'=(u',v')$ there is an edge $e=(f^{-1}(u), f^{-1}(v)) \in E$ such that $\beta(e) = \beta'(e')$.	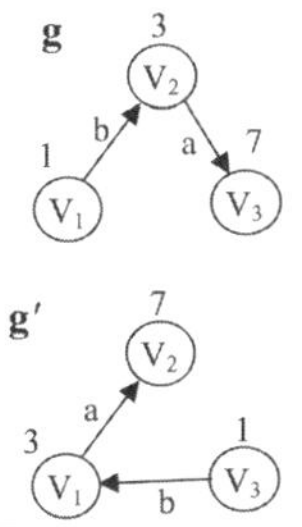 $f:$ (V_1,V_3) (V_2,V_1) (V_2,V_3)

Fig. 1. Definitions of graph and graph isomorphism

	Definition	Example
Sub-Graph Isomorphism	Let g and g' be graphs. If $f_{sub}: V \to V''$ is a graph isomorphism between graphs g and g'', and g'' is a subgraph of graph g', i.e. $g'' \subseteq g'$, then f_{sub} is called a *subgraph isomorphism* between g and g'.	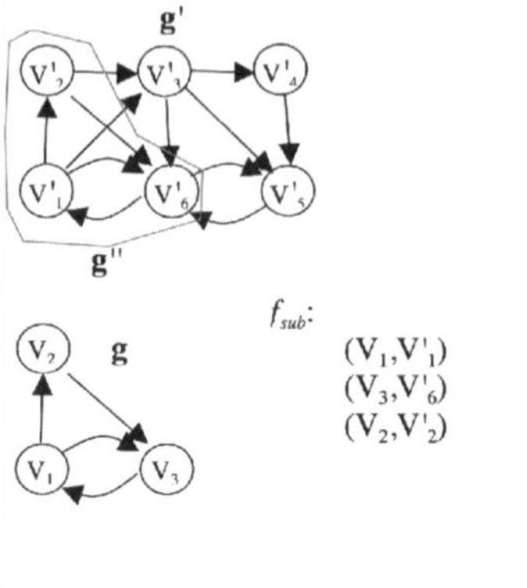 f_{sub}: (V_1, V'_1) (V_3, V'_6) (V_2, V'_2)
Monomorphism	Let g and g' be graphs. A monomorphism is an injective mapping $f_{mon}: V \to V'$ such that: - $\alpha(v) = \alpha'(f_{mon}(v))$ $\forall v \in V$ - for any edge $e=(u,v) \in E$ there is an edge $e' = (f_{mon}(u), f_{mon}(v)) \in E'$ such that $\beta(e) = \beta'(e')$.	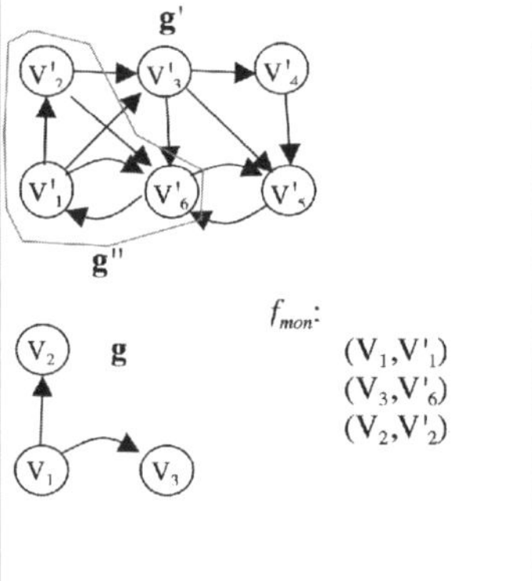 f_{mon}: (V_1, V'_1) (V_3, V'_6) (V_2, V'_2)

Fig. 2. Definitions of subgraph isomorphism and monomorphism

	Definition	Example
Maximum Common Subgraph	Let g and g' be graphs. A *Common Subgraph* of g and g', $CS(g, g')$, is a graph g'' such that there are subgraph isomorphisms between g'' and g and between g'' and g'. We call g'' a *Maximum Common Subgraph* of g and g', $MCS(g,g')$, if there is no other common subgraph of g and g' that has more nodes than g''.	$MCS_1 \equiv \{(V_1, V'_1), (V_2, V'_2)\}$ $MCS_2 \equiv \{(V_2, V'_2), (V_3, V'_3)\}$

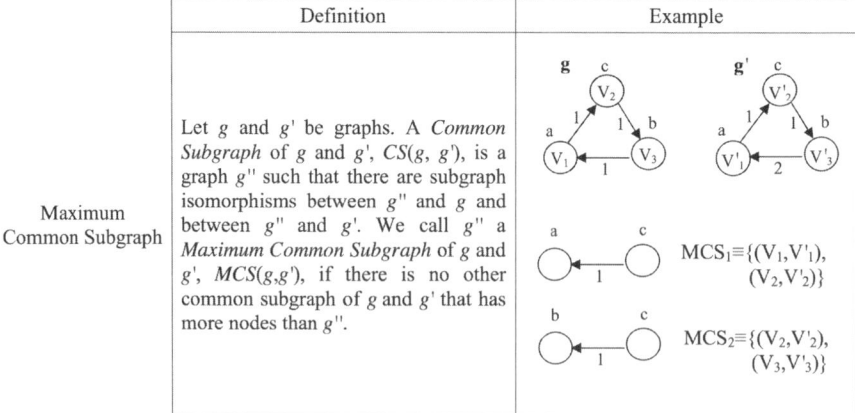

Fig. 3. Definition of maximum common subgraph

It has been demonstrated that the MCS problem is equivalent to the determination of the maximum *clique* (i.e., fully connected subgraph) in a so-called *association graph*, encoding the possible mappings between the nodes of the two graphs being matched; hence, many authors formulate the graph-matching algorithm in terms of clique detection. A generalization of MCS is weighted graph matching (WGM), where the edges of the graphs have a weight, and the goal is to find the common subgraph with the largest total weight.

Because of its strict requirements, isomorphism is not very used in PR applications, where it is customary that the graphs representing different instances of a same pattern have some structural differences due to noise or occlusions or to other causes. Subgraph isomorphism, monomorphism, and MCS are generally used in PR for finding an object, represented by a graph, as a part of a larger model graph (prototype), or for detecting the parts shared by two objects, structurally represented by graphs. Although, exact graph matching has exponential time complexity in the worst case. However, in many PR applications the actual computation time can be still acceptable, because of two factors: first, the kinds of graphs encountered in practice are usually different from the worst cases for the algorithms. Second, node and edge attributes can be used very often to reduce dramatically the search time.

2.2 A Commented Bibliography on Algorithms and Techniques for Exact Graph Matching

The first attempts for reducing the computational complexity of graph matching were aimed to define algorithms devised for special kinds of graphs. Among them, we find algorithms for some common graph topologies, as trees (special cases of graphs), proposed by Aho et al. in 1974 [2], planar graphs by Hopcroft and Wong in 1974 [3], and bounded valence graphs by Luks in 1982 [4]). Despite the historical relevance, this family of graph-matching algorithms can be used only in specific applicative areas, where the graphs being matched always have a same predefined structure.

Most of the algorithms for exact graph matching are based on some form of tree search with backtracking. The basic idea is that a partial match (initially empty) is iteratively expanded by adding to it new pairs of matched nodes; the pair is chosen using some necessary conditions that ensure its compatibility with the constraints imposed by the matching type with respect to the nodes mapped so far, and usually using also some heuristic condition to prune as early as possible unfruitful search paths. Eventually, either the algorithm finds a complete matching, or it reaches a point where the current partial mapping cannot be further expanded because of the matching constraints. In this latter case the algorithm backtracks, i.e., undoes the last additions until it finds a partial matching for which an alternative extension is possible. If all the possible mappings that satisfy the constraints have already been tried, the algorithm halts. Several different implementation strategies of this kind of algorithm have been employed, differing in the order the partial matches are visited. Probably the simplest is *depth-first search* that requires less memory than others and lends itself very well to a recursive formulation; it is alsoknown as *branch and bound*.

A nice property of such algorithms is that they can be very easily adapted to take into account the attributes of nodes and edges in constraining the desired matching, with no limitations on the kind of attributes that can be used. This is very important for PR applications where often attributes play a key role in reducing the computational time of the matching. The first important algorithm of this family is due to Ullmann, in 1976 [5], still widely used today. Also the approach proposed by Schmidt and Druffel in 1976 [6] adopts the same strategy, with the addition of a preprocessing that creates an initial partition of the graph nodes on the basis of the distance matrix, to reduce the search space. Another interesting monomorphism algorithm based on backtracking has been proposed by Ghahraman et al. in 1980 [7]; it prune the search space, using a so-called *netgraph* obtained from the Cartesian product of the nodes of two graphs being matched. Monomorphisms between these two graphs correspond to particular subgraphs of the netgraph. A major drawback of the algorithm is that the netgraph is represented using a matrix of size $N^2 \times N^2$, where N is the number of nodes of the largest graph. Consequently, only small graphs can be reasonably dealt with.

A more recent algorithm for both isomorphism and subgraph isomorphism is the VF algorithm [8, 9]. The authors define a heuristic that is based on the analysis of the sets of nodes adjacent to the ones already considered in the partial mapping. This heuristic is fast to compute leading in many cases to a significant improvement over Ullmann's and other algorithms, as shown in [10, 11]. Successively, the authors propose a modification of the algorithm [12, 13], called VF2, that reduces the memory requirement from $O(N^2)$ (that compares favorably with other algorithms) to $O(N)$ with respect to the number of nodes in the graphs, thus making the algorithm particularly interesting for working with large graphs. One of the most recent tree search methods for isomorphism has been proposed by Larrosa and Valiente in 2002 [14]; the authors reformulate graph isomorphism as a constraint satisfaction problem (CSP), a problem that has been studied very deeply in the framework of discrete optimization and operational research. Thus the authors apply to graph matching some heuristics derived from the CSP literature.

The backtracking approach has been applied also to problems different from graph isomorphism and subgraph isomorphism. For instance, Durand et al. [15] have used this approach to solve the maximal clique detection problem. Probably the most interesting matching algorithm that is not based on tree search is *Nauty*, developed by McKay in 1981 [16]. The algorithm deals only with the isomorphism problem, and is regarded by many authors as the fastest isomorphism algorithm available today. It uses some results coming from group theory for constructing the *automorphism* group of each of the input graphs. From them, a *canonical labeling* is derived, so that two graphs can be checked for isomorphism by simply verifying the equality of their canonical forms. The equality verification can be done in $O(N^2)$ time, but the construction of the canonical labeling can require an exponential time in the worst case. In the average case this algorithm has quite impressive performance, although in [11, 17] it has been verified that under some conditions it can be outperformed by other algorithms like the above mentioned VF2. Furthermore, it does not lend itself very well to exploit node and edge attributes of the graphs, that in many PR applications can provide an invaluable contribution to reduce the matching time.

Some matching algorithms are specifically aimed at reducing the cost of matching one input graph against a large library of graphs, suitably preprocessed. Messmer and Bunke proposed a very impressive algorithm in 1997 [18,19]. The algorithm, that deals with isomorphism and subgraph isomorphism, in a preprocessing phase builds a decision tree from the graph library. Using this decision tree, an input graph can be matched against the whole library in a time that is $O(N^2)$ with respect to the input graph size. An extension to MCS is presented in a paper by Shearer et al. in 1997 [20], further improved in [21].

Other two recent papers, by Lazarescu et al. in 2000 [22] and by Irniger and Bunke in 2001 [23], proposed the use of decision trees for speeding up the matching against a large library of graphs. In these cases, the decision tree is not used to perform the matching process, but only for quickly filtering out as many library graphs as possible, applying then a complete matching algorithm only to the remaining ones.

2.3 An Introduction to Inexact Graph-Matching Problems

The stringent constraints imposed by exact matching are in some circumstances too rigid for the comparison of two graphs. In many applications, the observed graphs are subject to deformations due to several causes: intrinsic variability of the patterns, noise in the acquisition process, presence of nondeterministic elements in the processing steps leading to the graph representation, are among the possible reasons for having actual graphs that differ somewhat from their ideal models.

So the matching process must accommodate the differences by relaxing, to some extent, the constraints that define the matching type. Usually, in these algorithms the matching between two nodes that do not satisfy the edge-preservation requirements of the matching type is not forbidden. Instead, it is penalized by assigning to it a cost that may take into account other differences (e.g., among the corresponding node/edge attributes). So the algorithm must find a mapping that minimizes the matching cost.

Optimal inexact matching algorithms always find a solution that is the global minimum of the matching cost so implying that if an exact solution exists, it will be found. Hence they can be seen as a generalization of exact matching algorithms. Optimal algorithms face the problem of graph variability, and they do not necessarily provide an improvement of the computation time, usually resulting fairly more expensive than their exact counterparts.

Approximate or *suboptimal* matching algorithms, instead, only ensure to find a local minimum of the matching cost, generally not very far from the global one. Even if an exact solution exists, they may not be able to find it and for some applications this may not be acceptable, but the suboptimality of the solution is abundantly repaid by a shorter, usually polynomial, matching time.

A significant number of inexact graph-matching algorithms base the definition of the matching cost on an explicit model of the errors (deformations) that may occur (i.e., missing nodes, etc.), assigning a possibly different cost to each kind of error.

These algorithms are often denoted as *error correcting* or *error tolerant*. Another way of defining a matching cost is to introduce a set of *graph edit operations* (e.g., node insertion, node deletion, etc.), each assigned a cost; the cheapest sequence of operations needed to transform one of the two graphs into the other is computed, and called *graph edit cost*.

Some of the inexact matching methods also propose the use of the matching cost as a measure of dissimilarity of the graphs, e.g., for selecting the most similar in a set of graphs, or for clustering. In some cases, the cost formulation verifies the mathematical properties of a distance function (e.g., the triangular inequality); then we have a *graph distance* that can be used to extend to graphs some of the algorithms defined in metric spaces. Of particular interest is the *graph edit distance*, obtained if the graph edit costs satisfy some constraints (e.g., the cost of node insertion must be equal to the cost of node deletion).

Some papers demonstrates equivalences holding between the graph edit distance and relevant graph-matching problems, as the graph isomorphism and subgraph isomorphism and MCS [24–28].

2.4 A Commented Bibliography on Algorithms and Techniques for Inexact Graph Matching

Tree search with backtracking can also be used for inexact matching. In this case the search is usually directed by the cost of the partial matching obtained so far, and by a heuristic estimate of the matching cost for the remaining nodes. This information can be used either to prune unfruitful paths in a branch and bound algorithm, or also to determine the order in which the search tree must be traversed, as in the A* algorithm. In this latter case, if the heuristic provides a close estimate of the future matching cost, the algorithm finds the solution quite rapidly; but if this is not the case, the memory requirement is considerably larger than for the branch and bound algorithm.

The first tree-based inexact algorithm is due to Tsai and Fu, in 1979 [29], and in an extended version in 1983 [30]. The paper introduces a formal definition of error-correcting graph matching of attributed relational graphs (ARG), based on the introduction of a graph edit cost, and defines a search method ensuring to find the optimal solution. A more recent paper by Wong et al. in 1990 [31] proposes an improvement of the heuristic of Tsai and Fu for error-correcting monomorphism, taking into account also the future cost of edge matching.

A similar approach is used in a paper by Sanfeliu and Fu in [32–34], where the definition of a true graph edit distance is attempted, and a suboptimal method, working in a polynomial time, for the distance computation is introduced. In a paper of 1980, Gharaman et al. [35], propose an optimal inexact graph monomorphism algorithm that is based on the use of branch and bound together with a heuristic derived from the netgraph.

Interesting early papers are due to Shapiro and Haralick in 1981 [36] and later in 1985 [37], with algorithms for finding the optimal error-correcting homomorphism and for evaluating the distance between two hypergraphs.

Among the more recent proposals based on tree search we can cite the algorithm using A*, for different purposes: Dumay et al. in 1992 [38], for evaluating a graph distance and Berretti et al. in their 2000 and 2001 papers [39–41], for finding the largest matching between two sets of nodes forming a *bipartite graph*, with the constraint that each node must be used at most once. A* search appears also in a recent paper by Gregory and Kittler in 2002 [42], where a fast, simple heuristic is used that takes into account only the future cost of unmatched nodes. The authors assume that at least for small graphs the less accurate estimate of the future cost is abundantly repaid by the time savings obtained in computing a less complicated heuristic.

Another recent inexact algorithm has been proposed by Cordella et al. in two papers of 1996 and 1997 [43,44]. This algorithm deals with deformations by defining a transformation model in which under appropriate conditions a subgraph can be collapsed into a single node. The transformation model is contextual, in the sense that a given transformation may be selectively allowed depending on the attributes of neighboring nodes and edges.

Along the same lines, Serratosa et al. in 1999 [45,46] present an inexact matching method that also exploits some form of contextual information. The authors define a distance between function described graphs (FDG) that are ARGs enriched with additional information relative to the joint probability of the nodes in order to model with one FDG a set of observed ARGs. As in the case of exact approach, efficiently inexact matching algorithms have been proposed for dealing with special, restricted classes of graphs, as planar graphs and region adjacency graphs (RAGs). For planar graphs, Rocha and Pavlidis [47] present an optimal algorithm for error-correcting homomorphism, while in a paper by Wang and Abe (1995) [48], a distance between RAGs is proposed, and is computed using a suboptimal algorithm. More recently, Llados et al. in a 2001 paper [49] define a graph edit distance for RAGs using edit operations that are devised to model common distortions in image segmentation; the distance is computed using an optimal algorithm based on branch and bound.

The matching methods examined so far rely on a formulation of the matching problems directly in terms of graphs. A radically different approach is to cast graph matching, that is inherently a discrete optimization problem, so as to use one of the many continuous, nonlinear optimization algorithms. The found solution needs to be converted back from the continuous domain into the initial discrete problem by a process that may introduce an additional level of approximation. Nevertheless, in many application contexts this approach is very appealing because of its extremely reduced computational cost that is usually polynomially dependent (and with a low exponent) on the size of the graphs.

The first family of methods based on this approach uses *relaxation labeling*. One of the pioneering works is due to Fischler and Elschlager in 1973 [50]. The basic idea is that each node of one of the graphs can be assigned one label out of a discrete set of possible labels, that determines which node of the other graph it corresponds to. During the matching process, for each node there is a vector of the probabilities of each candidate label, dynamically re-evaluated until the process converges to a stable solution. At this point, for each node the label having the maximum probability is chosen. In 1989 Kittler and Hancock [51] provide a probabilistic framework for relaxation labeling, in which the update rules previously used for the probabilities are given a theoretical motivation. In 1995, Christmas et al. [52] propose a method, based on the theoretical framework of Kittler and Hancock, that is able to take into account during the iteration process both node and edge attributes. Wilson and Hancock, in 1997 [53], extended the probabilistic framework by introducing a Bayesian consistency measure, that can be used as a graph distance. An extension of this method has been proposed by Huet and Hancock in 1999 [54]. This method also takes into account edge attributes in the evaluation of the consistency measure.

Myers et al. [55] in 2000 propose a new matching algorithm that introduces the definition of a Bayesian graph edit distance, approximated by considering independently the *supercliques* of the graphs, so as to perform the computation in polynomial time. Finally, in a recent paper (2001), Torsello and Hancock [56] propose the use of relaxation labeling also for computing an edit distance between trees.

A recent method by Luo and Hancock [57] is based on a probabilistic model of matching: the nodes of the input graph play the role of observed data while the nodes of the model graph act as hidden random variables; the matching is then found by using the expectation–maximization (EM) algorithm [58].

A different family of methods is based on a formulation of the problem as a WGM problem that permits the enforcement of two-way constraints on the correspondence. It consists in finding a matching, usually expressed by means of a matching matrix M, between a subset of the nodes of the first graph and a subset of the nodes of the second graph. The edges of the graphs are labeled with weights, that are real numbers, usually between 0 and 1. The desired matching must optimize a suitably defined goal function. Usually the problem is transformed into a continuous one by allowing M elements to have continuous values so making the WGM problem a quadratic optimization problem. An important limitation of this approach, from the perspective of PR applications, is that nodes cannot have attributes and edges cannot have other attributes than their weight. This restriction imposes a severe limit on the use of the semantic information often available in real applications.

Among the first papers based on this formulation is the work by Almohamad and Duffuaa in 1993 [59]. In this paper the quadratic problem is linearized and solved using the simplex algorithm [60]. The approximate, continuous solution found this way is then converted back into discrete form using the so-called Hungarian method [60] for the assignment problem. Rangarajan and Mjolsness [61], in 1996, proposed a method based on Lagrangian relaxation networks in which the constraints on the rows and on the columns of the matching matrix are satisfied separately and then equated through a Lagrange multiplier. Also in a 1996 paper, Gold and Rangarajan [62] present the graduated assignment graph-matching (GAGM) algorithm. In this algorithm a technique known as *graduated nonconvexity* is employed to avoid poor local optima. Another approach is based on a theorem by Motzkin and Straus that establishes a close relation between the clique problem and continuous optimization. Namely, they proves that all the maximum cliques of a graph correspond to maxima of a well-defined quadratic functional. In 1997, Bomze [63] proposed a modified functional for which the correspondence holds in both senses.

The papers by Pelillo and Jagota in 1995 [64, 65] propose a matching method based on the above cited theorem and an implementation where the quadratic problem is solved by means of relaxation networks [66]. In [67] a unified framework for relational matching based on the Bomze functional is presented. In 1999, Pelillo et al. [68] introduced a technique to reduce the MCS problem between trees to a clique problem and then solved it using replicator equations. Branca et al. [69] proposed in 1999 an extension of the framework defined by Pelillo [67] that is able to deal with a weighted version of the clique problem.

Several other inexact matching methods based on continuous optimization have been proposed in the recent years, as the fuzzy graph matching (FGM) by Medasani et al. [70, 71], that is a simplified version of WGM based on fuzzy logic. Another recent approach, proposed by van Wyk et al. in 2002 [72, 73] is based on the theory of the so-called reproducing Kernel Hilbert spaces (RKHS) for casting the matching problem into a system identification problem; this latter is then solved by constructing a *RKHS interpolator* to approximate the unknown mapping function.

Spectral methods are based on the following observation: the eigenvalues and the eigenvectors of the adjacency matrix of a graph are invariant with respect to node permutations. Hence, if two graphs are isomorphic, their adjacency matrices will have the same eigenvalues and eigenvectors. Unfortunately, the converse is not true: we cannot deduce from the equality of eigenvalues/eigenvectors that two graphs are isomorphic. However, since the computation of eigenvalues/eigenvectors is a well-studied problem, that can be solved in polynomial time, there is a great interest in their use for graph matching. An important limitation of these methods is that they are purely structural, in the sense that they are not able to exploit node or edge attributes, that often, in PR applications, convey information very relevant for the matching process. Further, some of the spectral methods are actually able to deal only with real weights assigned to edges by using an adjacency matrix with real-valued elements.

The pioneering work on spectral methods is the paper by Umeyama, in 1988 [74], proposing an algorithm for the weighted isomorphism between two graphs. It uses the eigendecomposition of adjacency matrices of the graphs to derive a simple expression of the orthogonal matrix that optimizes the objective function, under the assumption that the graphs are isomorphic. From this expression he derives a method for computing the optimal permutation matrix when the two graphs are isomorphic, and a suboptimal permutation matrix if the graphs are nearly isomorphic. In 2001, Xu and King [75], propose a solution to the weighted isomorphism problem, by approximating the permutation matrix with a generic orthogonal matrix. An objective function is defined using Principal Component Analysis and then gradient descent is used to find the optimum of this function.

In 2001 Carcassoni and Hancock [76] propose a spectral method that is based on the use of spectral features to define clusters of nodes that are likely to be matched together in the optimal correspondence; the method uses hierarchical matching by first finding a correspondence between clusters and then between the nodes in the clusters. Another method that combines a spectral approach with the idea of clustering has been presented by Kosinov and Caelli in 2002 [77]: a vector space, called the *graph eigenspace*, is defined using the eigenvectors of the adjacency matrices, and the nodes are projected onto points in this space and a clustering algorithm is used to find nodes of the two graphs that are to be put in correspondence.

A method that is partly related to spectral techniques has been proposed in 2001 by Shokoufandeh and Dickinson [78]. The authors use the eigenvalues to associate to each node of a Directed Acyclic Graph a topological signature vector (TSV) that is related to the structure of the subgraph made of the descendants of the node. These TSV are used both for a quick indexing in a graph database and for the actual graph-matching algorithm. This latter is based on the combination of a greedy search procedure and of bipartite graph matching.

Finally, we must say that other heuristic approaches to inexact graph matching have been proposed: at least in principle, any of the heuristic techniques that have been used for combinatorial problems or for continuous global optimization problems can be adapted to some approximate form of graph matching. With no presumption of completeness, we can cite here, as examples, simulated annealing (Jagota et al. [79]) and tabu search (Gendreau et al. [80]; Williams et al. [81]).

3 Application Taxonomy

In the last decade several applications of graph matching in PR and machine vision have been reported in the literature. As regards the role of such applications within the global context of the containing work, we can recognize two situations. In a first type of works, that we could name *application-driven papers*, the main concern of the authors is solving an applicative problem. They present their graph-based techniques as a solution that is as effective or more effective than other, nongraph-based methods, for solving the problem at hand. Application-driven papers are of course the most interesting ones for an audience involved in deciding whether a graph-based technique is more or less suitable for a given problem, since they usually provide a comparison, either theoretical or experimental (or both), between their proposal and other approaches of different kind to the same problem. A second type of works, say *technique-driven papers*, instead is more centered around the presentation of a novel graph-based algorithm or technique that could be potentially applicable in several situations, and make use of an applicative problem to provide a performance benchmark of the proposed technique in comparison to other, often similar, methods. Since the application is not the main concern, in these papers the authors do not investigate thoroughly the advantages of a graph-based method over a different kind of approach. Nevertheless, these paper provide very useful insight to a slightly different audience: that is, the researchers that have already chosen to cast their problem in a graph framework and are now looking for the best performing technique or algorithm known to solve a problem of that kind. In this review, we will present both application-driven and technique-driven papers, but will focus mainly on the first type of works, providing only a shallow overview of the second type.

We have grouped the applications of graph matching according to the topic within the PR and machine vision fields. Namely, we have individuated five broad areas that cover the vast majority of the applications:

- 2D and 3D image analysis
- Document processing
- Biometric identification
- Image databases
- Video analysis

Among the applications to 2D and 3D image analysis, we have found both low-level problems such as *edge detection* and *stereomatching*, and middle/high-level problems such as *automatic navigation*, *robotic vision*, and *object recognition*. *Handwritten recognition*, *OCR*, and *symbol recognition* are the most relevant document processing applications addressed in the literature, while *face recognition and authentication*, *facial expression recognition*, *hand posture recognition*, *ear recognition*, and *fingerprint recognition* are examples of biometric identification applications. In the field of image databases, *indexing* and *retrieval* have been considered. *Retrieval* from video databases, *annotation* of video databases, *object tracking* and *motion estimation* are the typical applications in the context of video analysis.

While most graph-matching applications fall in the previously outlined categories, there are also a few, isolated works (mainly in the fields of biology and biomedicine) that do not fall neatly into one of the above-mentioned areas. We have presented some of these papers as *miscellaneous applications*.

In the following subsections we will provide details about each application areas, discussing the peculiarity of how graph-based methods have been applied in their context, highlighting (wherever it is possible) recurring patterns of usage and correspondence between the problem, the representation, and the matching technique.

3.1 2D and 3D Image Analysis

Among the papers that address 2D and 3D image analysis problems with a graph-matching technique, a significant number lies in the technique-driven category we have previously defined. For the 2D image analysis, this is the case of papers that report applications in the object recognition field [31, 73, 82–86], of papers that address the shape recognition [56, 68, 77, 87–89] or the scene recognition [90, 91] problem, and of papers that work on SAR images [53, 92]. As regards 3D image analysis, papers that report results on robotic vision [34], stereomatching [52, 93] object matching [57], object recognition [94–97], and object reconstruction [98] applications can be cited.

However, there are several interesting application-driven papers that we will now examine with more detail. In the field of 2D image analysis, the problems faced by the application-driven papers are three: *object recognition* (by Meth and Chellappa [99], Li and Lee [100], Belongie and Malik [101]), *shape recognition* (Sebastian et al. [102]) and *visual inspection* (Koo and Yoo [103]).

In object recognition the goal is to find all the occurrences, within an image, of a distinguished set of objects. Usually objects of interest belong to different classes (having different shapes) and neither their number nor their positions within the image are known; the image often contains also background elements (possibly complex) that should not be detected by the system. In some cases, the objects of interest may be partially occluded by other objects or by background elements. The basic idea of graph-based object recognition is to decompose the whole image into smaller parts, obtaining a graph representation describing those parts and their relations, and then to look for subgraphs of this large graph that correspond to the shapes of the objects of interest, by means of some kind of inexact matching algorithm. The above-mentioned papers differ in the adopted representations, ranging from low-level [99] to middle-level representations [100, 101], and also in matching techniques (error-correcting subgraph isomorphism with a similarity measure [99], inexact matching with a neural approach [100], weighted bipartite matching [101]); Shape recognition is very similar to object recognition, differing for the fact that only shape information is available (and not, say, color or texture information), and usually the image is not cluttered with background elements. If the shapes are simply connected (i.e., they not contain holes), they can be represented using a tree instead of a fully general graph, and the matching can be performed using error-correcting tree isomorphism [102]. Visual inspection is also similar to the object recognition problem, with the important difference that a model of which objects are expected to be in the image and which should be their positions is known; indeed, the purpose of visual inspection is actually to spot any difference with respect to the expected situation. For this reason, exact matching methods can be more appropriate for this problem [103].

The distribution of the matching algorithms and of the graph representations used within this applicative area is shown in Fig. 4.

Entering into details, in the field of 2D object recognition, Meth and Chellappa in [99] work on SAR images. They use a low-level representation: a node of the graph is associated to each pixel of the image. The node labels depend on the so-called topographical primal sketch (TPS). A TPS assigns to each pixel a label that is invariant under monotonic transformations of the grey levels. This is obtained by fitting a local two dimensional cubic surface on the image for estimating the intensity surface around each pixel. On the basis of the derivatives of this surface, one of the following six labels is given to the pixel: peak, pit, ravine, ridge, saddle, "no zero crossing." Two graph-matching techniques are proposed, the first one is based on a distance measure between node labels, while the second one is based on a similarity measure between features associated to node labels. In both cases, the test and the model image are first registered with respect to the node labels position. The first matching technique calculates a cost based on the relative distance between nodes with the same label in the test and in the model image. The second one associates a feature vector (by calculating the second derivative extrema, the directions of the second derivative extrema and the gradient) to nodes that have a certain label; on the basis of these feature vectors a similarity measure is computed. Results are reported on 81 images belonging to three different categories.

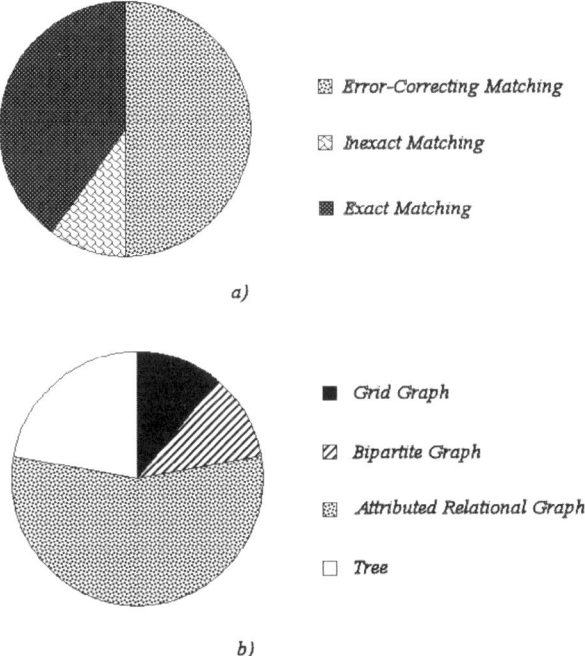

Fig. 4. Distribution of (**a**) the matching algorithms and (**b**) the graph representations used within applications in the 2D and 3D image analysis field

In [101] Belongie and Malik define a new middle-level shape descriptor, that they call *shape context*, for measuring shape similarity. Given an image, the edges are extracted and a certain number of uniformly spaced points, say N, on these edges is selected. A compact descriptor for each sample point is obtained by computing a coarse histogram of the relative coordinates of the remaining points, in a log-polar coordinate system. All the N histograms are flattened and concatenated so as to obtain the so-called shape context of the image. In addition to this representation, another one based on the local appearance, in particular on the tangent angle calculated for each of the N points, is also used. So, if a node of a graph is associated to each of these points, the cost of matching two nodes relative to points on two images can be expressed by taking into account two contributions, one relative to the difference between histograms and the other one relative to the tangent angles dissimilarity. The object recognition problem is then viewed as a weighted bipartite matching problem that can be solved with the Hungarian method. Results on the same database used in [104] are presented and also on other silhouette image databases. Furthermore, the authors suggest that their method can be also used for the retrieval from an image database, as it provides a similarity measure between 2D objects.

In the paper by Li and Lee [100], a graph represents a 2D scene that is described by using a polygonal approximation. The nodes of the graph are the vertices of the polygon and the edges represent the sides. The angle subtended by each vertex is the node attribute, while the distance between two nodes is used as edge attribute. Given such a graph (called *scene graph*), in order to cope with distortions and occlusions, the authors propose to divide it into smaller pieces, called *subscene graphs*. Then, an inexact subscene graph matching is performed between each subscene graph and a model graph, by using an Hopfield neural net. The correct match for the complete scene graph can be obtained from the statistics of the matching results between each subscene graph and the model graph. In the paper, tests are made on images representing 2D hand tools.

Sebastian et al. [102] propose a system to recognize object shapes on the basis of their silhouette. They represent each object using a shock tree, which is derived from a thinning of the shape. For the matching, they propose the definition of a tree edit distance, in which the edit costs are not fixed arbitrarily but are derived analytically from a small set of hypotheses related to the cost of deforming a silhouette. This distance is computed by means of an error-correcting tree isomorphism algorithm based on dynamic programming.

Finally, Koo and Yoo [103] address the problem of visual inspection by using an high-level representation scheme. They consider Printed Circuit Board images, that are represented by means of a tree. Images are first binarized, then the binarized image is partitioned into nonoverlapping regions (*blobs*) each one made up of adjacent pixels having the same value. A node is associated to each blob; a tree is constructed by adding an edge between two nodes if the blobs they represent are spatially included into each other and have different pixel values. The root of the tree is the blob that contains the outer boundary pixels of the image. The tree obtained from a test image is compared with the tree derived from a defect-free image by means of the tree isomorphism algorithm proposed in [2]. If the two trees are not isomorphic a defect is detected and the inspection process stops. Otherwise, additional polygonal-boundary information are extracted and a second matching step is performed. In particular, a tolerance zone is defined and the proposed algorithm through a polygonal-boundary matching function checks if such tolerance is respected between pairs of matched nodes.

In the field of the 3D image analysis applications, Branca et al. [69] addressed the problem of automatic navigation, while Bauckhage et al. [105] and Olatunbosun et al. [106] the recognition of 3D objects, and Fuchs and Le Men [107, 108] the 3D object reconstruction.

The 3D object recognition problem is quite similar to the 2D version; the main differences are the need to take into account the changes in the object appearance due to a different point of view, and the increased importance of the occlusion phenomenon. These differences lead to the use of graph-matching algorithms more tolerant to structural changes.

The 3D object reconstruction is aimed at deriving the three-dimensional structure of a scene from 2D images. Under some constraints on the objects being reconstructed, it can been faced with an approach that reduces this problem to object recognition, by defining a set of 3D structural primitives whose occurrence can be recognized in the image. Automatic navigation consists in the detection of still or moving objects (such as obstacles a vehicle has to avoid) in a 3D scene, usually represented by means of a pair of stereo images. The problem is different from object recognition since the shape of the objects is not known a priori. However, it turns out that this problem can be faced with techniques very similar to the ones used for 3D object recognition. In fact, by finding for each part of one of the two images the corresponding part in the other (which is similar to what an object recognizer does), the distance of each part from the camera can be estimated.

The 3D applications mainly use ARGs [69, 105, 107, 108] for representing objects, and two different matching techniques: either perform a MCS detection (by means of a maximal clique search on the association graph) [69, 106] or employ an error-correcting subgraph isomorphism algorithm [105, 107, 108]. With more detail, Branca et al. [69] presents an application of graph matching to autonomous navigation. In particular, the detection of ground floor obstacles and of moving objects are considered. Relational graphs are used for object representation, where the object features extracted by means of the Moravec interest operator are the nodes and the edge linking them are weighted by projective invariant values. Given two graphs obtained from two different images acquired by a TV camera mounted on a mobile vehicle, the goal is to determine, into the association graph, the maximal clique of nodes that are mutually compatible according to the similarity imposed by the invariant relations encoded into the edges. The nodes of this clique will belong to the same object and this permit to detect into a given image the features that belong to an obstacle, or to individuate the feature pertaining to a moving object. In the paper an algorithm for finding the maximum edge weighted clique in an high-order association graph is presented, based on an optimization procedure that use the Motzkin–Straus theorem.

As regards 3D object recognition, Bauckhage et al. present in [105] a system that uses graphs to recognize mechanical assemblies in a dynamic construction environment. In particular, the main objective of their project is to develop a robot that assembles parts from a wooden construction-kit for children, made up of bolts, rings, bars, and cubes. So, given an assembly described by a graph, they use a graph-matching technique for recognizing if that assembly is already present in the knowledge base of the robot. In the negative case it will be added to the database. They introduce the *mating feature* graph for representation. The nodes of this graph represent mating features or subparts of an assembly and are labeled with the type of the subpart. Nodes connected with a pair of edges represent subparts that belong to the same object, while single directed edges between nodes represent the fact that the corresponding subparts are attached to the same bolt. If an object is connected to several bolts, the nodes that correspond to these bolts are linked by a bidirectional edge; this edge is labeled with a value that indicates the angle between the bolts. For matching two mating feature graphs, they use the error-correcting matching procedure presented in [109]. They also propose an application of the mating feature graph to the 3D reconstruction of assembled objects.

Another approach is proposed in Olatunbosun et al. [106], where a special kind of RAGs, called color region adjacency graph (CRAG), is used for representing 3D objects. Graph nodes are the segmented regions, using the coordinates of their centroids as node attributes, and edges represent connections between regions. By using the line length ratio, i.e., the ratio of the distance between a pair of nodes into the model image and a pair of nodes in a test image, and the line angles, i.e., angles between three nodes into the model and the test image, an association graph where the nodes are provisionally matched is built. Then the Bron–Kerbosh [110] algorithm is used to find the maximal clique on this association graph: an high clique value imply high similarity between the model image and a test image. The authors also propose to reduce the computational complexity of the maximal clique search method, by adopting a model-based approach. In order to recognize an object, a test images is first filtered, by eliminating from it all the color regions that do not belong to the model CRAG. So, the maximal clique search is performed on a smaller association graph.

Finally, Fuchs and Le Men in [107] and [108] use graph matching in the field of 3D building reconstruction from aerial stereopairs. In particular, in [108] the 3D object extraction problem is addressed, while in [107] the goal is the reconstruction of the structure of the roofs. They use a model driven strategy: the models used are ARGs, where each nodes represent a 3D feature (a 3D line segment, or a 3D planar region, or a facade of a building), while each edge encodes a geometric property (such as parallelism, orthogonality, and so on) between nodes. The building reconstruction is based on the computation of a subgraph isomorphism between a model and a graph built on a set of 3D features derived from the images. As regards the matching procedure, in [108] they use the error-correcting subgraph isomorphism detection presented in [109], with an estimation of the subgraph distance based on a stochastic heuristic, while in [107] propose a modification of the algorithm proposed in [109] in order to take benefit of an external information (e.g., an user input or a precomputed information). If the correspondence between some nodes of the model and some nodes of the input data is already known before the matching, the search space of the matching problem can be pruned by integrating the external information in the error-correcting subgraph isomorphism algorithm.

3.2 Document Processing

Among the various document processing applications, OCR, handwritten recognition, string recognition, symbol and graphic recognition have been addressed in the literature by using graph-matching techniques.

These problems are relatively similar to each other, entailing the recognition of small elements having a definite meaning within a printed or handwritten document. The number of different categories (classes) to be considered varies from ten to several hundreds, and also the shape variability of the elements belonging to a same class can range from reasonably small (e.g., for high-resolution printed characters) to very high (for handwritten characters or symbols). As regards the strategy adopted to face the problem, entities to be recognized (characters, symbols, or graphics) are usually decomposed into geometric primitives, which are in most cases approximated as thin lines (also called *strokes*), since in handwriting and in printed scripts, thickness does not convey useful information. This decomposition is then represented as a graph, and the recognition process is performed as a graph matching with model graphs corresponding to the different classes of characters, symbols, or graphics to be recognized. The proposed approaches differ in complexity of the geometric primitives, in the way the decomposition is translated into a graph, and in the kind of graph matching performed. A problem that is strongly related is the construction of such model graphs from a set of examples, which is usually performed by means of algorithms that involve a graph matching as one of their step.

The distribution of the matching algorithms and of the graph representations used within the document processing field is shown in Fig. 5. Now we will examine with more detail each of the problems belonging to this field.

Since OCR and handwritten recognition are among the most classical PR problems, and many of large datasets are available, they are often used as test cases in technique-driven papers. This is the case of the papers by Sanfeliu and Fu [34], Foggia et al. [43, 111, 112], Chan [113] and Rangarajan and Mjolness [61]. But also several application-driven papers have been written on the handwritten recognition problem (both offline and online) or on the optical character recognition problem: this is the case of the papers by Lee and Liu [114] and Suganthan and Yan [115], and Liu et al. [116], Lu et al. [117], Chen and Lieu [118], and Rocha and Pavlidis [47].

Independently on the main focus of the considered papers, both printed and handwritten characters are typically described by ARGs [34, 43, 47, 111, 112, 115, 117, 118]. Two description schemes have been used (1) the nodes of the graph represent

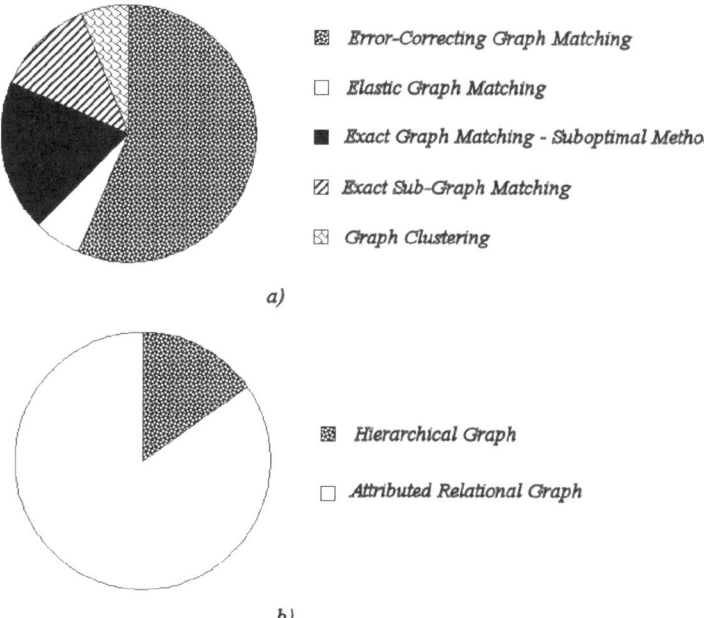

Fig. 5. Distribution of (**a**) the matching algorithms and of (**b**) the graph representations used within applications in the document processing field

the structural primitives in which a character can be decomposed after a thinning process and the edges represent the relations between them these primitives or (2) the nodes are the junctions between strokes (singular points) and edges represent the primitives into which characters are decomposed. Authors dealing with Latin characters and Arabic digits [34, 43, 47, 111, 112] use circular arcs and segments as primitives, while straight line segments or strokes are typically used in case of Chinese characters [115–118]. As regards the representation, a quite different approach is proposed in [114], where the authors propose an architecture for the recognition of handwritten Chinese character that integrates the feature extraction, the segmentation, and the recognition phase. The feature extraction phase is performed by means of Gabor filters; such features are used to segment characters using an optimization module based on a genetic algorithm. Finally, elastic graph matching is used in the recognition phase. Besides this paper, other authors mainly use error-correcting graph matching for dealing with the high variability of handwritten characters. Another feature that is peculiar to this kind of applications is the graph size: graphs describing characters are typically made up of few nodes.

As regards technique-driven papers on handwritten recognition, in [34] and [111] the authors use respectively handwritten characters and handwritten digits to validate a distance measure between ARGs in the framework of error-correcting graph matching. In the same framework, a matching algorithm using subgraph transformations is applied to handwritten characters [43]. Chinese characters are used in [113] as test case for a learning algorithm that build templates starting from fuzzy-attribute graphs; while in [61] the authors present a suboptimal method for exact graph matching, based on a lagrangian relaxation network, using handwritten digits for testing. Finally, handprinted digits are used as application of a graph learning algorithm [112].

As regards the recognition of Chinese characters, both offline and online approaches are present in the literature. In [115] the matching between input graph and model graph for offline Chinese character recognition is performed by means of an Hopfield network (presented in [119]) that is specifically devised to allow the segments of a broken stroke of an input character to be matched to a stroke of the model graph.

The recognition of online handwritten Chinese character addressed by graph matching has the additional problem of a significant computational cost due to the large number of categories. Therefore for developing an online recognition system it is mandatory to find an adequate structural representation together with matching algorithms that can efficiently address this recognition problem. To this aim, some authors [117, 118] used a sort of hierarchical graphs to represent a character. Such graphs have two layers: nodes and edges in the first layer represent high-level components and relations between them; while in the second layer each component is described by a graph in which nodes and edges represent the strokes of that component and their relations. In [116] a Chinese character is described with a complete relational graph (CRG), where each node describe one of the segments in which a stroke obtained from a pen down–pen up movement on a digitizer can be decomposed. In order to reduce matching time, a suboptimal solution is proposed. The problem of matching CRGs is transformed into a two-layer assignment problem and solved with the Hungarian method. Within the OCR field, in [47] an error-correcting subgraph-matching algorithm is used. It allows a multiple-to-one matching from a set of feature (a path) of an input graph to a feature of a model graph (prototype), on the basis of a set of predefined transformations. These graph transformations regard straightening of strokes, rewriting of strokes into arcs, insertion and deletion of features, and attribute transformations. Having associated a cost to each transformation, the matching procedure for each input graph selects the prototype that gives rise to the matching with the minimum cost. It is worth noting that prototypes are manually defined, without using a specific learning procedure.

A quite peculiar approach is proposed in [120], where the OCR problem is addressed with an ad hoc matching defined between the so-called graph embeddings. A graph embedding, used in this paper for representing characters, is a labeled graph where each node is labeled with its coordinates in the x–y plane.

The handwritten digit string recognition problem has been addressed by [121]. Starting from an input image and after a thinning process, the authors construct a graph whose nodes are the branches or the ending points of the thinned image and whose edges represent lines of the thinned image. The input graph is then submitted to a segmentation process by using a set of heuristic rules. It gives rise to a number of separate symbols, called *blocks*. The recognition procedure consists in matching the input blocks with the prototype graphs of the digits, by applying a set of transformations to each input block. The matching is therefore an error-correcting graph–subgraph isomorphism. As transformations, the combination of two nodes into one, the transformation of a loop to an edge and the deletion of edges or nodes are considered.

Finally, in the field of symbol and graphics recognition fall the paper of Llads et al. [49, 122], Changhua et al. [123], Cordella et al. [8] and Jiang et al. [124]. While the last two papers are devoted to exploit the performance of an exact subgraph-matching algorithm in detecting component parts within technical drawings [8] and of a graph-clustering algorithm [124], respectively, the others have their main focus on the application domain.

As in case of character recognition, almost all the approaches use ARGs for representing symbols or graphical drawings; as already said, in [8] an exact subgraph-matching algorithm is used, while other authors employs different kinds of error-correcting subgraph-matching algorithms for recognition. The main difference with respect to the case of character recognition is in the number of the nodes of graphs representing maps, diagrams, or technical drawings, that can be up to some hundreds or even thousands.

In [49] the problem of finding a model graph, that represents a prototype symbol, as a subgraph of an input graph, that represents a drawing, is addressed. To do this, a two-level graph representation for graphical symbols is used. In the first level, a vectorized document is approximated by graphs whose nodes represent characteristic points (i.e., junctions, end or corner points, and so on) and whose edges approximate the segments between them. In the second level, data is organized in terms of RAGs. The RAG nodes represent the regions, i.e., minimal closed loops of the first level graphs, and the edges are the neighboring relations between regions. Symbols are then recognized by means of an inexact subgraph-matching procedure that computes the minimum distance from a model RAG to an input RAG. This distance is considered to be the weighted sum of the costs of edit operations to transform one RAG into another one.

In [122] the authors try to identify building blocks in a hand-drawn floor plan. After a scanning and a vectorization process, drawings are described by means of ARGs. An inexact subgraph isomorphism algorithm based on discrete relaxation is used for matching the obtained ARG against model graphs representing the building elements. In order to speed up the process, a straight line Hough transform is also used. It allows the detections of regions filled with parallel straight lines, such as walls that are typically characterized by hatching patterns.

Finally in [123] graphical hand-sketched symbols are represented through ARGs and a similarity measure calculated using the A* algorithm is used for recognition.

3.3 Biometric Identification

Graph-based techniques have been widely used within the context of biometric applications, mainly with reference to identification problems implemented by means of elastic graph-matching procedures. Rarely, in this application area, graph-learning algorithms have been used.

Among all the biometric identification problems, a key role is played by *face authentication*, *face recognition*, and *fingerprint recognition*. Moreover, there are other applications based on facial images, as *facial expression recognition* and *face pose estimation*, as well as other probably less-known applications, as *hand posture recognition* and *ear recognition*. In all these problems, the goal is to compare a graph representation obtained from a sample image of some biometric trait of an individual with a model graph. This comparison has to take into account the possibility of severe distortion of the sample graph with respect to the model, due to the extreme variability in the appearance of biometric traits. In the case of authentication, there is only one model graph, and the problem is to decide whether the model and the sample correspond to the same person. In biometric recognition problems, instead, there are several models (corresponding to different persons, but also possibly to different gestures of a single person) and the system has to identify the person (or the gesture) shown in the sample. A characteristic that is common to many applications of this category, is that the reliability of the identification is extremely important, since the cost of errors is significantly larger than, for example, that of document processing applications.

In almost all cases, papers falling in this area are mainly application-driven, so using rather standard graph-based techniques; sometimes minor adjustments to classical algorithms have been introduced so as to take into account peculiarities of the problem at hand. Generally, the graphs used for describing the patterns are made of tens of nodes and have a rather simple structure (sometimes regular ones, as the grids of nodes used for applications dealing with facial images). The attributes of the nodes of the graphs are rather complex, and frequently are given by feature vectors made of many components, while the edge attributes are simpler and typically represent distances between given points in the original image. For these characteristics, the matching technique that is most commonly used is elastic graph matching.

The distribution of the matching algorithms and of the graph representations used in the Biometric Identification field is summarized in Fig. 6.

In the areas of face authentication and face recognition, graph matching has been used in the systems proposed by Van Der Malsburg, Wiskott et al. [125–127], by Lim and Reinders [128], by Kotropoulos, Pitas, and Tefas [129, 130], by Duc et al. [131] and by Lyons et al. [132]. All these approaches use a graph, in particular a labeled rectangular grid, as an intermediate representation level for representing a face. In this grid, each node of the graph is associated to a specific facial landmark, called *fiducial point*. The labels associated to the nodes are of two different types: those based on *Gabor coefficients* [125–128, 131, 132], the so-called *jets*, and those made up of a vector of features evaluated on small areas of interest in the input image by means of multiscale dilation–erosion techniques [129, 130]. The face identification process is carried out by standard elastic graph-matching algorithms. The grid representing the input face is compared with the ones representing face models. During the matching process the feature vectors associated to matched nodes are used to calculate a distance, so as to evaluate an overall distance between the two compared input graphs. The matching procedure is elastic in the sense that it copes with deformations, rotations, or scale variations in the areas of interest of the input image.

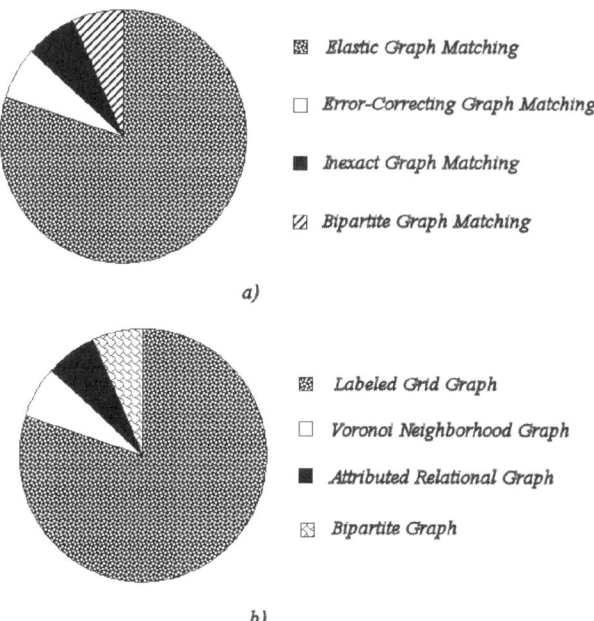

Fig. 6. Distribution of (**a**) the matching algorithms and of (**b**) the graph representations used within applications in the biometric identification field

In more details, one of the simplest description schemes is the one proposed in [128], where the authors describe a face image by a graph made of four nodes representing prefixed landmark points of the face as eyes, the nose, and the mouth. Each node is labeled with a jet, while an edge of the graph is associated an attribute representing the distance existing between the points of the images relative to the nodes it connects. In this paper the elastic graph-matching procedure is specifically tailored for dealing with affine transformations on the considered images in the neighbors of the landmark points, and the authors denote their matching algorithm as *affine graph matching*. The algorithm is used for localizing a face within an image, and this task is accomplished by maximizing the similarity measure proposed in [127]; they take into account only the magnitude value of the jets, and use a genetic algorithm for exploring the search space more efficiently.

In the papers by Van der Malsburg, Wiskott et al. [125–127] faces are described by a larger graph, in particular a rectangular graph (a *grid graph*) where each node label is associated to a vector of Gabor wavelet complex coefficients. In [125] only the magnitude of these coefficients is used in the recognition process; while in [127] the addition of the phase of the coefficients allowed to achieve a more accurate location of the landmark points within the considered image. Moreover, in the latter paper, a new data structure, called *bunch graph*, is introduced for dealing with generalized representations of faces. A face bunch graph (FBG) is a sort of prototype of a set of images. As the previous graphs, it has a grid structure, and each node is devoted to represent the homologous nodes (fiducial points) of the represented graphs. The term *bunch* is used to denote the set of jets referring to the same fiducial point, and associated to a node of a FBG. The FBGs used to represent the images are obtained by an elastic graph-matching procedure, described in more details in [126]. The latter paper also explores the possibility of determining facial attributes, as sex, presence or absence of glasses or beard by using FBGs.

The magnitude of Gabor coefficients as features associated to grid nodes have been also used by Duc et al. [131] and Lyons et al. [132] in combination with techniques based on discriminant analysis. In particular in [131], after the elastic graph-matching phase, the authors use a local discriminant analysis on the feature vectors associated to grid node to verify the correct identity of the input face. In [132], instead, the authors use discriminant analysis before the matching. In particular, they submit the feature vectors to a principal component analysis so as to reduce the dimensionality of the feature space. They also present results on sex, race, and expression recognition.

Instead of using Gabor coefficients, Pitas et al. in [129] associate to each node of the grid a feature vector obtained by applying a multiscale dilation–erosion operator to the input image; they also propose a variant of the elastic graph matching, called morphological elastic graph matching (MEGM) that uses in the elastic graph-matching procedure the feature vectors obtained by morphological operators. The use of such operators is justified by considering that the computation of Gabor coefficients is time consuming while dilation and erosions can be computed in a very fast way. Moreover, dilations and erosions deal with local minima or maxima in an image and revealed to provide an effective characterization of facial features.

In a more recent paper [130] the same authors describe a method to improve the recognition performance of MEGM. In particular they propose to estimate the best coefficients for weighting the similarity values associated to the grid nodes by means of discriminant analysis techniques and support vector machines.

Among the other applications dealing with face images, papers by Wang et al. [133] and Hong et al. [83] make use of graph-matching techniques in the context of *facial expression recognition* while Elagin et al. [134] use graph matching for *pose estimation*.

In particular, as usual in this application area, Hong et al. [83] use grid graphs, labeled with jets, for representing faces and rather standard elastic graph-matching algorithm for recognizing seven face expressions: neutrality (that means no expression), happiness, sadness, anger, disgust, fear, and surprise.

Only three expressions are instead considered by Wang et al. in [133]: happiness, surprise, and anger. Indeed, their main goal is rather different and is aimed to estimate the changes of face expression from sequences of facial images. To this concern, the correspondence between images relative to successive frames is viewed as an elastic matching, even if the authors call it "labeled graph-matching problem." In detail, 19 nodes are used to represent a face image. Each node is labeled using a template matrix of the 17×17 pixels (in gray levels) around each node, while to each edge is associated a measure of the distance between the nodes it links. The graph matching is carried out by minimizing a cost function that takes into account both the template similarity and the topological information.

In the framework of the pose estimation problem, Elagin et al. [134] use graphs with 16 nodes to represent a face. Each node is associated to a facial landmark, as the pupils, the tip of the nose, the mouth angles, and so on. Also in this case, the nodes are labeled with Gabor coefficients, while the labels of the edges represent the distances between the points of the image associated to the nodes. Five different orientations are considered for the pose. As in [127] a bunch graph is used to represent set of faces, and so a bunch graph-matching procedure is used in order to perform the estimation.

The use of graph matching in the context of hand posture recognition, is described in the paper by Triesch and von der Malsburg [135]. The authors employ a description and recognition scheme similar to those typically utilized in the field of face recognition. In fact, Gabor coefficient as graph labels and elastic graph matching for recognition are used. In addition to conventional Gabor jets, a *color-Gabor* jet is introduced. It measures the similarity of each pixel to the skin color and together with the Gabor jet constitutes the so-called *compound jet*. The elastic graph-matching procedure is also modified in order to cope with this compound jet. After describing each hand by graphs made up of 15 nodes manually placed at anatomically significant points, 12 different hand postures are recognized.

Another biometric system is the one proposed by Burge and Burger [136], that make use of features extracted by ear images for subject identification. They consider 300×500 pixels images, acquired using a CCD camera. Also in this case a middle level representation is used; after the localization of the ear within the images, an edge extraction based on the Canny operator is performed, followed by a curve extraction. On the basis of the regions delimited by the obtained curves, a Voronoi neighborhood graph is constructed. The identification process is accomplished by a subgraph error-correcting graph matching between the model graph and the input graph. To this aim, the authors propose a matching procedure that specifically takes into account the possibility of broken curves into the input graph. This procedure tries to merge neighboring curves if their Voronoi regions indicate that they are part of the same underlying feature.

Finally, fingerprint recognition by means of graph matching, has been addressed in the papers by Maio and Maltoni [137] by Fan et al. [138] and by Neuhaus and Bunke [93]. This latter is a technique driven paper, while the other two are application driven. They use different approaches both for representing fingerprint and for recognizing them.

The first paper [137] uses ARG for describing fingerprints. The original fingerprint image is first processed in order to calculate a directional image. Then the directional image is segmented into regions, and each region is represented by a node of the graph. Each node has an attribute that measures the area of the region it represents, while each edge has three attributes: the phase difference between the average directions, the distance between the centroids, the length of the boundary between the regions represented by the two nodes it links. For the recognition phase an inexact graph matching is proposed, based on a branch and bound search within the space state.

On the other hand, Fan et al. [138] use bipartite graphs for representing the sample fingerprint image and template fingerprints. A fingerprint image is preprocessed in order to extract clusters of feature points (minutiae). A set of 24 attributes is then calculated for each feature point cluster and is associated to a node of the graph. The feature point clusters of a test image are the set of the left nodes of a fuzzy bipartite weighted graph while the feature point clusters of the template fingerprint are the right nodes. Fingerprint verification is then treated as a fuzzy bipartite graph-matching problem.

3.4 Image Databases

Another field in which graph-based techniques have been successfully employed is the one of image databases. Typical applications involving this kind of databases are indexing and retrieval: few papers addressed both the aspect [41,139], while the most part [42,54,104,140–147] investigated only the retrieval problem. Among all these papers, in [144] Hlaoui and Wang use a simple image database only for testing the performance of a new error-correcting matching algorithm with edit operations.

The indexing and retrieval problems are very similar from a conceptual point of view, but their different requirements in terms of performance and accuracy have brought to the use of different techniques and algorithms. In both cases, the goal is to find the images in the database that are similar to a given query image. While this can bear some resemblance to a recognition problem, there is an important conceptual difference: the images in the database are not partitioned in a set of fixed, nonoverlapping classes, to which the unknown class of the query image belongs.

Instead, the images have to be considered relevant to the query only on the basis of a vaguely defined perceptual similarity; there is no clear-cut, exact desired response for the system. Furthermore, the number of images in the database can be really large, imposing strong constraints on the performance of the algorithm. In the retrieval problem it is usually desired that the result images are provided in an order that reflects a similarity scoring, to allow the user to choose interactively the one that fits his needs. This mandates for a matching technique that yields some sort of cost or distance, such as error-correcting graph-matching techniques. As regards the indexing problem, the focus is to obtain a fast screening of the images before performing a retrieval operation, to reduce the search time. So it is not required to provide a distance measure, and it is acceptable if some images that are not relevant to the query are returned in the result set (the converse is not true, i.e., it is not acceptable if indexing excludes strongly relevant images). The main concerns for indexing are how fast it performs, and how many nonrelevant images it is able to filter out.

The distribution of the matching algorithms and of the graph representations used in the image databases field is shown in Fig. 7.

A peculiarity of this applicative context is that there is little agreement on the choice of the kind of graph representation to be used for the images. In most cases images are represented by ARGs [41, 139, 140, 143], but also RAGs [42], directed ordered acyclic graphs [141], shock graphs [104], dual graphs [147], pyramidal

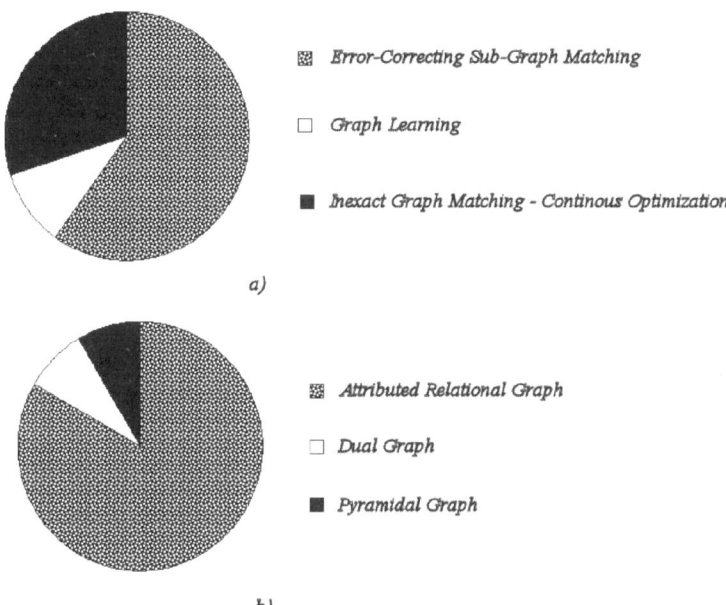

Fig. 7. Distribution of (**a**) the matching algorithms and of (**b**) the graph representations used within applications in the image databases field

graphs [142], and n nearest-neighbor graphs [54, 145, 146] are used. As regards the matching phase, mainly error-correcting subgraph isomorphism algorithms are used. In some cases, however, learning techniques have also been employed [141].

Among the papers that address both the indexing and the retrieval problem, Berretti et al. in [41] propose the use of a metric indexing scheme for managing the organization of large archives of ARGs with a common size. In particular, the indexing is performed using m-trees. They also propose a new algorithm for retrieval, combining the A* search with an original look ahead estimate. The estimate is derived as the optimal solution of a weighted assignment, which relaxes the optimal look-ahead problem so as to remove its basic factor of exponential complexity. This sort of minimal simplification results in an extremely well-informed estimate which can still be computed in polynomial time. The database used for testing the approach is composed of about 1,000 images, coming from paintings of the library of a web-museum. For each image of the database 10 further images are generated, synthetically changing color and color positions. In the system they proposed, for querying the database, an user can both select an example image or submit a query by sketch by drawing a set of colored regions and by arranging them in order to represent the expected appearance of the searched images. All the images are modeled with ARGs having a fixed number of nodes, namely eight. Each node come from the clustering of the color histogram in the $L*u*v*$ color space and the node attributes encode the triple of normalized coordinates of the average color of the cluster. For any two objects corresponding to different regions in the user sketch, the edge attribute encodes the relationship between the regions themselves.

In the paper by Petrakis and Faloutsos [139] ARGs that model medical images are reduced to a vectorial representation, so enabling R-tree indexing, under the assumption that all the graphs contain a set of anchor entities with predefined labels. Non-anchor entities are also allowed, but their number determines a linear degradation in the efficiency of the index. In addition to this indexing technique, the authors propose a subgraph isomorphism algorithm with a distance measure for retrieval. In particular, given an iconic query, all the images under a suitably chosen threshold are selected. As regards the representation, each image is segmented into regions, each one represented by a node. Size, roundness, and orientation of each region have been chosen as node attributes.

Among the papers that mainly address the retrieval problem, Cho and Yoo in [140] use graphs whose nodes represent objects of the image, while the edges encode spatial relations between objects. An object is characterized by its color, the ratio between its area and the whole image area, the ratios between the x-coordinate and the width and the ratio between the y-coordinate and the height of the image. The attribute edge can assume one of the eight possible spatial relations between two objects (N, NE, E, SE, S, SW, W, NW). They also define the *prime edge graph*, obtainable from a graph by deleting edges that are unnecessary for representing the structure of the image. The matching is realized with a subgraph isomorphism algorithm that makes use of a similarity measure.

In [143] Folkers et al. propose an exact subgraph isomorphism with a bottom-up strategy. They also define a similarity measure for pruning some isomorphism checks. The proposed measure takes into account both the contextual and the spatial similarity between ARGs. In their description scheme, once again nodes represent the symbols of the image and the edge the relationships between them.

In the papers by Hancock and Huet [54,145,146], the aim is to retrieve 2D images from large databases. In their description scheme, a set of line-patterns are represented by means of a special type of ARG, i.e., a N-nearest-neighbor graph. In a N-nearest-neighbor graph, each node represent a line structure segmented from a 2D image. For each node n of the graph exactly N edges are created, the ones that link n to the nodes representing the N line segments having the closest distances from the line represented by n itself. Distances between lines are computed by considering distances between their centers.

In [145] the authors use six nearest-neighbor graphs. The line orientation and the line length constitute the attributes of the nodes, while the measure of the relative position and of the relative orientation of two lines whose representing nodes are linked by an edge are the attribute of that edge. The proposed matching is of inexact type; in particular a fuzzy variant of the Hausdorff distance that use only the values of the edge attributes is proposed for comparing graphs. For each image graphs of 3–400 nodes are considered. A first screening of a possible query result is made by considering only the histograms of the edge attributes, that are compared using the Bhattacharyya distance. Then, the fuzzy version of the Hausdorff distance is employed on the N-nearest images that are found, for refining the search. In [54] the node attributes are two normalized histograms, the one of the relative angles and the one of the relative lengths with respect to the remaining line segments in the pattern. The matching process is realized by means of a Bayesian graph-matching algorithm that utilize a two-step process. Firstly, a correspondence matches between the nodes in the a query pattern and each of the patterns in the database is established. This is made by maximizing an a posteriori measurement probability. In particular, the authors use an extension of the graph-matching technique reported by Wilson and Hancock in [53]; in order to minimize the computational overheads associated with establishing correspondence matches only edge information are used. Once the maximum a posteriori probability correspondence matches have been established for each pattern in the database, the pattern which has maximum matching probability is selected. This is made by using the Bhattacharyya distance for comparing the histogram attributes of the matched nodes.

In [146] Huet et al. present an application of the image retrieval for verifying similarities among different technical drawings representing patents. They use ARGs obtained as six-nearest-neighbor graph from the line drawings, using the same description model of [145]. The matching is of inexact type, and is realized by means of the fuzzy variant of the Hausdorff distance presented in [145].

Among the other representation schemes employed in the literature, in [42] Gregory and Kittler utilizes RAGs. Images are segmented so that a RAG can be built. Each pixel in the image is represented as a 5D vector, where the first three dimensions are the RGB color values for the pixel and the last two dimensions are the pixel coordinates. This feature space is then clustered and to every pixel a label corresponding to the cluster which it has been classified to is given. The region labels correspond to homogeneous color regions within the image. A connected component analysis stage ensures that only to connected pixels may be assigned the same label. At this point each obtained region is represented by a node, whose attributes are the number of pixels and the average values of the red, green, and blue pixels within the region it represent. The segmentation is further improved by merging adjacent nodes which have a small number of pixels, or "similar" feature space representation. The database used for the testing phase is made up of flag images, that give rise to graphs of about 15 nodes. The matching is performed by using an error-correcting subgraph isomorphism with edit operations and the A* procedure.

On the other hand, in [104] Sharvit et al. use shock graphs for representing images. The shock graph is directly extracted from the image on the basis of the symmetries exhibited by the image itself. As regards the matching procedure, they use a WGM that is a variant of the method presented in [62]. For testing, they employ a database consisting of binary shapes, and match grayscale images of isolated objects and user-drawn sketches against this database. The resulting shock graphs are made up of few nodes.

Finally, in [147] Park et al. propose the use of dual graphs for representing images. In particular, an ARG called modified color adjacency graph (MCAG) is used for indexing and a spatial variance graph (SVG) is used to disambiguate different images having equal MCAG representations. In a MCAG each node represents a bin of the quantized RGB color histogram. Node attributes are then the pixel count of each RGB chromatic component, while the edge attributes encodes spatial adjacency (based on 8-connectivity) between two color regions. The average number of nodes of a MCAG is about 100. On the other hand, each node of the SVG graph has as attribute the within-class variance relative to the pixels of the node it represents, while each edge attribute encodes the between-class variance. Graph matching is performed by defining a similarity measure directly obtainable by the adjacency matrices of the graphs.

Finally, in [141] a learning technique for facing the retrieval problem is proposed by De Mauro et al. Database images are described by means of RAG that are successively transformed into directed ordered acyclic graphs (DOAG). This transformation becomes necessary because it is more difficult to process undirected graphs than directed ones. The task of learning the search criteria for visual retrieval is accomplished by means of a Recursive Neural Network that map DOAGs into vectors. This net learn to map DOAG representing similar images into near vectors. Then, the retrieval problem is reformulated as the one of finding the N-nearest neighbors of the vector into which the net transform the DOAG of the query image.

3.5 Video Analysis

Among the video analysis problems addressed by using graph-based techniques, retrieval from video databases [21, 142, 148], annotation of video databases [149], object tracking [150–153], and motion estimation [154] have been proposed.

Retrieval from video databases is similar to retrieval from static image databases from a conceptual point of view; the main differences are the considerably larger size of the databases and the possibility to exploit information about the motion of parts of the scene to improve the retrieval performance. The other problems, instead, are rather peculiar of video analysis. In particular, their common aspect is that they are focused on extracting some kind of information from the sequence of the frames composing a video. This implies a comparison between successive frames, and the need to establish a correspondence between regions of two frames representing the same object or the same part of it. In motion estimation, the goal is to measure the velocity of moving elements of the scene. In object tracking, which can be considered as an evolution of motion estimation, where the application should be able to follow the motion of an object and compute its trajectory, distinguishing the different objects presents in the scene. A further evolution consists in the recognition, on the basis of the object trajectories, of events that bear a specific meaning within the context of the application: this gives rise to the possibility of automatic annotation of the video sequence, allowing a user to perform retrieval with classic, keyword-based search.

Since these problems are quite different each other, it should not be amazing the fact that very different kinds of graphs have been used. In particular, ARGs [21, 148], pyramidal graphs [142, 153], bipartite graphs [150, 152], multivalued neighborhood graphs [151], and medial graphs [154] are used. Obviously, also the matching techniques proposed in the various paper are quite different. The distribution of the matching algorithms and of the graph representations used in this applicative area is summarized in Fig. 8.

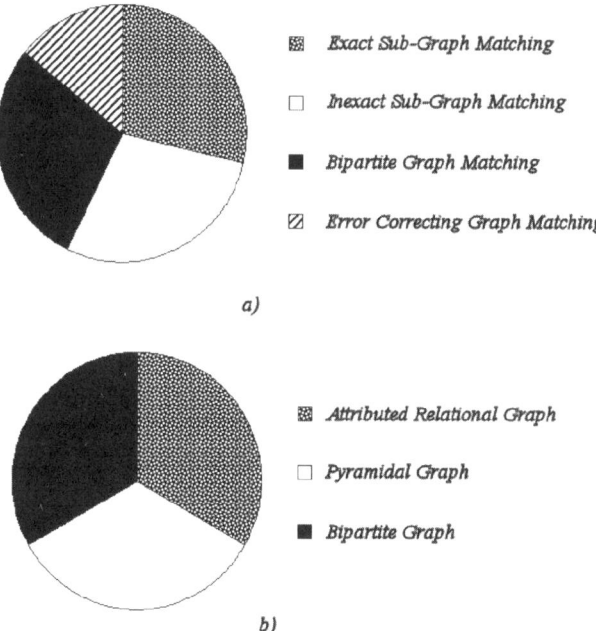

Fig. 8. Distribution of (**a**) the matching algorithms and of (**b**) the graph representations used within applications in the video analysis field

In the framework of the retrieval from video databases, Shearer et al. [21, 148] proposed two different approaches that do not make particular hypotheses on the nature of the video at hand, where Doulamis et al. [142] propose a system specifically tailored for the retrieval of people images in a video database. Furthermore, in the first two papers ARGs are used for representing video frames, while the last one propose the use of pyramidal graphs.

Entering into details, in [21] Shearer et al. describe a new algorithm to solve the largest common subgraph problem. Such algorithm significantly reduces the computational complexity of detection of the largest common subgraph between a known database of models, and a query given online. This approach can be fruitfully applied to video databases. In fact, when searching a video database, we are typically interested in the largest subpicture match that can be found. So, the largest common subgraph method will find the largest subpicture in common between a query image and a database of video frames. As regard the representation, ARGs are used. The authors consider each frame of the video and decompose it into objects. Then, graph nodes represent objects, while the edges are labeled with one of five categories (*Disjoint – Meets – Contains – Belongs to – Overlaps*) that represent the relationships between two objects. The proposed retrieval procedure is realized by using a decision tree algorithm based on a decision tree constructed using the adjacency matrix representation for the model graphs.

A different approach is presented in [148], where a modified version of an algorithm presented by Bunke and Messmer in [18] is proposed. It is able to cope with dynamically changing graphs. Such graphs can be employed for representing videos: the sequence of images that make a video can be represented by means of an initial graphs that represent the initial image and a sequence of graph edit operation that represent the successive images. As in the previous paper, for each image the nodes of the graph represent objects, while the edges encode the spatial relations between objects. An experimental evaluation of the algorithm is also presented, by using query graphs with 9 nodes against models having 4–10 nodes. In particular, the application of this algorithm consists in querying a video database with a sequence of frames. Each query frame is built starting from a number of object labels that can be spatially arranged by the user. The system transforms these query frames into a graph representing the initial frame and into a set of edit operations. Then, it uses the proposed matching algorithm in order to find the video sequence that match the sequence of selected frames.

In [142] Doulamis et al. propose a system for extracting people images from MPEG-coded videos. After a segmentation phase in which objects such as the face, the human body, and the background are extracted from each frame, graphs are used for representing these objects and their spatial relationships. As attribute of the nodes, the average color and the texture of an object, as well as its size and location within the scene are considered. The authors make use of two different types of graphs, one with edge attributes, that encode the direction and the orientation between two objects, and another one without edge attributes. Moreover, in order to enhance the querying flexibility of their system, they also propose a further decomposition of each node into other graphs, so giving rise to a pyramidal graph representation of the visual content. As an example, the human face can be considered as an object containing the regions of eyes, mouth, and lips, each having their own properties. As regards the problem of retrieval from a video database, they do not clarify in the paper what type of graph-matching technique they use.

A quite peculiar approach to the problem of retrieval from databases is the one presented by Ozer et al. in [149]. The aim of this work is to annotate images and/or videos where a particular object of interest (OOI) is present. So, a simple textual query can be performed to extract images of OOI from a preprocessed database. As an example, they consider cars in video and image libraries, that they describe using ARGs. In case of video sequences, the feature points of an object are tracked and then grouped together according to their moving directions and distances. The object extraction is performed by means of a color image segmentation technique combined with an edge detector algorithm. Since an object usually contains several subobjects (in this case wheels, windows, lights, etc. of a car) a hierarchical segmentation scheme is also proposed. Three different views of a car are considered – front view, rear view, and side view. The three subgraphs relative to these views are joined together to form a unique graph representing all the possible views of the object. As attributes of the nodes, Hu moments and the compactness of the segmented regions are considered. Given two adjacent regions represented by two nodes, the ratio of the areas, the ratio of the perimeters, the relative position and orientation, and the overlapping area between two adjacent regions are the attribute of the edge that links those nodes. As regards the graph-matching procedure, they propose an inexact subgraph matching with a matching cost based on the attribute values, using a depth-first search with a brute force approach.

The papers by Chen et al. [150], Gomila and Mayer [151], and Conte et al. [152, 153] exploit the use of graph matching for object tracking in video sequences. They use different middle level representations and also different matching techniques.

In Conte et al. [152] the definition and the performance assessment of a tracking method devised for video-surveillance applications are presented. The tracking problem is factorized into two subproblems: the first is the definition of a suitable measure of similarity between regions in adjacent frames. Provided with this measure, the second subproblem is the search for an optimal matching between the regions appearing in the frames. As regards the first subproblem (the definition of a similarity measure), several different metrics are proposed, jointly used during the detection phase, according to a sort of signal fusion approach. The subproblem of the optimal matching has been instead formulated in a graph-theoretic framework, and then reduced to a weighted bipartite graph matching, for which a standard algorithm has been used.

Chen et al. [150] apply a shape contour extraction and a shadow deletion to each frame. Therefore they obtain the silhouette of each object within the scene. To each object a probability distribution is associated, that takes into account the intensity values of the area within the object contour. To model the multiobject tracking problem a bipartite graph is used. Each node represents an object and has as attributes its position, its intensity distribution, and the dimension of its enclosing bounding box. The two classes of nodes in the bipartite graph are the so-called *profile nodes* and *object nodes* that correspond to the objects in the past and the present frame respectively. A bipartite matching algorithm is used to find the best match among nodes of the two successive frames in order to resolve the identities of the objects. If there are unmatched nodes, it implies that new objects have been detected and so new profiles will be created for tracking them within the successive frames.

On the other hand, Gomila and Mayer [151] segment the image of each frame on the basis of the color information and represents the segmented image with a multivalued neighborhood graph. Node attributes measure the intrinsic features of the region they represents, while edge attributes represent relational constraints between nodes. Matching graphs relative to two successive frames permits to follow the objects along the video sequence. In order to cope with different segmentation of the same object in two successive frames, split and merge operation are performed on the images before the matching. The proposed matching algorithm is an error correcting one using the relaxation labeling.

Conte et al. [153] use a multiresolution graph pyramid for representing objects at different levels of detail. They use a hierarchical graph-matching procedure to deal with partial occlusions of the objects being tracked. The advantage of their approach is that it uses a fast, coarse grained, weighted bipartite graph matching as long as there are no occlusions in the scene. When two tracked objects come to overlap, a more refined subgraph isomorphism procedure is used to distinguish the parts of the occluding objects, possibly recurring to a finer level or detail until a reasonable solution is found.

Finally, Salotti and Laachfoubi in [154] present an application of motion estimation in aerial videos. Given an aerial video, their aim is to estimate the shift of the part of the image that represents the smoke, in order to collect information for preventing fires. They use *topographic graphs* (that are similar to medial graphs) for describing aerial images. Each frame of the video is segmented on the basis of the color information and the smoke area is described by means of a topographic graph. The shift estimation is performed by means of an inexact matching procedure that defines a cost function for matching nodes of two topographic graphs relative to successive images. These cost function examines only shifts in a small square window centered on each node, since it is reasonable that the move of the smoke is not too fast.

3.6 Miscellanea

Besides the application areas detailed in previous sections, there are other application-driven papers that are not strongly related to each other, neither on the basis of the problem they face, nor on the basis of the adopted approach (besides, obviously, the fact that they are based on graph matching). For the sake of completeness, we present here some of these works, although in these cases we cannot individuate any sort of common scheme in the exploitation of graph matching. The papers we have chosen deal with biomedical [38, 133, 155, 156] and the

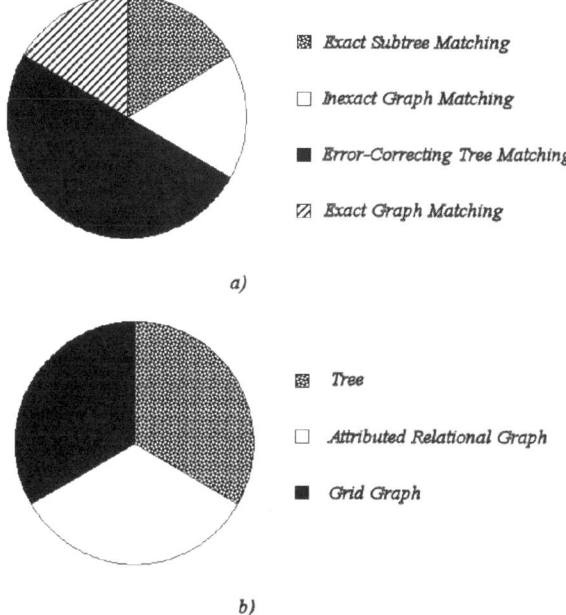

Fig. 9. Distribution of (**a**) the matching algorithms and of (**b**) the graph representations used within applications in the miscellanea field

biological [157, 158] applications (see also Fig. 9 for the distribution of the matching algorithms and the graph representations used in this case).

Namely, the paper by Wang et al. [133] is technique driven, and presents an algorithm for finding the largest approximately common substructure of two trees. In the experimental results, the authors discuss its application for finding motifs in multiple RNA secondary structures. On the contrary, both the biomedical application described in [38, 155] have their main focus on the application context and address the problem of the correct identification of coronary arteries (artery labeling) starting from medical images, and are based on prior knowledge about the expected coronary arterial tree (CAT) structure and attributes. They use different graph-matching techniques to realize the labeling.

In the paper by Dumay et al. [38], the authors start from an arteriogram image and project a geometric model of the artery against the image. From this projected model, an ARG made up of about ten nodes is constructed: the nodes of the graph represents arterial segments and have as attributes the position, the mean diameter, and the orientation of the segments, while edges represent the parental relationship (parent–child and grandparent–child) between segments. Starting from the anatomy of a left coronary tree of normal functioning hearts, an inexact graph-matching procedure is used in order to assign anatomic labels to the node of an input image. Since missing branch and/or false structures can corrupt the input image, a cost function is defined in order to cope with transformations (substitution, insertion, and deletion of a node and/or an edge) of the input graph. An A* algorithm is used to perform the state space search.

Also the paper by Charnoz et al. [156] faces a somewhat analogous problem: the matching of several CAT-images of the intrahepatic vascular system of a same patient, acquired at different times. The authors propose an error-correcting tree matching algorithm that is robust with respect to topological modifications.

In the paper by Haris et al. [155], the authors use ARGs to represent the CAT. Starting from the input image, the CAT is detected by constructing an approximation of its centerline and borders. This results in a directed acyclic graph representing the CAT. The attribute of each node of this graph are the position of the artery element it represents, the direction of the artery and its approximate width. Given the input graph and a 3D CAT model which encapsulates the expected anatomic and geometric structure of a normal human CAT, a graph-matching algorithm assigns the appropriate labels to the input CAT using weighted maximal cliques on the association graph corresponding to the two given graphs. So the labeling problem is reformulated as one of finding the best maximal clique of the association graph.

A biological application is the identification of diatoms described by Fischer et. al. [157] and Ambauen et al. [158]. Diatoms are unicellular algae found in water and in other places where there is sufficient humidity and light for allowing photosynthesis. The technique used for describing diatom images is the same used in the face recognition field. A middle-level representation based on labeled grid graphs is used. On each image a rectangular 16×8 grid is superimposed and each node of the graph is associated to a rectangle of the image. Each node is labeled with 13 features derived from the gray-level co-occurrence matrices and from the Gabor coefficients. In [157] the matching procedure can be seen as a simple form of error-correcting graph matching. A dissimilarity measure is evaluated between two grid graphs as the sum of the distance between the feature vectors associated to the nodes. Moreover, in order to cope with geometric distortions, also translations of the nodes are allowed and a specific cost is introduced into the dissimilarity measure in order to weigh such translations. In [158] a more complex matching algorithm is proposed, based on the addition of new edit operations to the classical set of deletion, insertion, and mutation.

4 Performance Comparison

By reviewing the wide literature in the field of graph matching, it appears evident that the habit of proposing more and more new algorithms is prevailing against the need of assessing the performance of the existing ones in an objective way. The characterization of the graph-matching algorithms proposed up to now would instead allow potential users to predict the performance of an algorithm – at least to some degree – and could thus lead to substantial savings in system development time.

Starting from these considerations, the IAPR-TC15 community in the second GbR workshop held in 1999 (see the TC15website at the address: http://

`www.iapr-tc15.unisa.it`) declared the need for a serious benchmarking activity in the context of graph-matching algorithms. According to the proposal made in [159], such activity could start with exact graph-matching algorithms, by considering different kinds of morphism and by suitably defining the number of graphs to be matched, the size of the input graphs and the graph structure. Even if only exact graph-matching algorithms are considered, it is also worth noting that the matching problem may have, as pointed out in [160]:

- No solution (e.g., if the graphs are not isomorphic, or if there is no subgraph isomorphic to the given graph)
- One solution
- More than one solution (e.g., a square mesh matches its versions rotated by 90, 180, and 360°)

Generally, not only a fast solution for the second case is required, since a small disturbance caused by noise may transform it into the first case. Then, it is relevant the time needed by an algorithm for finding out that there is no perfect solution (*nonmatching* time). Nonmatching times are also crucial factors in the context of graph database filtering, as noted in [161]. Finally, the third case concerns algorithms that can find one or all the possible solutions to the given matching problem.

If we try to understand the reasons why up to now only a small number of serious attempts [11,162–164] has been made for comparing graph-matching algorithms, we can easily recognize that one of the main difficulties is the fact that only in the last few years some standard databases of graphs specifically designed for this purpose have been made available. The creation of a graph database, in fact, is definitely not a simple task, since several issues have to be faced [11]. Generally speaking, two approaches can be followed for generating a database; a first way is to start from graphs obtained from real data, otherwise the database can be obtained synthetically. Although the first approach allows us to obtain rather realistic graphs, it is generally more expensive as it requires the collection of real data and the selection of the set of algorithms to be used for obtaining graphs from data. In this case the graphs are dependent on both the domain under consideration and the preprocessing algorithm used, reducing significantly the generality of the database and its reusability in other contexts. On the contrary, the artificial generation of graphs is not only simpler and faster than collecting graphs from real applications, but also allows us to control the variation of several critical parameters of the underlying graph population, such as the average number of nodes, average number of edges per node, number of different labels, and so on.

By following the latter approach, three proposal have been recently made in the scientific community, in order to provide standard graph databases. The first two [164, 165] gave rise to databases of synthetically generated graphs explicitly devised for benchmarking (sub)graph isomorphism algorithms and MCS algorithms, respectively, while the third proposal [166] is based on the generation of a database of artificial images – by using a set of attributed plex grammars – and of their corresponding graph representations.

4.1 Graph Databases

The choice of the kind of graphs to be included in the first two cited database derived from an analysis of the graphs mainly used by members of the IAPR-TC15 community. Both databases are structured in pairs of graphs. In the first database, two categories of pairs of graphs have been introduced, namely pairs made of isomorphic graphs and pairs made of graphs in which the second graph is a subgraph of the first one. In the second database each pair of graphs has a MCS of at least two nodes.

In both cases, each category of pairs is made up of graphs that are different for structure and size. In particular, the following kinds of graphs have been considered (see [165] for a detailed discussion about their properties and the parameters characterizing them):

- *Randomly Connected Graphs*
- *Regular Meshes*, with different dimensionality: 2D, 3D and 4D
- *Irregular Meshes*
- *Bounded Valence Graphs*
- *Irregular Bounded Valence Graphs*

Randomly connected graphs are graphs in which the edges connect nodes without any structural regularity (see Fig. 10a). They can model applications in which objects (represented by nodes) can establish relations (represented by edges) with any other objects (not only the surrounding ones) independently of the relative positions. This hypothesis typically occurs in the middle and high processing levels of an image processing task.

Regular meshes (see Fig. 10b) have been introduced for simulating applications dealing with regular structures as those operating at the lower levels of an image processing task; while *Irregular mesh-connected* graphs (see Fig. 10c) can be used for simulating the behavior of graph-matching algorithms in presence of slightly distorted meshes. *Bounded valence graphs* (see Fig. 10d) model applications in which each object establish a fixed number of relations with other object, not necessarily with those belonging to its neighborhood. In order to introduce some irregularities in these kind of graphs, *Irregular bounded valence graphs* have been introduced too (see Fig. 10e).

So, in the first database (hereinafter denoted as *ISO-DB*) a total of 72,800 pairs of graphs have been generated: 18,200 pairs of isomorphic graphs and 54,600 pairs for which a subgraph isomorphism exists. Each kind of graphs has pairs of different size, ranging from few dozens to about 1,000 nodes (i.e., small and medium size graphs according to the classification presented in [159]). For each size and kind of graphs 100 different pairs have been generated. Moreover, in case of graph subgraph isomorphism, pairs in which the two graphs have three different size ratios have been generated.

The graphs composing the whole database have been distributed on a CD during the third IAPR-TC15 Workshop on Graph-Based Representations in Pattern Recognition and are also publicly available on the web at the URL: http://amalfi.dis.unina.it/graph.

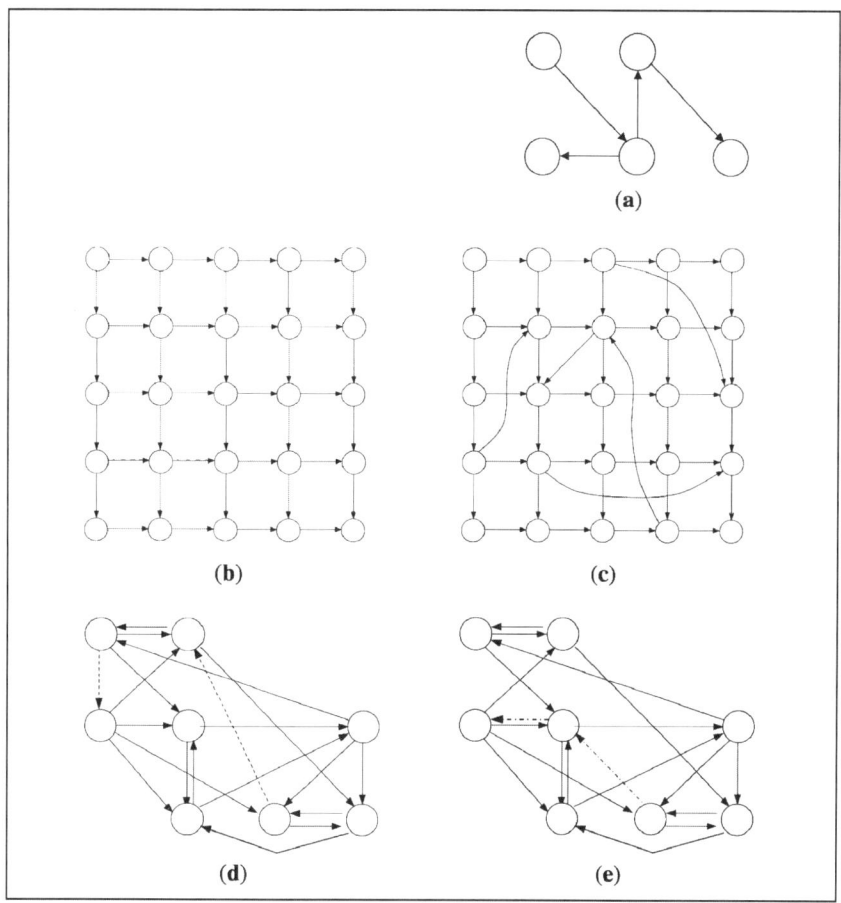

Fig. 10. (a) An example of a randomly connected graph. (b) A 2D regular mesh with size 5×5 and (c) an irregular mesh obtained by adding five further edges to the graph (b). For each added edge, the starting and the ending nodes are randomly determined according to an uniform distribution. (d) A bounded valence graph with a valence equal to 5; (e) an irregular bounded valence graph, obtained from the graph (d) by moving the two dashed edges

For the second database (from now on denoted as *MCS-DB*), a total of 81,400 pairs of labeled graphs have been generated. The choice of the labeling comes from the need of restricting the number of possible node or edge pairings because of the complexity of the MCS problem. For each of the above-mentioned kind of graphs, pairs of graphs having different sizes N, ranging from 10 to 100, have been included in the database. Moreover, for each value of N, five different sizes of the MCS have been taken into account and 100 pairs of graphs have been generated for each size. As regards the labeling, the authors proposal was to generate random values for the attributes, since any other choice would imply assumptions about an application

124   D. Conte et al.

dependent model of the represented graphs. The whole database is publicly available on the web at the URL: http://amalfi.dis.unina.it/graph.

The third database comes from an image generation method (downloadable from the URL: http://www.artificial-neural.net/) based on a combination of attribute grammars and plex grammars [167]. A plex grammar is a generic mechanism allowing to specify a number of rules that describe how an image is built up from simpler constituents. Because the rules can be recursive, a potentially infinite set of images can be described by a finite number of rules.

According to the authors' proposal [166], an image generated according to the previous method can be simply converted into a graphical representation. An example referring to the image of a policeman is reported in Fig. 11. The nodes of the graph correspond to regions in the image, while the edges represent spatial relations between the regions. A number of different attributes can be computed for each node and each edge. Examples of node attributes are color, center of gravity, and size of a region; while edge attributes represent geometric relations between regions (e.g., angle and distance of centers of gravity).

Since the rules of the attributed plex grammar are defined by the user, there are no restrictions on the underlying domain and there is not a predefined graphical representation of an image. In fact, through a number of parameters the user can choose the representation that is most suitable for the problem at hand. It is then possible to create image databases, together with their graphical representations, for various kinds of PR problems involving graphs. Exact and inexact graph matching, supervised and unsupervised learning of graphical representations from examples and graph clustering are actual examples.

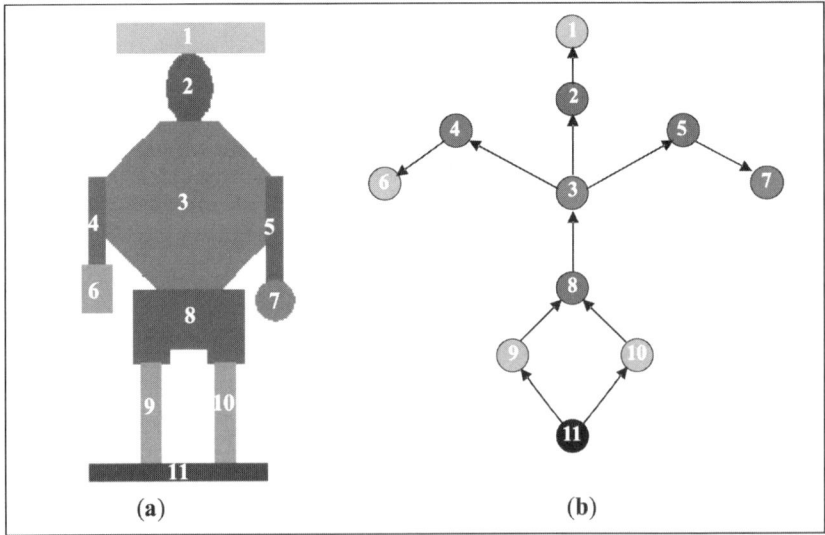

Fig. 11. (a) The image of a policeman generated according to the method proposed in [166] and (b) its representation as a graph. Labels associated with each node are not shown

Even if some authors [141, 163, 168] used the above-described image database for testing their graph-based approaches, no paper reporting a serious comparison of algorithms on graphs generated from it has been presented so far. On the other hand, there are some papers presenting benchmarking activities using graphs extracted from both the *ISO-DB* and the *MCS-DB*.

4.2 Benchmarking Activities

One of the first attempts to compare graph-matching algorithms has been made in [162]. Five inexact graph-matching algorithms have been considered and compared with respect to both the speed of algorithms, capacity for classification and suitability for different kind of graphs. Algorithms were tested in two real-world classification problems. Nevertheless, since the size of the graphs used in the tests ranges from three to nine nodes, the current usefulness of the obtained results is quite limited.

More recently, three works have tried to follow the indications of the IAPR TC 15 community for carried out a noteworthy activity in the context of graph-matching algorithms.

In particular, in [11], four exact graph-matching algorithms have been compared with respect to the times needed for finding a match on pairs of isomorphic graphs extracted from the *ISO-DB*. In particular the Ullmann's algorithm [5], the algorithm by Schmidt and Druffel [6], the VF2 algorithm [13] and Nauty [16] have been considered. As it could be expected, the authors conclude that an algorithm performing definitively better than all the others does not exist. In particular, for randomly connected graphs, the Nauty algorithm is the better if the graphs are quite dense and/or of quite large size. For smaller and quite sparse graphs, on the contrary, VF2 performs better. On more regular graphs, i.e., on 2D meshes, VF2 is definitely the best algorithm: in this case the Nauty algorithm is even not able to find a solution for graphs bigger than few dozens of nodes. In case of bounded valence graph, if the valence is small, VF2 is always the best algorithm, while for bigger values of the valence the Nauty algorithm is more convenient if the size of the graphs is small.

Two of the above-described algorithms, namely VF2 and Ullmann, have been also extensively compared in [161] with respect to their *nonmatching* times, i.e., the time needed for declaring that there is not a match between two given graphs. The comparison has been carried out on graphs extracted from the *ISO-DB*; both the cases of pairs of graphs with the same number of nodes and with a different number of nodes have been considered.

According to the tests reported in the paper, the *nonmatching* times obtained by the Ullmann's algorithm are almost always smaller than those achieved by the VF2 algorithm when pairs of graphs with the same number of nodes are taken into account. Only in case of very regular graphs, i.e., high-dimensional meshes, the nonmatching times of the two algorithms are practically identical. This behavior is substantially confirmed in case of graphs with different number of nodes, even if in this case the *nonmatching* times of VF2 are smaller for regular bounded valence graphs with at least 70 nodes and for high-dimensional meshes.

Finally, in [164] the matching times of three MCS algorithms, namely those proposed by McGregor [169], by Balas and Yu [170], and by Durand et al. [15], have been compared by using pairs of graphs extracted from the *MCS-DB*.

According to the authors, also in this case it does not exist an algorithm that is definitively better than the others. In particular, for randomly connected graphs, the McGregor algorithm is the best one if the graphs are quite sparse and/or of quite small size. For larger and quite dense graphs, on the contrary, the Durand et al. algorithm performs better. If the MCS has a more regular structure, i.e., on mesh-like MCS, the McGregor algorithm is in most cases the best algorithm; the Durand et al. algorithm performs better only for small and dense graphs. On the contrary, when the irregularity degree grows up, the Balas–Yu algorithm performs better for large and dense graphs. In case of bounded valence graphs, the McGregor algorithm is the best one if the valence is small, while for larger values of the valence the Durand et al. algorithm is more convenient when the graphs to be matched are dense. If the graphs are both dense and large, the Balas Yu algorithm is the best one. Finally, when an irregularity degree is added to the bounded valence graphs, the Durand–Pasari algorithm performs better in most cases, even if the Balas–Yu algorithm is still winning for large graphs.

5 Conclusions

In this paper we have presented a comprehensive review of graph-matching methods used in PR and computer vision applications, highlighting the relationships between the application domain and specific problem on one side, and the adopted graph representation and matching technique on the other side. An evaluation of the proposed methods and tools for assessing the performance of a graph-matching algorithm completed our work.

All together, these two parts provide useful information to applied researchers for deciding which graph-based technique best fits their needs.

References

1. Conte D, Foggia P, Sansone C, Vento M (2004) Thirty years of graph matching in pattern recognition. International Journal of Pattern Recognition and Artificial Intelligence 18(3):265–298
2. Aho AV, Hopcroft JE, Ullman JD (1974) The Design and Analysis of Computer Algorithms. Addison Wesley, Reading, MA
3. Hopcroft JE, Wong J (1974) Linear time algorithm for isomorphism of planar graphs. Proceedings of the Sixth Annual ACM Symposium on Theory of Computing, pp. 172–184
4. Luks EM (1982) Isomorphism of graphs of bounded valence can be tested in polynomial time. Journal of Computer and Systems Sciences 25:42–65
5. Ullman JR (1976) An algorithm for subgraph isomorphism. Journal of the Association for Computer Machinery 23:31–42

6. Schmidt DC, Druffel LE (1976) A fast backtracking algorithm to test directed graphs for isomorphism using distance matrices. Journal of the ACM 23:433–445
7. Ghahraman DE, Wong AKC, Au T (1980) Graph monomorphism algorithms. IEEE Transactions on Systems, Man and Cybernetics 10(4):189–197
8. Cordella LP, Foggia P, Sansone C, Vento M (2000) Fast graph matching for detecting CAD image components. Proceedings of 15th International Conference on Pattern Recognition, pp. 1034–1037
9. Cordella LP, Foggia P, Sansone C, Tortorella F, Vento M (1998) Graph matching: a fast algorithm and its evaluation. Proceedings of 14th International Conference on Pattern Recognition, pp. 1582–1584
10. Cordella LP, Foggia P, Sansone C, Vento M (1999) Performance evaluation of the VF graph matching algorithm. Proceedings of the International Conference on Image Analysis and Processing, pp. 1172–1177
11. De Santo M, Foggia P, Sansone C, Vento M (2003) A large database of graphs and its use for benchmarking graph isomorphism algorithms. Pattern Recognition Letters 24(8):1067–1079
12. Cordella LP, Foggia P, Sansone C, Vento M (2001) An improved algorithm for matching large graphs. Proceedings of the Third IAPR-TC15 Workshop on Graph-based Representations in Pattern Recognition, pp. 149–159
13. Cordella LP, Foggia P, Sansone C, Vento M (2004) A (sub)graph isomorphism algorithm for matching large graphs. IEEE Transactions on Pattern Analysis and Machine Intelligence 26(10):1367–1372
14. Larrosa J, Valiente G (2002) Constraint satisfaction algorithms for graph pattern matching. Mathematical Structural in Computer Science 12:403–422
15. Durand PJ, Pasari R, Baker JW, and Tsai CC (1999) An efficient algorithm for similarity analysis of molecules. Internet Journal of Chemistry 2
16. McKay BD (1981) Practical graph isomorphism. Congressus Numerantium 30:45–87
17. Foggia P, Sansone C, Vento M (2001) A performance comparison of five algorithms for graph isomorphism. Proceedings of the Third IAPR-TC15 Workshop on Graph-based Representations in Pattern Recognition, pp. 188–199
18. Bunke H, Messmer BT (1997) Recent advances in graph matching. International Journal of Pattern Recognition and Artificial Intelligence 11(1):169–203
19. Messmer BT, Bunke H (1999) A decision tree approach to graph and subgraph isomorphism detection. Pattern Recognition 32(12):1979–1998
20. Shearer K, Bunke H, Venkatesh S, Kieronska S (1997) Efficient graph matching for video indexing. Computing. Supplement 12:53–62
21. Shearer K, Bunke H, Venkatesh S (2001) Video indexing and similarity retrieval by largest common subgraph detection using decision trees. Pattern Recognition 34(5):1075–1091
22. Lazarescu M, Bunke H, Venkatesh S (2000) Graph matching: fast candidate elimination using machine learning techniques. Proceeding of the Joint IAPR International Workshops SSPR and SPR, pp. 236–245
23. Irniger C, Bunke H (2001) Graph matching: filtering large databases of graphs using decision trees. Proceedings of the Third IAPR-TC15 Workshop on Graph-based Representations in Pattern Recognition, pp. 239–249
24. Bunke H (1997) On a relation between graph edit distance and maximum common subgraph. Pattern Recognition Letters 18(8):689–694
25. Bunke H, Shearer K (1998) A graph distance metric based on the maximal common subgraph. Pattern Recognition Letters 19(3):255–259

26. Bunke H (1999) Error correcting graph matching: On the influence of the underlying cost function. IEEE Transactions on PAMI 21(9):917–922
27. Fernndez ML, Valiente G (2001) A graph distance metric combining maximum common subgraph and minimum common supergraph. Pattern Recognition Letters 22(6):753–758
28. Wallis WD, Shoubridge P, Kraetz M, Ray D (2001) Graph distances using graph union. Pattern Recognition Letters 22(6):701–704
29. Tsai WH, Fu KS (1979) Error-correcting isomorphisms of attributed relational graphs for pattern analysis. IEEE Transactions on Systems, Man and Cybernetics 9(12): 757–768
30. Tsai WH, Fu KS (1983) Subgraph error-correcting isomorphisms for syntactic pattern recognition. IEEE Transactions on Systems, Man and Cybernetics 13(1):48–61
31. Wong AKC, You M, Chan SC (1990) An algorithm for graph optimal monomorphism. IEEE Transactions on Systems, Man and Cybernetics 20(3):628–638
32. Eshera MA, Fu KS (1984) A similarity measure between attributed relational graphs for image analysis. Proceedings of the Seventh International Conference on Pattern Recognition, pp. 75–77
33. Eshera MA, Fu KS (1984) A graph distance measure for image analysis. IEEE Transactions on Systems, Man and Cybernetics 14(3):398–408
34. Sanfeliu A, Fu KS (1983) A distance measure between attributed relational graphs for pattern recognition. IEEE Transactions on Systems, Man and Cybernetics 13(3): 353–363
35. Ghahraman DE, Wong AKC, Au T (1980) Graph optimal monomorphism algorithms. IEEE Transactions on Systems, Man and Cybernetics 10(4):181–188
36. Shapiro LG, Haralick RM (1981) Structural descriptions and inexact matching. IEEE Transactions on Pattern Analysis and Machine Intelligence 3(5):504–519
37. Shapiro LG, Haralick RM (1985) A metric for comparing relational descriptions. IEEE Transactions on Pattern Analysis and Machine Intelligence 7(1):90–94
38. Dumay ACM, van der Geest RJ, Gerbrands JJ, Jansen E, Reiber JHC (1992) Consistent inexact graph matching applied to labelling coronary segments in arteriograms. Proceedings of the International Conference on Pattern Recognition, Conference C, pp. 439–442
39. Berretti S, Del Bimbo A, Vicario E (2000) A look-ahead strategy for graph matching in retrieval by spatial arrangement. International Conference on Multimedia and Expo, pp. 1721–1724
40. Berretti S, Del Bimbo A, Vicario E (2000) The computational aspect of retrieval by spatial arrangement. Proceedings of 15th International Conference on Pattern Recognition, pp. 1047–1051
41. Berretti S, Del Bimbo A, Vicario E (2001) Efficient matching and indexing of graph models in content-based retrieval. IEEE Transactions on Pattern Analysis and Machine Intelligence 23(10):1089–1105
42. Gregory L, Kittler J (2002) Using graph search techniques for contextual colour retrieval. Proceeding of the Joint IAPR International Workshops SSPR and SPR, pp. 186–194
43. Cordella LP, Foggia P, Sansone C, Vento M (1996) An efficient algorithm for the inexact matching of ARG graphs using a contextual transformational model. Proceedings of the 13th International Conference on Pattern Recognition, pp. 180–184
44. Cordella LP, Foggia P, Sansone C, Vento M (1997) Subgraph trasformations for inexact matching of attributed relational graphs. Computing. Supplement 12:43–52

45. Serratosa F, Alquezar R, Sanfeliu A (1999) Function-described graphs: A fast algorithm to compute a sub-optimal matching measure. Proceedings of the Second IAPR-TC15 Workshop on Graph-based Representations in Pattern Recognition, pp. 71–77
46. Serratosa F, Alquezar R, Sanfeliu A (2000) Efficient algorithms for matching attributed graphs and function-described graphs. Proceedings of 15th International Conference on Pattern Recognition, pp. 867–872
47. Rocha J, Pavlidis T (1994) A shape analysis model with applications to a character recognition system. IEEE Transactions on Pattern Analysis and Machine Intelligence 16(4):393–404
48. Wang C, Abe K (1995) Region correspondence by inexact attributed planar graph matching. Proceedings of the Fifth International Conference on Computer Vision, pp. 440–447
49. Llados J, Marti E, Villanueva JJ (2001) Symbol recognition by error-tolerant subgraph matching between region adjacency graphs. IEEE Transactions on Pattern Analysis and Machine Intelligence 23(10):1137–1143
50. Fischler M, Elschlager R (1973) The representation and matching of pictorial structures. IEEE Transactions on Computers (22):67–92
51. Kittler J, Hancock ER (1989) Combining evidence in probabilistic relaxation. International Journal of Pattern Recognition and Artificial Intelligence 3:29–51
52. Christmas WJ, Kittler J, Petrou M (1995) Structural matching in computer vision using probabilistic relaxation. IEEE Transactions on Pattern Analysis and Machine Intelligence 17(8):749–764
53. Wilson RC, Hancock ER (1997) Structural matching by discrete relaxation. IEEE Transactions on Pattern Analysis and Machine Intelligence 19(6):634–648
54. Huet B, Hancock ER (1999) Shape recognition from large image libraries by inexact graph matching. Pattern Recognition Letters 20:1259–1269
55. Myers R, Wilson RC, Hancock ER (2000) Bayesian graph edit distance. IEEE Transactions on Pattern Analysis and Machine Intelligence 22(6):628–635
56. Torsello A, Hancock ER (2001) Computing approximate tree edit-distance using relaxation labeling. Proceedings of the Third IAPR-TC15 Workshop on Graph-based Representations in Pattern Recognition, pp. 125–136
57. Luo B, Hancock ER (2001) Structural graph matching using the EM algorithm and singular value decomposition. IEEE Transactions on Pattern Analysis and Machine Intelligence 23(10):1120–1136
58. Dempster AP, Laird AP, and Rubin DB (1977) Maximum-likelihood from incomplete data via the EM algorithm. Journal of the Royal Statistical Society Series B 39:1–38
59. Almohamad HA, Duffuaa SO (1993) A linear programming approach for the weighted graph matching problem. IEEE Transactions on Pattern Analysis and Machine Intelligence 15(5):522–525
60. Lawler ES (2001) Combinatorial Optimization: Networks and Matroids. Dover, New York
61. Rangarajan A, Mjolsness ED (1996) A Lagrangian relaxation network for graph matching. IEEE Transactions on Neural Networks 7(6):1365–1381
62. Gold S, Rangarajan A (1996) A graduated assignment algorithm for graph matching. IEEE Transactions on Pattern Analysis and Machine Intelligence 18(4):377–388
63. Bomze IM (1997) Evolution towards the maximum clique. Journal of Global Optimization 10:143–164
64. Pelillo M (1995) Relaxation labeling networks for the maximum clique problem. Journal of Artificial Neural Networks 2:313–328

65. Pelillo M, A Jagota (1995) Feasible and infeasible maxima in a quadratic program for maximum clique. Journal of Artificial Neural Networks 2:411–420
66. Rosenfeld A, Hummel RA, Zucker SW (1976) Scene labelling by relaxation operations. IEEE Transactions on Systems, Man and Cybernetics 6(6):420–433
67. Pelillo M (1998) A unifying framework for relational structure matching. Proceedings of 14th International Conference on Pattern Recognition, pp. 1316–1319
68. Pelillo M, Siddiqi K, Zucker SW (1999) Matching hierarchical structures using association graphs. IEEE Transactions on Pattern Analysis and Machine Intelligence 21(11):1105–1120
69. Branca A, Stella E, Distante A (1999) Feature matching by searching maximum clique on high order association graph. Proceedings of the 10th International Conference on Image Analysis and Processing, pp. 642–658
70. Medasani S, Krishnapuram R (1999) A fuzzy approach to content-based image retrieval. Proceedings of the IEEE International Conference on Fuzzy Systems, pp. 1251–1260
71. Medasani S, Krishnapuram R, Choi YS (2001) Graph matching by relaxation of fuzzy assignments. IEEE Transactions on Fuzzy Systems 9(1):173–182
72. van Wyk BJ, van Wyk MA (2002) Non-bayesian graph matching without explicit compatibility calculations. Proceeding of the Joint IAPR International Workshops SSPR and SPR, pp. 74–82
73. van Wyk BJ, van Wyk MA, Hanrahan HE (2002) Successive projection graph matching. Proceeding of the Joint IAPR International Workshops SSPR and SPR, pp. 263–271
74. Umeyama S (1988) An eigendecomposition approach to weighted graph matching problems. IEEE Transactions on Pattern Analysis and Machine Intelligence 10(5):695–703
75. Xu L, King I (2001) A PCA approach for fast retrieval of structural patterns in attributed graphs. IEEE Transactions on Systems, Man and Cybernetics, Part B 31(5):812–817
76. Carcassoni M, Hancock ER (2001) Weighted graph-matching using modal clusters. Proceedings of the Third IAPR-TC15 Workshop on Graph-based Representations in Pattern Recognition, pp. 260–269
77. Kosinov S, Caelli T (2002) Inexact multisubgraph matching using graph eigenspace and clustering models. Proceedings of the Joint IAPR International Workshops SSPR and SPR, pp. 133–142
78. Shokoufandeh A, Dickinson S (2001) A unified framework for indexing and matching hierarchical shape structures. In: Arcelli C, Cordella LP, Sanniti di Baja G (eds), Workshop on Visual Form, LNCS 2059, pp. 67–84
79. Jagota A, Pelillo M, Rangarajan A (2000) A new deterministic annealing algorithm for maximum clique. Proceedings of the International Joint Conference on Neural Networks 6:505–508
80. Gendreau M, Salvail L, Soriano P (1993) Solving the maximum clique problem using a tabu search approach. Annals of Operations Research 41:385–403
81. Williams ML, Wilson RC, Hancock ER (1999) Deterministic search for relational graph matching. Pattern Recognition 32(7):1255–1271
82. Eshera MA, Fu KS (1986) An image understanding system using attributed symbolic representation and inexact graph-matching. IEEE Transactions on Pattern Analysis and Machine Intelligence 8(5):604–618
83. Hong P, Wang R, Huang T (2000) Learning patterns from images by combining soft decisions and hard decisions. Proceedings of the 2000 IEEE Computer Society Conference on Computer Vision and Pattern Recognition, pp. 79–83
84. Jiang X, Bunke H (1998) Marked subgraph isomorphism of ordered graphs. Proceeding of the Joint IAPR International Workshops SSPR and SPR, pp. 122–131

85. Seong D, Kim HS, Park KH (1993) Incremental clustering of attributed graphs. IEEE Transactions on Systems, Man and Cybernetics 23(5):1399–1411
86. Zhang H, Yan H (1999) Graphic matching based on constrained Voronoi diagrams. Proceedings of the Fifth International Symposium on Signal Processing and its Applications, pp. 431–434
87. Luo B, Robles-Kelly A, Torsello A, Wilson RC, Hancock ER (2001) Clustering shock trees. Proceedings of the Third IAPR-TC15 Workshop on Graph-based Representations in Pattern Recognition, pp. 217–228
88. Suganthan PN, Teoh EK, Mital D (1995) Pattern recognition by graph matching using the potts MFT neural networks. Pattern Recognition 28(7):997–1009
89. Torsello A, Hancock ER (2002) Learning structural variations in shock trees. Proceeding of the Joint IAPR International Workshops SSPR and SPR, pp. 113–122
90. Perchant A, Boeres C, Bloch I, Roux M, Ribeiro C (1999) Model-based scene recognition using graph fuzzy homomorphism solved by genetic algorithm. Proceedings of the Second IAPR-TC15 Workshop on Graph-based Representations in Pattern Recognition, pp. 61–70
91. Boeres MC, Ribeiro CC, Bloch I (2004) A Randomized Heuristic for Scene Recognition by Graph Matching. In: Ribeiro CC, Martins SL (eds) Experimental and Efficient Algorithms: Third International Workshop, LNCS 3059, pp. 100–113
92. Wilson RC, Hancock ER (1999) Graph matching with hierarchical discrete relaxation. Pattern Recognition Letters 20(10):1041–1052
93. Neuhaus M, Bunke H (2003) An error-tolerant approximate matching algorithm for attributed planar graphs and its application to fingerprint classification. In: Fred A, Caelli T, Camphilho A (eds), SSPR 2004, LNCS 3138
94. Jia J, Abe K (1998) Automatic generation of prototypes in 3D structural object recognition. Proceedings of the Fourteenth International Conference on Pattern Recognition, pp. 697–700
95. Shasha D, Wang JTL, Zhang K, Shih FY (1994) Exact and approximate algorithms for unordered tree matching. IEEE Transactions on Systems, Man and Cybernetics 24(4):668–678
96. Sanfeliu A, Serratosa F, Alquezar R (2004) Second-order random graphs for modeling sets of attributed graphs and their application to object learning and recognition. International Journal of Pattern Recognition and Artificial Intelligence 18(3):375–396
97. Robles-Kelly A, Hancock ER (2005) Graph edit distance from spectral seriation. IEEE Transactions on Pattern Analysis and Machine Intelligence 27(3):365–378
98. Englert R, Cremers AB, Seelmann-Eggebert J (1997) Recognition of polymorphic pattern in parameterized graphs for 3D building reconstruction. Computing. Supplement 12:11–20
99. Meth R, Chellappa R (1996) Target indexing in synthetic aperture radar imagery using topographic features. Proceedings of the IEEE International Conference on Acoustics, Speech, and Signal Processing, pp. 2152–2155
100. Li WJ, Lee T (2000) Object recognition by sub-scene graph matching. IEEE International Conference on Robotics and Automation, pp. 1459–1464
101. Belongie S, Malik J (2000) Matching with shape contexts. Proceedings of IEEE Workshop on Content-based Access of Image and Video Libraries, pp. 20–26
102. Sebastian TB, Klein PN, Kimia BB (2005) Recognition of shapes by editing their shock graphs. IEEE Transactions on Pattern Analysis and Machine Intelligence 26(5):550–571

103. Koo JH, Yoo SI (1998) A structural matching for two-dimensional visual pattern inspection. Proceedings of the IEEE International Conference on Systems, Man and Cybernetics, pp. 4429–4434
104. Sharvit D, Chan J, Tek H, Kimia BB (1998) Symmetry-based indexing of image databases. Proceedings of the IEEE Workshop on Content-Based Access of Image and Video Libraries, pp. 56–62
105. Bauckhage C, Wachsmuth S, Sagerer G (2001) 3D assembly recognition by matching functional subparts. Proceedings of the Third IAPR-TC15 Workshop on Graph-based Representations in Pattern Recognition, pp. 95–104
106. Olatunbosun S, Dowling GR, Ellis TJ (1996) Topological representation for matching coloured surfaces. Proceedings of the International Conference on Image Processing, pp. 1019–1022
107. Fuchs F, Le Men H (2000) Efficient subgraph isomorphism with 'A Priori' knowledge. Proceeding of the Joint IAPR International Workshops SSPR and SPR, pp. 427–436
108. Fuchs F, Le Men H (1999) Building reconstruction on aerial images through multiprimitive graph matching. Proceedings of the Second IAPR-TC15 Workshop on Graph-based Representations in Pattern Recognition, pp. 21–30
109. Messmer BT, Bunke H (1998) A new algorithm for error-tolerant subgraph isomorphism detection. IEEE Transactions on Pattern Analysis and Machine Intelligence 20(5):493–504
110. Bron C, Kerbosch J (1973) Finding all cliques of an undirected graph. Communications of ACM 16(9):575–577
111. Foggia P, Sansone C, Tortorella F, Vento M (1999) Definition and validation of a distance measure between structural primitives. Pattern Analysis and Applications 2: 215–227
112. Foggia P, Genna R, Vento M (2001) Symbolic vs. connectionist learning: an experimental comparison in a structured domain. IEEE Transactions on Knowledge and Data Engineering 13(2):176–195
113. Chan KP (1996) Learning templates from fuzzy examples in structural pattern recognition. IEEE Transactions on Systems, Man and Cybernetics, Part B 26(1):118–123
114. Lee RST, Liu JNK (1999) An oscillatory elastic graph matching model for recognition of offline handwritten Chinese characters. Third International Conference on Knowledge-Based Intelligent Information Engineering Systems, pp. 284–287
115. Suganthan PN, Yan H (1998) Recognition of handprinted Chinese characters by constrained graph matching. Image and Vision Computing 16(3):191–201
116. Liu JZ, Ma K, Cham WK, Chang MMY (2000) Two-layer assignment method for online Chinese character recognition. IEEE Proceedings Vision, Image and Signal Processing 147(1):47–54
117. Lu SW, Ren Y, Suen CY (1991) Hierarchical attributed graph representation and recognition of handwritten Chinese characters. Pattern Recognition 24:617–632
118. Chen LH, Lieh JR (1990) Handwritten character recognition using a 2-layer random graph model by relaxation matching. Pattern Recognition 23:1189–1205
119. Suganthan PN, Teoh EK, Mital D (1995) A self organizing Hopfield network for attributed relational graph matching. Image and Vision Computing 13(1):61–73
120. Pavlidis T, Sakoda WJ, Shi H (1995) Matching graph embeddings for shape analysis. Proceedings of the Third International Conference on Document Analysis and Recognition, pp. 729–733
121. Filatov A, Gitis A, Kil I (1995) Graph-based handwritten digit string recognition. Proceedings of the Third International Conference on Document Analysis and Recognition, pp. 845–848

122. Llados J, Lopez-Krahe J, Marti E (1996) Hand drawn document understanding using the straight line Hough transform and graph matching. Proceedings of the 13th International Conference on Pattern Recognition, pp. 497–501
123. Changhua L, Bing Y, Weixin X (2000) Online hand-sketched graphics recognition based on attributed relational graph matching. Proceedings of the Third World Congress on Intelligent Control and Automation, pp. 2549–2553
124. Jiang X, Munger A, Bunke H (1999) Synthesis of representative symbols by computing generalized median graphs. Proceedings of the International Workshop on Graphics Recognition GREC'99, pp. 187–194
125. Lades M, Vorbruggen JC, Buhmann J, Lange J, von der Malsburg C, Wurz RP, Konen W (1993) Distortion invariant object recognition in the dynamic link architecture. IEEE Transactions on Computers 42(3):300–311
126. Wiskott L, Fellous JM, Kruger N, von der Malsburg C (1997) Face recognition by elastic bunch graph matching. IEEE Transactions on Pattern Analysis and Machine Intelligence 19(7):775–779
127. Wiskott L (1997) Phantom faces for face analysis. Proceedings of the International Conference on Image Processing, pp. 308–311
128. Lim R, Reinders MJT (2001) Facial landmarks localization based on fuzzy and gabor wavelet graph matching. The Tenth IEEE International Conference on Fuzzy Systems, pp. 683–686
129. Kotropoulos C, Tefas A, Pitas I (2000) Frontal face authentication using morphological elastic graph matching. IEEE Transactions on Image Processing 9(4):555–560
130. Tefas A, Kotropoulos C, Pitas I (2001) Using support vector machines to enhance the performance of elastic graph matching for frontal face authentication. IEEE Transactions on Pattern Analysis and Machine Intelligence 23(7):735–746
131. Duc B, Fischer S, Bigun J (1999) Face authentication with Gabor information on deformable graphs. IEEE Transactions on Image Processing 8(4):504–516
132. Lyons MJ, Budynek J, Akamatsu S (1999) Automatic classification of single facial images. IEEE Transactions on Pattern Analysis and Machine Intelligence 21(12):1357–1362
133. Wang M, Iwai Y, Yachida M (1998) Expression recognition from time-sequential facial images by use of expression change model. Proceedings of the Third IEEE International Conference on Automatic Face and Gesture Recognition, pp. 354–359
134. Elagin E, Steffens J, Neven H (1998) Automatic pose estimation system for human faces based on bunch graph matching technology. Proceedings of the Third IEEE International Conference on Automatic Face and Gesture Recognition, pp. 136–141
135. Triesch J, von der Malsburg C (2001) A system for person-independent hand posture recognition against complex backgrounds. IEEE Transactions on Pattern Analysis and Machine Intelligence 23(12):1449–1453
136. Burge M, Burger W (2000) Ear biometrics in computer vision. Proceedings of 15th International Conference on Pattern Recognition, pp. 822–826
137. Maio D, Maltoni D (1996) A structural approach to fingerprint classification. Proceedings of the 13th International Conference on Pattern Recognition, pp. 578–585
138. Fan KC, Liu CW, Wang YK (1998) A fuzzy bipartite weighted graph matching approach to fingerprint verification. Proceedings of the IEEE International Conference on Systems, Man and Cybernetics, pp. 4363–4368
139. Petrakis GM, Faloutsos C (1997) Similarity searching in medical image databases. IEEE Transaction on Knowledge and Data Engineering 9(3):435–447

140. Cho SJ, Yoo SJ (1998) Image retrieval using topological structure of user sketch. IEEE International Conference on Systems, Man and Cybernetics, pp. 4584–4588
141. de Mauro C, Diligenti M, Gori M, Maggini M (2003) Similarity learning for graph-based image representations. Pattern Recognition Letters 24(8):1115–1122
142. Doulamis N, Doulamis A, Kollias S (1999) Efficient content-based retrieval of humans from video databases. Proceedings of the International Workshop on Recognition, Analysis, and Tracking of Faces and Gestures in Real-Time Systems, pp. 89–95
143. Folkers A, Samet H, Soffer A (2000) Processing pictorial queries with multiple instances using isomorphic subgraphs. Proceedings of the 15th International Conference on Pattern Recognition, pp. 51–54
144. Hlaoui A, Wang S (2002) A new algorithm for graph matching with application to content-based image retrieval. Proceeding of the Joint IAPR International Workshops SSPR and SPR, pp. 291–300
145. Huet B and Hancock ER (1998) Fuzzy relational distance for large-scale object recognition. Proceedings of IEEE Conference on Computer Vision and Pattern Recognition, pp. 138–143
146. Huet B, Kern NJ, G Guarascio, B Merialdo (2001) Relational skeletons for retrieval in patent drawings. Proceedings of the International Conference on Image Processing, pp. 737–740
147. Park IK, Yun ID, Lee SU (1997) Models and algorithms for efficient color image indexing. Proceedings of the IEEE Workshop on Content-Based Access of Image and Video Libraries, pp. 36–41
148. Shearer K, Venkatesh S, Bunke H (2001) Video sequence matching via decision tree path following. Pattern Recognition Letters 22(5):479–492
149. Ozer B, Wolf W, Akansu AN (1999) A graph based object description for information retrieval in digital image and video libraries. Proceedings of the IEEE Workshop on Content-Based Access of Image and Video Libraries, pp. 79–83
150. Chen HT, Lin H, Liu TL (2001) Multi-object tracking using dynamical graph matching. Proceedings of the 2001 IEEE Computer Society Conference on Computer Vision and Pattern Recognition, pp. 210–217
151. Gomila C, Meyer F (2001) Tracking objects by graph matching of image partition sequences. Proceedings of the Third IAPR-TC15 Workshop on Graph-Based Representations in Pattern Recognition, pp. 1–11
152. Conte D, Foggia P, Guidobaldi C, Limongiello A, Vento M (2004) An object tracking algorithm combining different cost functions. In: Campilho A, Kamel M (eds) ICIAR 2004, LNCS 3212, pp. 614–622
153. Conte D, Foggia P, Jolion JM, Vento M (2006) A graph-based, multi-resolution algorithm for tracking objects in presence of occlusions. Pattern Recognition 39(4):562–572
154. Salotti M, Laachfoubi N (2001) Topographic graph matching for shift estimation. Proceedings of the Third IAPR-TC15 Workshop on Graph-based Representations in Pattern Recognition, pp. 54–63
155. Haris K, Efstradiatis SN, Maglaveras N, Pappas C, Gourassas J, Louridas G (1999) Model-based morphological segmentation and labeling of coronary angiograms. IEEE Transactions on Medical Imaging 18(10):1003–1015
156. Charnoz A, Agnus V, Malandain G, Soler L, Tajine M (2005) Tree matching applied to vascular system. In: Brun L, Vento M (eds) Graph-based Representations in Pattern Recognition, LNCS 3434, pp. 183–192
157. Fischer S, Gilomen K, Bunke H (2002) Identification of diatoms by grid graph matching. Proceeding of the Joint IAPR International Workshops SSPR and SPR, pp. 94–103

158. Ambauen R, Fischer S, Bunke H (2003) Graph Edit Distance with Node Splitting and Merging, and Its Application to Diatom Identification. In: Hancock ER, Vento M (eds) Graph Based Representations in Pattern Recognition, LNCS 2726, pp. 95–106
159. Bunke H, Vento M (1999) Benchmarking of graph matching algorithms, Proceedings of Second IAPR TC-15 GbR Workshop, Haindorf, pp. 109–114
160. Kropatsch W (2001) Benchmarking graph matching algorithms – A complementary view. Proceedings of Third IAPR – TC15 Workshop on Graph-Based Representations, Italy, pp. 170–175
161. Irniger C, Bunke H (2003) Theoretical analysis and experimental comparison of graph matching algorithms for database filtering. In: Hancock ER, Vento M (eds) Graph Based Representations in Pattern Recognition, LNCS 2726, pp. 118–129
162. Kalviainen H, Oja E (1990) Comparisons of attributed graph matching algorithms for computer vision. In: Proceedings of STEP-90, Finnish Artificial Intelligence Symposium, pp. 354–368
163. Bunke H, Foggia P, Guidobaldi C, Vento M (2003) Graph clustering using the weighted minimum common supergraph. In: Hancock ER, Vento M (eds) Graph Based Representations in Pattern Recognition, LNCS 2726, pp. 235–246
164. Conte D, Guidobaldi C, Sansone C (2003) A comparison of three maximum common subgraph algorithms on a large database of labeled graphs. In: Hancock ER, Vento M (eds) Graph Based Representations in Pattern Recognition, LNCS 2726, pp. 130–141
165. Foggia P, Sansone C, Vento M (2001) A database of graphs for isomorphism and sub graph isomorphism benchmarking. Proceedings of the Third IAPR TC-15 Workshop on Graph-based Representations in Pattern Recognition, Ischia, May 23–25, pp. 176–187
166. Hagenbuchner M, Gori M, Bunke H, Tsoi AC, Irniger C (2003) Using attributed plex grammars for the generation of image and graph databases. Pattern Recognition Letters 24(8):1081–1087
167. Feder J, (1971) Plex languages. Information Science 3:225–241
168. Cordella LP, Foggia P, Sansone C, Vento M (2002) Learning structural shape descriptions from examples. Pattern Recognition Letters 23(12):1427–1437
169. McGregor JJ (1982) Backtrack search algorithm and the maximal common subgraph problem. Software – Practice and Experience 12(1):23–34
170. Balas E, Yu CS (1986) Finding a maximum clique in an arbitrary graph. SIAM Journal on Computing 15(4):1054–1068

Efficient Algorithms on Trees and Graphs with Unique Node Labels

Gabriel Valiente

Summary. There is a growing interest on trees and graphs with unique node labels in the field of pattern recognition, not only because graph isomorphism and related problems become polynomial-time solvable when restricted to them but also in the light of important practical applications in structural pattern recognition. Current algorithms for testing graph and subgraph isomorphism and computing the graph edit distance, a shortest edit script, a largest common subgraph, and a smallest common supergraph of two graphs with unique node labels, take time quadratic in the number of nodes in the graphs, and the same holds for similar problems on trees with unique node labels. In this paper, simple algorithms are presented for solving these problems in time linear in the number of nodes and edges in the trees or graphs. These new algorithms are based on radix sorting the sets of nodes and edges in the trees or graphs by node label and source and target node label, respectively, followed by a simultaneous traversal of the ordered sets of nodes and edges.

Key words: Graph matching, Trees, Graphs with unique node labels, Graph isomorphism, Subgraph isomorphism, Edit distance, Edit script, Largest common subgraph, Smallest common supergraph, Efficient algorithms

1 Introduction

Graph theoreticians and theoretical computer scientists have been suffering the so-called *graph isomorphism disease* (to establish the complexity of graph isomorphism and related problems) for several decades now [1,2] but, surprisingly, it is the restriction to graphs with unique node labels [3,4] what makes these problems polynomial-time solvable and with important practical applications in pattern recognition [3,5].

A graph with unique node labels [3] is just a directed graph with nodes labeled over an ordered alphabet such that no two nodes share the same label. Formally, let Σ_V be an ordered set of node labels, and let Σ_E be a set of edge labels. A graph is a four-tuple $G = (V, E, \alpha, \beta)$, where V is a finite set of nodes, $E \subseteq V \times V$ is a finite set of edges, $\alpha : V \to \Sigma_V$ is a node labeling mapping, and $\beta : E \to \Sigma_E$ is an edge labeling mapping. A graph $G = (V, E, \alpha, \beta)$ is a graph with unique node labels if $\alpha(v) \neq \alpha(w)$ for all $v, w \in V$ with $v \neq w$.

Graph matching [6] has been studied in the pattern recognition literature in various forms: graph isomorphism [7–10], subgraph isomorphism [9–11] [12, 13], largest common subgraph [14–18], smallest common supergraph [19,20], and graph edit distance [21–25]. However, a fundamental limitation for the practical application of graph matching in the field of pattern recognition, lies in the complexity of graph matching because subgraph isomorphism, largest common subgraph, smallest common supergraph, and graph edit distance are all NP-complete problems [26].

In the class of graphs with unique node labels, these problems become polynomial-time solvable, because they reduce to the computation of either set union or set intersection for the set of nodes and the set of edges in the graphs. For instance, computing a largest common subgraph of two graphs with unique node labels takes $O(n^2)$ time, where n is the number of nodes [3, 5].

In this paper, we show that the problems of testing graph and subgraph isomorphism and computing the graph edit distance, a shortest edit script, a largest common subgraph, and a smallest common supergraph of two trees or graphs with unique node labels can all be solved in optimal $O(n + m)$ time, where n is the number of nodes and m is the number of edges. The algorithms themselves are not complicated to implement, and they only require the use of standard data structures.

The rest of the paper is organized as follows. In Sect. 2, the notion of graph with unique node labels is recalled. Efficient algorithms for the problems of graph isomorphism, subgraph isomorphism, graph edit distance, shortest edit script, largest common subgraph, and smallest common supergraph on trees and graphs with unique node labels are presented in detail in Sect. 3. Finally, some conclusions are drawn in Sect. 4.

2 Trees and Graphs with Unique Node Labels

The class of graphs with unique node labels, introduced in [3], is characterized by the requirement of each node label being unique. Graphs with unique node labels find application in those problem domains in which objects are modeled by nodes with some property that can be used to uniquely identify them. Some applications of graphs with unique node labels, discussed in [4], include computer network monitoring (where each client, server, or router in a computer network is represented by a node, and an address uniquely identifies such a node in a computer network) and web document analysis (where each unique term that occurs in a document is represented by a node, and multiple occurrences of the same term are represented by the same node). Further application domains for trees and graphs with unique node labels include biochemical networks (where each biochemical reaction in the metabolic pathway of an organism is represented by a node, and multiple occurrences of the same biochemical reaction in the metabolism of an organism are represented by the same node; see [27]) and taxonomic classifications (where each group of species or species name labels a different node of a taxonomic tree; see [28]) in computational biology.

Definition 1. *Let Σ_V be an ordered set of node labels, and let Σ_E be a set of edge labels. A graph is a four-tuple $G = (V, E, \alpha, \beta)$, where V is a finite set of nodes, $E \subseteq V \times V$ is a finite set of edges, $\alpha : V \to \Sigma_V$ is a node labeling mapping, and $\beta : E \to \Sigma_E$ is an edge labeling mapping. A graph $G = (V, E, \alpha, \beta)$ is a graph with unique node labels if $\alpha(v) \neq \alpha(w)$ for all $v, w \in V$ with $v \neq w$.*

In the class of graphs with unique node labels, the problems of testing graph and subgraph isomorphism and computing the graph edit distance, a shortest edit script, a largest common subgraph, and a smallest common supergraph of two graphs become polynomial-time solvable, because they reduce to computation of either set union or set intersection for the set of nodes and the set of edges in the graphs. For instance, the algorithm given in [3,5] for computing a largest common subgraph of two graphs with unique node labels, can be stated in pseudocode form as follows.

Algorithm 1. Given two graphs $G_1 = (V_1, E_1, \alpha_1, \beta_1)$ and $G_2 = (V_2, E_2, \alpha_2, \beta_2)$ with unique node labels, compute a largest common subgraph $G = (V, E, \alpha, \beta)$ of G_1 and G_2.

1. $V := \emptyset$
2. **foreach** node $v_1 \in V_1$
     foreach node $v_2 \in V_2$
       if $\alpha_1(v_1) = \alpha_2(v_2)$ **then**
         $V := V \cup \{v\}$, where v is a new node
         $\alpha(v) := \alpha_1(v_1)$
       endif
     endfor
   endfor
3. $E := \emptyset$
4. **foreach** node $v \in V$
     let v_1 be the node of G_1 with $\alpha_1(v_1) = \alpha(v)$
     let v_2 be the node of G_2 with $\alpha_2(v_2) = \alpha(v)$
     foreach node $w \in V$
       let w_1 be the node of G_1 with $\alpha_1(w_1) = \alpha(w)$
       let w_2 be the node of G_2 with $\alpha_2(w_2) = \alpha(w)$
       if $(v_1, w_1) \in E_1$, $(v_2, w_2) \in E_2$ and $\beta_1(v_1, w_1) = \beta_2(v_2, w_2)$ **then**
         $E := E \cup \{(v, w)\}$
         $\beta(v, w) := \beta_1(v_1, w_1)$
       endif
     endfor
   endfor
5. **return** G

Computation of a largest common subgraph of two graphs with unique node labels using the previous algorithm takes $O(n^2)$ time, where n is the number of nodes in the graphs. A more efficient algorithm is presented in Sect. 3 that only takes $O(n+m)$ time, where n is the number of nodes and m is the number of edges in the graphs.

3 Efficient Algorithms on Trees and Graphs with Unique Node Labels

The problems of testing graph and subgraph isomorphism and computing the graph edit distance, a shortest edit script, the largest common subgraph, and the smallest common supergraph of two graphs with unique node labels, can be solved in time linear in the number of nodes and edges in the graphs, only if the sets of node labels can be sorted in time linear in the number of nodes and the sets of edge source and target node labels can also be sorted in time linear in the number of nodes and edges in the graphs. The procedure was first sketched in [29].

While sorting takes, in general, quasilinear time, there are at least two particular cases of much interest in pattern recognition for which nodes labels can be sorted in linear time. On the one hand, if node labels are small integers, as in [3], let k be a fixed, but arbitrary, constant. Since n integers in the range $\{1, \ldots, kn\}$ can be sorted in $O(n)$ time, by bucket sorting techniques, it follows that the sets of node labels and the sets of edge source and target node labels can be sorted in time linear in the number of nodes and edges in the graphs.

On the other hand, if node labels are strings, as in [5], let k be again a fixed, but arbitrary, constant. Since n strings of total length at most kn can be sorted in $O(n)$ time, by radix sorting techniques [30], it follows that the sets of node labels and the sets of edge source and target node labels can be sorted in time linear in the total length of the strings. In particular, if node labels are all short strings, of $O(1)$ length each.

All trees and graphs are assumed to be given in adjacency list representation in the rest of the paper.

3.1 Graph Isomorphism

Definition 2. *Two graphs $G_1 = (V_1, E_1, \alpha_1, \beta_1)$ and $G_2 = (V_2, E_2, \alpha_2, \beta_2)$ are isomorphic if there is a bijection $\mu : V_1 \to V_2$ such that, for every node $v_i \in V_1$, $\alpha_1(v_i) = \alpha_2(\mu(v_i))$ and for every pair of nodes $v_1, w_1 \in V_1$, $(v_1, w_1) \in E_1$ if and only if $(\mu(v_1), \mu(w_1)) \in E_2$ and $\beta_1(v_1, w_1) = \beta_2(\mu(v_1), \mu(w_1))$. In such a case, μ is a graph isomorphism of G_1 to G_2.*

The efficient computation of the isomorphism of two graphs $G_1 = (V_1, E_1, \alpha_1, \beta_1)$ and $G_2 = (V_2, E_2, \alpha_2, \beta_2)$ with unique node labels proceeds as follows. Sort V_1 and V_2 by node label and, during a simultaneous traversal [30] of the ordered sets of nodes, map each node $v_1 \in V_1$ to the only node $v_2 \in V_2$ such that $\alpha_1(v_1) = \alpha_2(v_2)$, that is, set $\mu(v_1) = v_2$. In a similar vein, sort E_1 and E_2 by source node label and target node label and then, during a simultaneous traversal of the ordered sets of edges, for each edge $e_1 \in E_1$ and $e_2 \in E_2$, say $e_1 = (v_1, w_1)$ and $e_2 = (v_2, w_2)$, with $\alpha_1(v_1) = \alpha_2(v_2)$ and $\alpha_1(w_1) = \alpha_2(w_2)$, check that $\beta_1(e_1) = \beta_2(e_2)$. Then, the node mapping $\mu : V_1 \to V_2$ obtained in the first stage is a graph isomorphism of G_1 to G_2 if and only if all nodes of V_1 were mapped and the latter test was successful for all edges of E_1.

Algorithm 2. Given two graphs $G_1 = (V_1, E_1, \alpha_1, \beta_1)$ and $G_2 = (V_2, E_2, \alpha_2, \beta_2)$ with unique node labels, compute a graph isomorphism μ of G_1 to G_2, if it exists.

1. sort V_1 and V_2 by node label
2. $isomorph := true$
3. **while** $V_1 \neq \emptyset$ and $V_2 \neq \emptyset$ and $isomorph$ **do**
       let v_1 and v_2 be the first element of V_1 and V_2, respectively
       if $\alpha_1(v_1) = \alpha_2(v_2)$ **then**
           $\mu(v_1) := v_2$
           $V_1 := V_1 \setminus \{v_1\}$
           $V_2 := V_2 \setminus \{v_2\}$
       else
           $isomorph := false$
       endif
   endwhile
4. sort E_1 and E_2 by target node label
   sort E_1 and E_2 by source node label
5. **while** $E_1 \neq \emptyset$ and $E_2 \neq \emptyset$ and $isomorph$ **do**
       let (v_1, w_1) and (v_2, w_2) be the first element of E_1 and E_2
       if $\mu(v_1) = v_2$ and $\mu(w_1) = w_2$ and $\beta_1(v_1, w_1) = \beta_2(v_2, w_2)$ **then**
           $E_1 := E_1 \setminus \{(v_1, w_1)\}$
           $E_2 := E_2 \setminus \{(v_2, w_2)\}$
       else
           $isomorph := false$
       endif
   endwhile
6. return $(\mu, isomorph)$

3.2 Subgraph Isomorphism

Definition 3. *A* subgraph isomorphism *of a graph $G_1 = (V_1, E_1, \alpha_1, \beta_1)$ into a graph $G_2 = (V_2, E_2, \alpha_2, \beta_2)$ is an injection $\mu : V_1 \rightarrow V_2$ such that, for every node $v_i \in V_1, \alpha_1(v_i) = \alpha_2(\mu(v_i))$ and for every pair of nodes $v_1, w_1 \in V_1$ with $(v_1, w_1) \in E_1, (\mu(v_1), \mu(w_1)) \in E_2$ and $\beta_1(v_1, w_1) = \beta_2(\mu(v_1), \mu(w_1))$. In such a case, μ is a subgraph isomorphism of G_1 into G_2.*

The efficient computation of a subgraph isomorphism of a graph $G_1 = (V_1, E_1, \alpha_1, \beta_1)$ with unique node labels into another graph $G_2 = (V_2, E_2, \alpha_2, \beta_2)$ with unique node labels proceeds as follows. Sort V_1 and V_2 by node label and, during a simultaneous traversal [30] of the ordered sets of nodes, map each node $v_1 \in V_1$ to the only node $v_2 \in V_2$ such that $\alpha_1(v_1) = \alpha_2(v_2)$, that is, set $\mu(v_1) = v_2$. In a similar vein, sort E_1 and E_2 by source node label and target node label and then, during a simultaneous traversal of the ordered sets of edges, for each edge $e_1 \in E_1$ and $e_2 \in E_2$, say $e_1 = (v_1, w_1)$ and $e_2 = (v_2, w_2)$, with $\alpha_1(v_1) = \alpha_2(v_2)$ and $\alpha_1(w_1) = \alpha_2(w_2)$, check that $\beta_1(e_1) = \beta_2(e_2)$. Then, the node mapping $\mu : V_1 \rightarrow V_2$ obtained in the first stage is a subgraph isomorphism of G_1 into G_2 if and only if all nodes of V_1 were mapped and the latter test was successful for all edges of E_1.

Algorithm 3. Given two graphs $G_1 = (V_1, E_1, \alpha_1, \beta_1)$ and $G_2 = (V_2, E_2, \alpha_2, \beta_2)$ with unique node labels, compute a subgraph isomorphism μ of G_1 into G_2, if it exists.

1. sort V_1 and V_2 by node label
2. **while** $V_1 \neq \emptyset$ and $V_2 \neq \emptyset$ **do**
   let v_1 and v_2 be the first element of V_1 and V_2, respectively
   if $\alpha_1(v_1) = \alpha_2(v_2)$ **then**
   $\mu(v_1) := v_2$
   $V_1 := V_1 \setminus \{v_1\}$
   endif
   $V_2 := V_2 \setminus \{v_2\}$
   endwhile
3. sort E_1 and E_2 by target node label
   sort E_1 and E_2 by source node label
4. **while** $E_1 \neq \emptyset$ and $E_2 \neq \emptyset$ **do**
   let (v_1, w_1) and (v_2, w_2) be the first element of E_1 and E_2
   if $\mu(v_1) = v_2$ and $\mu(w_1) = w_2$ and $\beta_1(v_1, w_1) = \beta_2(v_2, w_2)$ **then**
   $E_1 := E_1 \setminus \{(v_1, w_1)\}$
   endif
   $E_2 := E_2 \setminus \{(v_2, w_2)\}$
   endwhile
5. $isomorph := (V_1 = \emptyset \text{ and } E_1 = \emptyset)$
6. return $(\mu, isomorph)$

3.3 Graph Edit Distance

The edit operations of node and edge deletion, insertion, and substitution allow one to transform any given graph into any other graph. In the class of graphs with unique node labels, edge label substitution are allowed but node label substitutions are forbidden, because they may generate graphs with nonunique node labels [3,4].

A non-negative cost is assigned to each edit operation, the cost of a sequence of edit operations is given by the sum of the individual cost over all of the edit operations in the sequence, and the edit distance of two graphs is defined as the least cost over all sequences of edit operations that transform one graph into the other.

In practical applications, the cost of an edit operations is equal to 1 except for node substitutions, which have infinite cost. Under this assumption of unit cost, the edit distance coincides with the size of a largest common subgraph [21]. Therefore, under the assumption of unit cost, the algorithm for computing a largest common subgraph of two graphs with unique node labels presented below can also be used to compute the edit distance of two graphs with unique node labels.

3.4 Shortest Edit Script

Definition 4. *An* edit script *of a graph* $G_1 = (V_1, E_1, \alpha_1, \beta_1)$ *to a graph* $G_2 = (V_2, E_2, \alpha_2, \beta_2)$ *is a set S of edit operations that, if applied in the right order (essentially, inserting an edge only after having inserted the nodes incident with the* inserted edge), allow one to transform G_1 into G_2. An edit script S of G_1 to G_2 is shortest *if there is no edit script of G_1 to G_2 of smaller size than S.*

The efficient computation of a shortest edit script of two graphs $G_1 = (V_1, E_1, \alpha_1, \beta_1)$ and $G_2 = (V_2, E_2, \alpha_2, \beta_2)$ with unique node labels proceeds as follows. Sort V_1 and V_2 by node label and, during a simultaneous traversal [30] of the ordered sets of nodes, for each node $v_1 \in V_1$ such that there is no node $v_2 \in V_2$ with $\alpha_1(v_1) = \alpha_2(v_2)$, output the edit operation "delete node $\alpha_1(v_1)$" and for each node $v_2 \in V_2$ such that there is no node $v_1 \in V_1$ with $\alpha_1(v_1) = \alpha_2(v_2)$, output the edit operation "insert node $\alpha_2(v_2)$."

In a similar vein, sort E_1 and E_2 by source node label and target node label and then, during a simultaneous traversal of the ordered sets of edges, for each edge $e_1 \in E_1$ and $e_2 \in E_2$, say $e_1 = (v_1, w_1)$ and $e_2 = (v_2, w_2)$, with $\alpha_1(v_1) = \alpha_2(v_2)$ and $\alpha_1(w_1) = \alpha_2(w_2)$, if $\beta_1(e_1) \neq \beta_2(e_2)$, then output the edit operation "substitute edge $\alpha_1(v_1)$ to $\alpha_1(w_1)$ label $\beta_2(v_2, w_2)$." Also, for each edge $e_1 = (v_1, w_1) \in E_1$ such that there is no edge $e_2 = (v_2, w_2) \in E_2$ with $\alpha_1(v_1) = \alpha_2(v_2)$ and $\alpha_1(w_1) = \alpha_2(w_2)$, output the edit operation "delete edge $\alpha_1(v_1)$ to $\alpha_1(w_1)$" and for each edge $e_2 = (v_2, w_2) \in E_2$ such that there is no edge $e_1 = (v_1, w_1) \in E_1$ with $\alpha_1(v_1) = \alpha_2(v_2)$ and $\alpha_1(w_1) = \alpha_2(w_2)$, output the edit operation "insert edge $\alpha_2(v_2)$ to $\alpha_2(w_2)$."

3.5 Largest Common Subgraph

Definition 5. *A common subgraph of two graphs $G_1 = (V_1, E_1, \alpha_1, \beta_1)$ and $G_2 = (V_2, E_2, \alpha_2, \beta_2)$ is a graph G such that there exist subgraph isomorphisms of G into G_1 and into G_2. A common subgraph G of G_1 and G_2 is maximal if there is no subgraph isomorphism of G into any other common subgraph G' of G_1 and G_2, and it is largest if there is no common subgraph G' of G_1 and G_2 of larger size than G.*

The efficient computation of a largest common subgraph of two graphs $G_1 = (V_1, E_1, \alpha_1, \beta_1)$ and $G_2 = (V_2, E_2, \alpha_2, \beta_2)$ with unique node labels proceeds as follows. Let $G = (V, E, \alpha, \beta)$ be an empty graph and let $\gamma : V_1 \to V$ be an array of nodes indexed by the nodes of G_1. Sort V_1 and V_2 by node label and, during a simultaneous traversal [30] of the ordered sets of nodes, for each node $v_1 \in V_1$ and $v_2 \in V_2$ with $\alpha_1(v_1) = \alpha_2(v_2)$, add a new node v to G with $\alpha(v) = \alpha_1(v_1)$ and set $\gamma(v_1) = v$. In a similar vein, sort E_1 and E_2 by source node label and target node label and then, during a simultaneous traversal of the ordered sets of edges, for each edge $e_1 \in E_1$ and $e_2 \in E_2$, say $e_1 = (v_1, w_1)$ and $e_2 = (v_2, w_2)$, with $\alpha_1(v_1) = \alpha_2(v_2)$ and $\alpha_1(w_1) = \alpha_2(w_2)$, if $\beta_1(e_1) = \beta_2(e_2)$, then add a new edge $e = (v, w)$ to G with $\beta(e) = \beta_1(e_1)$, where $v = \gamma(v_1)$ and $w = \gamma(w_1)$.

Notice that graph and subgraph isomorphism can also be tested by just comparing the size of a largest common subgraph with the size of the given graphs.

3.6 Smallest Common Supergraph

Definition 6. *A common supergraph of two graphs $G_1 = (V_1, E_1, \alpha_1, \beta_1)$ and $G_2 = (V_2, E_2, \alpha_2, \beta_2)$ is a graph G such that there exist subgraph isomorphisms of G_1*

Algorithm 4. Given two graphs $G_1 = (V_1, E_1, \alpha_1, \beta_1)$ and $G_2 = (V_2, E_2, \alpha_2, \beta_2)$ with unique node labels, output a shortest edit script of G_1 and G_2.

1. sort V_1 and V_2 by node label
2. **while** $V_1 \neq \emptyset$ and $V_2 \neq \emptyset$ **do**
       let v_1 and v_2 be the first element of V_1 and V_2, respectively
       case $\alpha_1(v_1) < \alpha_2(v_2)$
           output "delete node $\alpha_1(v_1)$"
           $V_1 := V_1 \setminus \{v_1\}$
       case $\alpha_1(v_1) > \alpha_2(v_2)$
           output "insert node $\alpha_2(v_2)$"
           $V_2 := V_2 \setminus \{v_2\}$
       otherwise
           $V_1 := V_1 \setminus \{v_1\}$
           $V_2 := V_2 \setminus \{v_2\}$
       endcase
   endwhile
3. sort E_1 and E_2 by target node label
   sort E_1 and E_2 by source node label
4. **while** $E_1 \neq \emptyset$ and $E_2 \neq \emptyset$ **do**
       let (v_1, w_1) and (v_2, w_2) be the first element of E_1 and E_2
       case $\alpha_1(v_1) < \alpha_2(v_2)$ or $(\alpha_1(v_1) = \alpha_2(v_2)$ and $\alpha_1(w_1) < \alpha_2(w_2))$
           output "delete edge $\alpha_1(v_1)$ to $\alpha_1(w_1)$"
           $E_1 := E_1 \setminus \{(v_1, w_1)\}$
       case $\alpha_1(v_1) > \alpha_2(v_2)$ or $(\alpha_1(v_1) = \alpha_2(v_2)$ and $\alpha_1(w_1) > \alpha_2(w_2))$
           output "insert edge $\alpha_2(v_2)$ to $\alpha_2(w_2)$"
           $E_2 := E_2 \setminus \{(v_2, w_2)\}$
       otherwise
           if $\beta_1(v_1, w_1) \neq \beta_2(v_2, w_2)$ **then**
               output "substitute edge $\alpha_1(v_1)$ to $\alpha_1(w_1)$ label $\beta_2(v_2, w_2)$"
           endif
           $E_1 := E_1 \setminus \{(v_1, w_1)\}$
           $E_2 := E_2 \setminus \{(v_2, w_2)\}$
       endcase
   endwhile

---

and G_2 into G. A common supergraph G of G_1 and G_2 is minimal *if there is no subgraph isomorphism into G of any other common supergraph G' of G_1 and G_2*, and it is smallest *if there is no common supergraph G' of G_1 and G_2 of smaller size than G*.

The efficient computation of a smallest common supergraph of two graphs $G_1 = (V_1, E_1, \alpha_1, \beta_1)$ and $G_2 = (V_2, E_2, \alpha_2, \beta_2)$ with unique node labels proceeds as follows. Let $G = (V, E, \alpha, \beta)$ be an empty graph, and let $\gamma : V_1 \to V$ be an array of nodes indexed by the nodes of G_1. Sort V_1 and V_2 by node label and, during a simultaneous traversal [30] of the ordered sets of nodes, for each node $v_1 \in V_1$ and $v_2 \in V_2$ with $\alpha_1(v_1) = \alpha_2(v_2)$, add a new node v to G with $\alpha(v) = \alpha_1(v_1)$ and set

Algorithm 5. Given two graphs $G_1 = (V_1, E_1, \alpha_1, \beta_1)$ and $G_2 = (V_2, E_2, \alpha_2, \beta_2)$ with unique node labels, compute a largest common subgraph $G = (V, E, \alpha, \beta)$ of G_1 and G_2.

1. sort V_1 and V_2 by node label
2. $V := \emptyset$
3. **while** $V_1 \neq \emptyset$ and $V_2 \neq \emptyset$ **do**
    let v_1 and v_2 be the first element of V_1 and V_2, respectively
    case $\alpha_1(v_1) < \alpha_2(v_2)$
        $V_1 := V_1 \setminus \{v_1\}$
    case $\alpha_1(v_1) > \alpha_2(v_2)$
        $V_2 := V_2 \setminus \{v_2\}$
    otherwise
        $V := V \cup \{v\}$, where v is a new node
        $\alpha(v) := \alpha_1(v_1)$
        $\gamma(v_1) := v$
        $V_1 := V_1 \setminus \{v_1\}$
        $V_2 := V_2 \setminus \{v_2\}$
    endcase
   endwhile
4. sort E_1 and E_2 by target node label
   sort E_1 and E_2 by source node label
5. $E := \emptyset$
6. **while** $E_1 \neq \emptyset$ and $E_2 \neq \emptyset$ **do**
    let (v_1, w_1) and (v_2, w_2) be the first element of E_1 and E_2
    case $\alpha_1(v_1) < \alpha_2(v_2)$ or $(\alpha_1(v_1) = \alpha_2(v_2)$ and $\alpha_1(w_1) < \alpha_2(w_2))$
        $E_1 := E_1 \setminus \{(v_1, w_1)\}$
    case $\alpha_1(v_1) > \alpha_2(v_2)$ or $(\alpha_1(v_1) = \alpha_2(v_2)$ and $\alpha_1(w_1) > \alpha_2(w_2))$
        $E_2 := E_2 \setminus \{(v_2, w_2)\}$
    otherwise
        if $\beta_1(v_1, w_1) = \beta_2(v_2, w_2)$ **then**
            $E := E \cup \{(\gamma(v_1), \gamma(w_1))\}$
            $\beta(\gamma(v_1), \gamma(w_1)) := \beta_1(v_1, w_1)$
        endif
        $E_1 := E_1 \setminus \{(v_1, w_1)\}$
        $E_2 := E_2 \setminus \{(v_2, w_2)\}$
    endcase
   endwhile
7. return G

$\gamma(v_1) = v$ and $\gamma(v_2) = v$. Also, for each node $v_1 \in V_1$ such that there is no node $v_2 \in V_2$ with $\alpha_1(v_1) = \alpha_2(v_2)$, add a new node v to G with $\alpha(v) = \alpha_1(v_1)$ and set $\gamma(v_1) = v$, and for each node $v_2 \in V_2$ such that there is no node $v_1 \in V_1$ with $\alpha_1(v_1) = \alpha_2(v_2)$, add a new node v to G with $\alpha(v) = \alpha_2(v_2)$ and set $\gamma(v_2) = v$.

In a similar vein, sort E_1 and E_2 by source node label and target node label and then, during a simultaneous traversal of the ordered sets of edges, for each edge $e_1 \in E_1$ and $e_2 \in E_2$, say $e_1 = (v_1, w_1)$ and $e_2 = (v_2, w_2)$, with $\alpha_1(v_1) = \alpha_2(v_2)$ and $\alpha_1(w_1) = \alpha_2(w_2)$, if $\beta_1(e_1) = \beta_2(e_2)$, then add a new edge $e = (v, w)$ to G

with $\beta(e) = \beta_1(e_1)$, where $v = \gamma(v_1)$ and $w = \gamma(w_1)$. Also, for each edge $e_1 = (v_1, w_1) \in E_1$ such that there is no edge $e_2 = (v_2, w_2) \in E_2$ with $\alpha_1(v_1) = \alpha_2(v_2)$ and $\alpha_1(w_1) = \alpha_2(w_2)$, add a new edge $e = (v, w)$ to G with $\beta(e) = \beta_1(e_1)$, where $v = \gamma(v_1)$ and $w = \gamma(w_1)$, and for each edge $e_2 = (v_2, w_2) \in E_2$ such that there is no edge $e_1 = (v_1, w_1) \in E_1$ with $\alpha_1(v_1) = \alpha_2(v_2)$ and $\alpha_1(w_1) = \alpha_2(w_2)$, add a new edge $e = (v, w)$ to G with $\beta(e) = \beta_2(e_2)$, where $v = \gamma(v_1)$ and $w = \gamma(w_1)$.

4 Conclusion

Graph matching encompasses a series of related problems with important practical applications in combinatorial pattern matching, pattern recognition, chemical structure search, computational biology, and other areas of engineering and life sciences. In the class of trees and graphs with unique node labels, these problems become polynomial-time solvable and current algorithms for testing graph and subgraph isomorphism and computing the graph edit distance, a shortest edit script, a largest common subgraph, and a smallest common supergraph of two graphs with unique node labels, take time quadratic in the number of nodes in the graphs, and the same holds for similar problems on trees with unique node labels.

The main contribution of this paper is the development of a simple technique for performing set-theoretical operations on the nodes and edges of two trees or graphs with unique node labels. The technique is based on radix sorting the sets of nodes and edges in the trees or graphs by node label and source and target node label, respectively, followed by a simultaneous traversal of the ordered sets of nodes and edges.

Application of this technique to graph matching resulted in simple algorithms for testing graph and subgraph isomorphism and computing the graph edit distance, a shortest edit script, a largest common subgraph, and a smallest common supergraph of two trees or graphs with unique node labels in time linear in the number of nodes and edges in the trees or graphs. The algorithms themselves, for which detailed pseudocode is given, are not complicated to implement, and they only require the use of standard data structures.

Acknowledgment

The research described in this paper was partially supported by BBSRC grant BB/C004310/1, by the Spanish CICYT, project GRAMMARS (TIN2004-07925-C03-01), and by the Japan Society for the Promotion of Science through Long-term Invitation Fellowship L05511 for visiting JAIST (Japan Advanced Institute of Science and Technology).

Algorithm 6. Given two graphs $G_1 = (V_1, E_1, \alpha_1, \beta_1)$ and $G_2 = (V_2, E_2, \alpha_2, \beta_2)$ with unique node labels, compute a smallest common supergraph $G = (V, E, \alpha, \beta)$ of G_1 and G_2.

1. sort V_1 and V_2 by node label
2. $V := \emptyset$
3. **while** $V_1 \neq \emptyset$ and $V_2 \neq \emptyset$ **do**
   let v_1 and v_2 be the first element of V_1 and V_2, respectively
   case $\alpha_1(v_1) < \alpha_2(v_2)$
     $V := V \cup \{v\}$, where v is a new node
     $\alpha(v) := \alpha_1(v_1)$
     $\gamma(v_1) := v$
     $V_1 := V_1 \setminus \{v_1\}$
   case $\alpha_1(v_1) > \alpha_2(v_2)$
     $V := V \cup \{v\}$, where v is a new node
     $\alpha(v) := \alpha_2(v_2)$
     $\gamma(v_2) := v$
     $V_2 := V_2 \setminus \{v_2\}$
   otherwise
     $V := V \cup \{v\}$, where v is a new node
     $\alpha(v) := \alpha_1(v_1)$
     $\gamma(v_1) := v$
     $\gamma(v_2) := v$
     $V_1 := V_1 \setminus \{v_1\}$
     $V_2 := V_2 \setminus \{v_2\}$
   endcase
   endwhile
4. sort E_1 and E_2 by target node label
   sort E_1 and E_2 by source node label
5. $E := \emptyset$
6. **while** $E_1 \neq \emptyset$ and $E_2 \neq \emptyset$ **do**
   let (v_1, w_1) and (v_2, w_2) be the first element of E_1 and E_2
   case $\alpha_1(v_1) < \alpha_2(v_2)$ or $(\alpha_1(v_1) = \alpha_2(v_2)$ and $\alpha_1(w_1) < \alpha_2(w_2))$
     $E := E \cup \{(\gamma(v_1), \gamma(w_1))\}$
     $\beta(\gamma(v_1), \gamma(w_1)) := \beta_1(v_1, w_1)$
     $E_1 := E_1 \setminus \{(v_1, w_1)\}$
   case $\alpha_1(v_1) > \alpha_2(v_2)$ or $(\alpha_1(v_1) = \alpha_2(v_2)$ and $\alpha_1(w_1) > \alpha_2(w_2))$
     $E := E \cup \{(\gamma(v_2), \gamma(w_2))\}$
     $\beta(\gamma(v_2), \gamma(w_2)) := \beta_1(v_2, w_2)$
     $E_2 := E_2 \setminus \{(v_2, w_2)\}$
   otherwise
     if $\beta_1(v_1, w_1) = \beta_2(v_2, w_2)$ **then**
       $E := E \cup \{(\gamma(v_1), \gamma(w_1))\}$
       $\beta(\gamma(v_1), \gamma(w_1)) := \beta_1(v_1, w_1)$
     endif
     $E_1 := E_1 \setminus \{(v_1, w_1)\}$
     $E_2 := E_2 \setminus \{(v_2, w_2)\}$
   endcase
   endwhile
7. return G

References

1. Gati, G.: Further annotated bibliography on the isomorphism disease. J. Graph Theory **3**(1) (1979) 95–109
2. Read, R.C., Corneil, D.G.: The graph isomorphism disease. J. Graph Theory **1**(4) (1977) 339–363
3. Dickinson, P.J., Bunke, H., Dadej, A., Kraetzl, M.: On graphs with unique node labels. In Hancock, E.R., Vento, M., eds.: Proceedings of Fourth IAPR International Workshop on Graph Based Representations in Pattern Recognition. Volume 2726 of Lecture Notes in Computer Science. Berlin Heidelberg New York, Springer (2003) 13–23
4. Dickinson, P.J., Bunke, H., Dadej, A., Kraetzl, M.: Matching graphs with unique node labels. Pattern Anal. Appl. **7**(3) (2004) 243–254
5. Schenker, A., Bunke, H., Last, M., Kandel, A.: Polynomial time complexity graph distance computation for web content mining. In Basu, M., Ho, T.K., eds.: Data Complexity in Pattern Recognition. Berlin Heidelberg New York, Springer (2006)
6. Conte, D., Foggia, P., Sansone, C., Vento, M.: Thirty years of graph matching in pattern recognition. Int. J. Pattern Recogn. Artif. Intell. **18**(3) (2004) 265–298
7. Jiang, X.Y., Bunke, H.: Including geometry in graph representations: A quadratic-time graph isomorphism algorithm and its applications. In: Proceedings of Sixth International Workshop on Structural and Syntactical Pattern Recognition. Volume 1121 of Lecture Notes in Computer Science. Berlin Heidelberg New York, Springer (1996) 110–119
8. Jiang, X.Y., Bunke, H.: Optimal quadratic-time isomorphism of ordered graphs. Pattern Recogn. **32**(7) (1999) 1273–1283
9. Messmer, B.T., Bunke, H.: Error-correcting graph isomorphism using decision trees. Int. J. Pattern Recogn. **12**(6) (1998) 721–742
10. Messmer, B.T., Bunke, H.: A decision tree approach to graph and subgraph isomorphism detection. Pattern Recogn. **32**(12) (1999) 1979–1998
11. Jiang, X.Y., Bunke, H.: Marked subgraph isomorphism of ordered graphs. In: Proceedings of Joint IAPR International Workshops on Structural and Syntactical Pattern Recognition and Structural Pattern Recognition. Volume 1451 of Lecture Notes in Computer Science. Berlin Heidelberg New York, Springer (1998) 122–131
12. Wong, A.K.C., You, M., Chan, S.C.: An algorithm for graph optimal monomorphism. IEEE Trans. Syst. Man Cybern. **20**(3) (1990) 628–636
13. Wong, E.K.: Model matching in robot vision by subgraph isomorphism. Pattern Recogn. **25**(3) (1992) 287–303
14. Akinniyi, F.A., Wong, A.K.C., Stacey, D.A.: A new algorithm for graph monomorphism based on the projections of the product graph. IEEE Trans. Syst. Man Cybern. **16**(5) (1986) 740–751
15. Bunke, H., Messmer, B.T.: Recent advances in graph matching. Int. J. Pattern Recogn. **11**(1) (1997) 169–203
16. Bunke, H., Shearer, K.: A graph distance metric based on the maximal common subgraph. Pattern Recogn. Lett. **19**(3–4) (1998) 255–259
17. Hidovic, D., Pelillo, M.: Metrics for attributed graphs based on the maximal similarity common subgraph. Int. J. Pattern Recogn. Artif. Intell. **18**(3) (2004) 299–313
18. Wallis, W.D., Shoubridge, P., Kraetz, M., Ray, D.: Graph distances using graph union. Pattern Recogn. Lett. **22**(6–7) (2001) 701–704
19. Bunke, H., Jiang, X.Y., Kandel, A.: On the minimum common supergraph of two graphs. Computing **65**(1) (2000) 13–25

20. Fernández, M.L., Valiente, G.: A graph distance measure combining maximum common subgraph and minimum common supergraph. Pattern Recogn. Lett. **22**(6–7) (2001) 753–758
21. Bunke, H.: On a relation between graph edit distance and maximum common subgraph. Pattern Recogn. Lett. **18**(8) (1997) 689–694
22. Bunke, H.: Error-tolerant graph matching: A formal framework and algorithms. In: Advances in Pattern Recognition. Volume 1451 of Lecture Notes in Computer Science. Berlin Heidelberg New York, Springer (1998) 1–14
23. Sanfeliu, A., Fu, K.S.: A distance measure between attributed relational graphs for pattern recognition. IEEE Trans. Syst. Man Cybern. **13**(3) (1983) 353–363
24. Shapiro, L.G., Haralick, R.M.: Structural descriptions and inexact matching. IEEE Trans. Pattern Anal. **3**(5) (1981) 504–519
25. Tsai, W.H., Fu, K.S.: Error-correcting isomorphism of attributed relational graphs for pattern analysis. IEEE Trans. Syst. Man Cybern. **9**(12) (1979) 757–769
26. Garey, M.R., Johnson, D.S.: Computers and Intractability: A Guide to NP-Completeness. New York, Freeman (1979)
27. Deville, Y., Gilbert, D., van Helden, J., Wodak, S.J.: An overview of data models for the analysis of biochemical pathways. Brief. Bioinform. **4**(3) (2003) 246–259
28. Page, R.D.M., Valiente, G.: An edit script for taxonomic classifications. BMC Bioinform. **6** (2005) 208
29. Valiente, G.: Comment to "Polynomial time complexity graph distance computation for web content mining". In Basu, M., Ho, T.K., eds.: Data Complexity in Pattern Recognition. Berlin Heidelberg New York, Springer (2006)
30. Valiente, G.: Algorithms on Trees and Graphs. Berlin Heidelberg New York, Springer (2002)

A Generic Graph Distance Measure Based on Multivalent Matchings

Sébastien Sorlin, Christine Solnon and Jean-Michel Jolion

Summary. Many applications such as information retrieval and classification, involve measuring graph distance or similarity, i.e., matching graphs to identify and quantify their common features.

Different kinds of graph matchings have been proposed, giving rise to different graph similarity or distance measures. Graph matchings may be *univalent* – when each vertex is associated with at most one vertex of the other graph – or *multivalent* – when each vertex is associated with a set of vertices of the other graph. Also, graph matchings may be *exact* – when all vertex and edge features must be preserved by the matching – or *error-tolerant* – when some vertex and edge features may not be preserved by the matching.

The first goal of this chapter is to propose a new graph distance measure based on the search of a best matching between the vertices of two graphs, i.e., a matching minimizing vertex and edge distance functions. This distance measure is generic in the sense that it allows both univalent and multivalent matchings and it is parameterized by vertex and edge distance functions defined by the user depending on the considered application. The second goal of this chapter is to show how to use this generic measure to model and to solve classical graph matching problems such as (sub-)graph isomorphism problem, error-tolerant graph matching, and nonbijective graph matching.

1 Introduction

In many applications such as information retrieval or classification, measuring object similarity is an important issue [1]. Measuring the similarity of two objects consists in identifying and quantifying their commonalities. A dual problem is to measure the distance of these two objects, i.e., identify and quantify their differences.

Graphs are often used to model structured objects, e.g., scene representation [2–5], design objects [6], molecule representations [7, 8], and web documents [9]. Vertices represent object components while edges represent binary relations between these components. Vertices and edges may be labeled by their features. For example, to represent an image by a graph, one usually associates a vertex with each region of the segmented image, and an edge with each couple of vertices corresponding to two adjacent regions. In order to better represent images, each vertex may be labeled by the size and the bounding box of its associated region and each edge may be labeled by a value representing how much two regions are connected (by means of the number of adjacent pixels) [2].

1.1 Graph Matchings and Distance Measures

Computing the distance/similarity of two graphs usually involves finding a "best" matching of the graph vertices (i.e., the one that most preserves vertex and edge features) and then quantifying this set of preserved features. Hence, graph distance measures are closely related to graph matching problems and the capacity of a measure to identify the commonalities of graphs depends on the kind of considered matching.

Graph matchings may be *univalent* – when each vertex is associated with at most one vertex of the other graph – or *multivalent* – when each vertex is associated with a set of vertices of the other graph. Also, graph matchings may be *exact* – when all vertex and edge features must be preserved by the matching – or *error-tolerant* – when some vertex and edge features may not be preserved by the matching.

Examples of univalent exact matchings are:

1. Graph isomorphism, that involves finding a bijection between the graph vertices that preserves all vertex and edge features of the graphs and that is used to prove graph equivalence
2. Subgraph isomorphism, that involves finding an injection from the vertices of the first graph to the vertices of a second graph that preserves all vertex and edge features of the first graph and that is used to prove graph inclusion

In many applications, we are looking for similar objects and not "identical" ones and error-tolerant matchings are needed. Examples of univalent error-tolerant matchings are:

1. Maximum common subgraph [10, 30], that looks for the largest matching (with respect to the number of matched vertices) that preserves all the edges of the matched vertices
2. Graph edit distance [10, 30] that looks for the minimum cost set of operations (i.e., vertex and edge insertion, deletion and relabeling) needed to transform the first graph into a graph that is isomorphic to the second graph

Many applications involve comparing objects described at different granularity levels and multivalent matchings are needed. Different graph distance/similarity measures based on multivalent error-tolerant graph matchings have been proposed:

1. Champin and Solnon [6] measure the similarity of design[ed] objects where one single component of an object may play the same role as that of a set of components of another object, depending on the granularity of object description. Therefore, the graph similarity measure is based on multivalent matchings where one vertex in a graph may be associated with a set of vertices of the other graph.
2. Boeres et al. [4] and Deruyver et al. [12] use graph matching to match an image to its model. In this application, the model has a schematic aspect easy to segment while the image is noised and usually over-segmented. Therefore, scene

recognition is better expressed as a multivalent matching problem where a set of vertices of the scene may be matched with a same vertex of the model.
3. Ambauen et al. [2] propose a new graph edit distance to overcome the problem of comparing over and under segmented images. This distance is based on multivalent matchings: two new edit operations – vertex splitting and merging – are introduced in order to merge or to split over- or under-segmented regions.

1.2 Motivation and Outline of the Chapter

Many different graph distance/similarity measures have been proposed in the literature [13, 14]. These measures are based on different definitions of a "best" matching between two graphs depending on the considered application. For example, the graph similarity measure of Boeres et al. [4] is specific to the recognition of brain images, and in this context specific constraints are added (e.g., all model vertices must be mapped and each image vertex must be mapped to exactly one model vertex). Therefore, it is difficult to use this measure in other applications.

Ambauen et al. defines [2] a more generic graph distance measure: the measure is parameterized by the cost of each possible operation and these costs can be chosen depending on the considered application. As in [4], this measure adds an image recognition specific constraint on the considered multivalent matching: the multivalent matching operations (vertex merging and splitting) must be nonoverlapping, i.e., if one wants to link two vertices u and v of one graph to another vertex u', one has to merge u and v and as a consequence, it will not be possible anymore to link u with a vertex v' without linking v to v'. If this constraint makes sense in a context of over-segmented regions, it is not a desirable property in all applications (in particular for the application of [6]). Also, graph distance measure of [2] is not generic enough to express all kinds of multivalent matching problems: for example, it cannot be used to model the problem described in [4].

In [15] Sorlin and Solnon prove that the similarity measure of Champin and Solnon [6] is generic, i.e., it can be used to compute many other similarity measures (including measures of [4] and [2]). However, if it is generic, it is not always straightforward to use. This measure deals with multilabeled graphs and the similarity of two multilabeled graphs is computed with respect to the set of identical labels that are associated by a mapping. These labels are discrete values, and each label is either recovered or lost by a mapping. However, in many applications and in particular in an image recognition context, one has to compare continuous values. For example, the size of a region of an image is a continuous value and in order to compare two regions, one has to compute the difference between their sizes. Furthermore, when two components are merged, one needs an operator to aggregate these continuous values (for example, the sum of the sizes or the average color of a set of merged regions). Finally, some constraints on matchings are difficult to express in [6]. For example, it is difficult to constrain a vertex to be linked to vertices having a given property only. To express these kinds of constraints on matchings, we show in [15] that one can label the graph vertices in such a way that the original matching can be reconstituted from the set of recovered labels. As a consequence, the similarity of [6] can be used to compute any other similarity measures based on a best graph matching, whatever the constraints on the matching are.

Our goal is to propose a generic graph distance measure, i.e., a unifying framework for all graph matchings and distance measures. This framework offers a better understanding of the different existing matchings and distance measures. It also allows us to define generic algorithms that can be used to compute any kind of graph distance/similarity measures. Indeed, many algorithms have been proposed for computing graph distance measures or solving graph matching problems. However, all these algorithms are dedicated to one problem and cannot be used to solve other kinds of graph matching problems.

Our generic distance has the same power of expression than the similarity measure of Champin and Solnon [6]. However, it is more flexible: it is based on a multivalent matching of the graph vertices like in [6] but it is parameterized by vertex and edge distance functions that can more easily deal with vertex and edge properties (such as labels, real values, etc.).

In Sect. 2, we introduce some definitions and notations needed to define our distance measure. In Sect. 3, we propose a new generic graph distance measure. In Sect. 4, we compare this measure with some classical graph matching problems. In Sect. 5, we prove that our distance and the graph similarity measure of Champin and Solnon [6] are equivalent in the sense that they have the same power of expression. We conclude in Sect. 6 with some computational issues.

2 Definitions and Notations

2.1 Graph

A *graph* is a pair $G = (V, E)$ such that:

1. V is a finite set of *vertices*
2. $E \subseteq V \times V$ is a set of oriented couples of vertices called *edges*

Given an edge $(u, v) \in E$, the vertices u and v are called the *endpoints* of the edge (u, v).

Partial Subgraph and Induced Subgraph

A graph $G' = (V', E')$ is a *partial subgraph* of a graph $G = (V, E)$ (noted $G' \subseteq_p G$) if and only if $V \subseteq V'$ and $E' \subseteq E \cap (V \times V')$.

A graph $G' = (V', E')$ is an *induced subgraph* of a graph $G = (V, E)$ (noted $G' \subseteq_i G$) if and only if $V \subseteq V'$ and $E' = E \cap (V' \times V')$. An induced subgraph $G' = (V', E')$ of a graph $G = (V, E)$ is the graph that contains all the edges of G having their endpoints into V'. As a consequence, an induced subgraph is always a partial subgraph of G (Fig. 1).

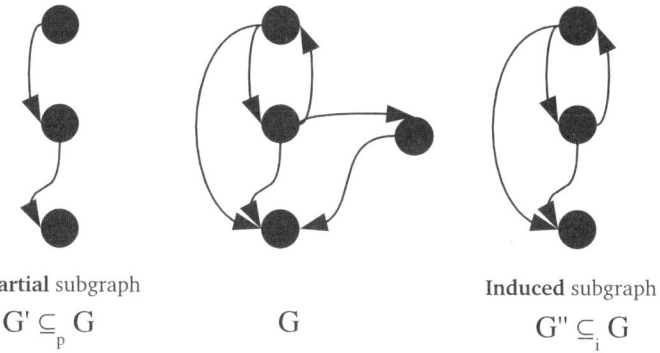

Fig. 1. Examples of a graph G, a partial subgraph G' of G, and an induced subgraph G'' of G

Graph Matching

Given two graphs $G = (V, E)$ and $G' = (V', E')$, a *multivalent matching* m between G and G' is a relation between V and V', i.e., $m \subseteq V \times V'$. Without loss of generality, we shall suppose that $V \cap V' = \emptyset$.

Given a matching m, we note $m(v)$ the set of vertices matched with a vertex v. More formally, we define:

$$\forall v \in V, m(v) \doteq \{v' \in V' | (v, v') \in m\}$$
$$\forall v' \in V', m(v') \doteq \{v \in V | (v, v') \in m\}$$

By extension, when the set of vertices matched with a vertex v is a singleton (i.e., $|m(v)| = 1$), we shall also use $m(v)$ to denote the single vertex that belongs to $m(v)$.

When there is no constraint on the matching, i.e., each vertex may be associated in m with 0, 1 or several vertices, the matching is said to be *multivalent*.

However, one may add constraints on the number of vertices a vertex may be matched with, thus defining matchings that are *partial functions*, *total functions*, *univalent matchings*, *injective matchings*, and *bijective matchings*. Given two graphs $G = (V, E)$ and $G' = (V', E')$, a matching $m \subseteq V \times V'$ is said to be:

1. A *partial function* from G to G' if m links each vertex of V to at most one vertex of G', i.e.:

$$\forall v \in V, |m(v)| \leq 1$$

2. A *total function* from G to G' if m links each vertex of V to exactly one vertex of G', i.e.:

$$\forall v \in V, |m(v)| = 1$$

3. A *univalent matching* between G and G' if m links each vertex of V and V' to at most one vertex, i.e.:

$$\forall v \in V, |m(v)| \leq 1 \land \forall v' \in V', |m(v)| \leq 1$$

4. An *injective matching* from G to G' if m links each vertex of V to a different vertex of V', i.e.:

$$\forall v \in V, |m(v)| = 1 \land \forall (u,v) \in V \times V, u \neq v \Rightarrow m(u) \neq m(v)$$

Another definition of an injective matching from G to G' is a matching m such that:

$$\forall v \in V, |m(v)| = 1 \land \forall v' \in V', |m(v')| \leq 1$$

5. A *bijective matching* between G and G' if m links each vertex of V (resp. V') to a different vertex of V' (resp. V), i.e.:

$$\forall v \in V, |m(v)| = 1 \land \forall (u,v) \in (V \times V), u \neq v \Rightarrow m(u) \neq m(v)$$
$$\forall v' \in V', |m(v')| = 1 \land \forall (u',v') \in (V' \times V'), u' \neq v' \Rightarrow m(u') \neq m(v')$$

Another definition of a bijective matching between G and G' is a matching m such that m links each vertex of V and V' to exactly one vertex, i.e.:

$$\forall v \in V, |m(v)| = 1 \land \forall v' \in V', |m(v')| = 1$$

Edges Matched by a Matching

Given a matching m of the vertices of two graphs $G = (V, E)$ and $G' = (V', E')$, an edge $(u, v) \in E$ is said to be matched with another edge $(u', v') \in E'$ if and only if $\{(u, u'), (v, v')\} \subseteq m$. By extension, we shall note $m(u, v)$ the set of edges matched with the edge (u, v) by the matching m, i.e.:

$$\forall (u,v) \in E, m(u,v) \doteq \{(u',v') \in E' | u' \in m(u), v' \in m(v)\}$$
$$\forall (u',v') \in E', m(u',v') \doteq \{(u,v) \in E | u \in m(u'), v \in m(v')\}$$

Subgraph Induced by a Matching

Given a matching m of two graphs $G = (V, E)$ and $G' = (V', E')$, the subgraph of G (resp. G') induced by m is noted $G_m = (V_m, E_m)$ (resp. $G'_m = (V'_m, E'_m)$) where V_m and E_m (resp. V'_m and E'_m) are the sets of vertices and edges of G (resp. G') matched with at least one vertex or edge of G' (resp. G), i.e.:

$$V_m = \{v \in V / m(v) \neq \emptyset\}, E_m = \{(u,v) \in E / m(u,v) \neq \emptyset\}$$
$$V'_m = \{v' \in V' / m(v') \neq \emptyset\}, E'_m = \{(u',v') \in E' / m(u',v') \neq \emptyset\}$$

Given a matching m of two graphs $G = (V, E)$ and $G' = (V', E')$, if the subgraph of G induced by m, $G_m = (V_m, E_m)$, is equal to G, then, m is an homomorphism between G and G', i.e., m is a function that links each edge of G to an edge of G' (Fig. 2).

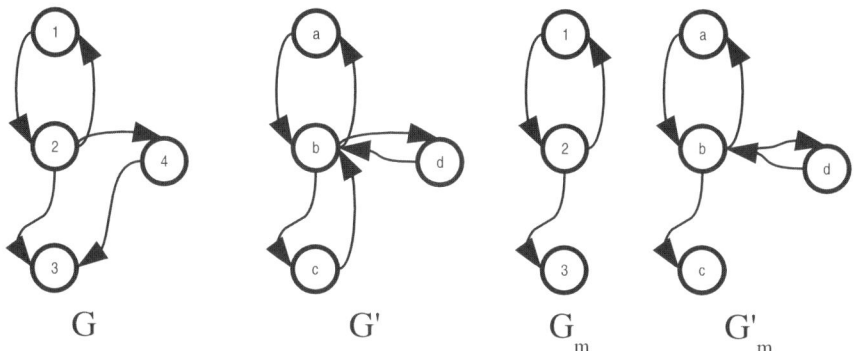

Fig. 2. Two graphs G and G' and their subgraphs induced by the matching $m = \{(1, a), (1, d), (2, b), (3, c)\}$

3 A New Graph Distance Measure

3.1 Vertex and Edge Distance Functions

The first step when computing the distance between two graphs is to match their vertices in order to identify their commonalities. We consider here multivalent graph matchings, i.e., each vertex of a graph may be matched with a – possibly empty – set of vertices of the other graph.

Given a matching m, one has to know for each vertex and each edge how much its properties are recovered by m. Therefore, we assume the existence of a vertex (resp. edge) distance function δ_{vertex} (resp. δ_{edge}) that gives for each vertex v (resp. edge (u, v)) of the two graphs and each set of vertices s_v (resp. set of edges s_e) of the other graph a real value from the interval $[0, +\infty[$ expressing the distance between v (resp. (u, v)) and the set s_v (resp. s_e). More formally, we assume the existence of the two following functions:

$$\delta_{vertex} : (V, \wp(V')) \cup (V', \wp(V)) \to [0, +\infty[$$
$$\delta_{edge} : (E, \wp(E')) \cup (E', \wp(E)) \to [0, +\infty[$$

Roughly speaking, the functions δ_{vertex} and δ_{edge} express the *local preferences* on the way to match a vertex and an edge. These functions depend on the considered application and are used to reflect both the similarity knowledge and constraints that a matching must satisfy.

Generally, the distance is equal to $+\infty$ if the vertex v (resp. the edge (u, v)) is not comparable with the set of vertices s_v (resp. the set of edges s_e), i.e., when it is not possible to match v (resp. (u, v)) with s_v (resp. s_e). The distance is equal to 0 when all the properties of v (resp. (u, v)) are recovered by the set s_v (resp. s_e).

For example, if we are looking for an univalent matching (i.e., each vertex is linked to at most one other vertex) that recovers a maximum number of vertices and edges, one can define the functions δ_{vertex} and δ_{edge} as follows:

$$\forall (v, s_v) \in (V \times \wp(V')) \cup (V' \times \wp(V)), \delta_{vertex}(v, s_v) = 1 \quad \text{if } s_v = \emptyset$$
$$= 0 \quad \text{if } |s_v| = 1$$
$$= +\infty \quad \text{otherwise}$$
$$\forall ((u, v), s_e) \in (E \times \wp(E')) \cup (E' \times \wp(E)), \delta_{edge}((u, v), s_e) = 1 \quad \text{if } s_e = \emptyset$$
$$= 0 \quad \text{if } |s_e| = 1$$
$$= +\infty \quad \text{otherwise}$$

3.2 Graph Distance

Given a matching $m \subseteq V \times V'$ of two graphs $G = (V, E)$ and $G' = (V'E')$ and two distance functions δ_{vertex} and δ_{edge}, the distance of these two graphs with respect to the matching m depends on the distance between each vertex (resp. edge) and the set of vertices (resp. edges) they are matched with, i.e.:

$$\delta_m(G, G') = \otimes(\{(v, \delta_{vertex}(v, m(v)))/v \in V \cup V'\} \cup \qquad (1)$$
$$\{((u, v), \delta_{edge}((u, v), m(u, v)))/(u, v) \in E \cup E'\})$$

where $\otimes$ is an application-dependent function which is used to aggregate the different vertex and edge distances. Roughly speaking, the function $\otimes$ is used to express the global preferences on the distances of the vertices and the edges of the graphs. The function $\otimes$ should be defined in such a way that the minimal distance between two graphs with respect to a matching is equal to 0 and if the distance between two graphs G and G' is equal to $+\infty$, the matching of these two graphs is not acceptable with respect to the considered application. In most cases, the function $\otimes$ is defined as a sum or a weighted sum of the distances of each component. However, in order to express more sophisticated distances, we do not restrict ourself to this particular case. For example, the function $\otimes$ may be defined in such a way that the distance between two graphs depends on the number of vertices that have at most one incoming or outgoing edge having a distance higher than a threshold.

Formula (1) defines the distance of two graphs with respect to a given matching m between the graph vertices. Now, we define the distance of two graphs G and G' as the distance induced by the best matching, i.e., the matching giving rise to a minimal distance:

$$\delta(G, G') = \min_{m \subseteq V \times V'} \delta_m(G, G') \qquad (2)$$

Finally, given two graphs G and G', a distance measure between G and G' is defined as a triple $\delta =< \delta_{vertex}, \delta_{edge}, \otimes >$ where δ_{vertex} is the vertex distance function, δ_{edge} the edge distance function, and $\otimes$ is the function used to aggregate the distances of all vertices and edges of the graphs.

Note that the word "distance" is used here in its common sense: the distance of two graphs is low when the two graphs share a lot of common properties and is equal to 0 (the minimum) when we can find a "perfect" matching of the two graphs (with respect to the considered application). In the general case, our distance measure does not have the mathematical properties of a classical distance measure and is not a metric. As a consequence, the distance between two graphs may have an infinite value, it may not respect the triangular inequality, nor be symmetric and the distance between a graph and itself may not be equal to 0. However, depending on the functions δ_{vertex}, δ_{edge}, and $\otimes$, our distance measure may be a metric.

3.3 Graph Similarity

We have chosen to define the distance of two graphs but distance and similarity measures are two dual concepts and we could use this graph distance measure to define a graph similarity measure of two graphs. For example, in many applications, the distance between two graphs G and G' is always lower or equal to the sum of the distance between each graph and the empty graph $G_\emptyset$ (i.e., $G_\emptyset = (\emptyset, \emptyset)$). As a consequence, we could define a graph similarity measure using this property:

$$sim(G, G') = 1 - \frac{\delta(G, G')}{\delta(G, G_\emptyset) + \delta(G', G_\emptyset)}$$

4 Equivalence with Other Graph Matchings and Distance/Similarity Measures

In this section, we show how our graph distance measure can be used to solve classical graph matching problems.

In this section, the function $\otimes$ is always defined by the function $\otimes_\Sigma$ that returns the sum of the distances of each vertex and each edge of the two graphs. More formally, we define $\otimes_\Sigma : (V \cup V' \cup E \cup E') \times [0, +\infty[\rightarrow [0, +\infty[$ by:

$$\otimes_\Sigma(S) = \sum_{(u,d)\in S} d + \sum_{((u,v),d)\in S} d$$

4.1 Exact Graph Matchings

In this section we show how to reformulate exact graph matching problems with our graph distance measure. For all these kinds of problems, we are looking for an univalent matching between the vertices of two graphs. As a consequence, the vertex and edge distance functions are defined in such a way that a multivalent matching always involves an infinite positive distance. Furthermore, as these problems are satisfaction problems, the objective is always to find a matching m such that $\delta_m(G, G') = 0$.

Graph Isomorphism

Problem Definition

Given two graphs that have the same number of vertices, the graph isomorphism problem consists in deciding if these two graphs are identical minor a renaming of their vertices. More formally, two graphs $G = (V, E)$ and $G' = (V', E')$ such that $|V| = |V'|$ are isomorphic if and only if there exists a bijective matching $m \subseteq V \times V'$ such that $(u, v) \in E \Leftrightarrow (m(u), m(v)) \in E'$.[1]

Measure Definition

To solve the graph isomorphism problem using our distance measure, we have to define vertex and edge distance functions such that these functions return 0 if the vertex or edge is matched with exactly one element and $+\infty$ otherwise (in order to forbid nonbijective matchings). More formally:

$$\forall v \in V \cup V', \forall s_v \subseteq V \cup V', \delta^{iso}_{vertex}(v, s_v) = 0 \quad \text{if } |s_v| = 1$$
$$= +\infty \quad \text{otherwise}$$
$$\forall (u, v) \in E \cup E', \forall s_e \subseteq E \cup E', \delta^{iso}_{edge}(u, v, s_e) = 0 \quad \text{if } |s_e| = 1$$
$$= +\infty \quad \text{otherwise}$$

$$\delta^{iso} = <\delta^{iso}_{vertex}, \delta^{iso}_{edge}, \otimes_{\sum}>$$

Theorem 1. *Given two graphs $G = (V, E)$ and $G' = (V', E')$, the two following properties are equivalent:*

1. *G and G' are isomorphic*
2. *$\delta^{iso}(G, G') = 0$*

Proof. (1) $\Rightarrow$ (2). By definition, if the two graphs are isomorphic, there exists a bijective matching $m \subseteq V \times V'$ such that $(u, v) \in E \Leftrightarrow (m(u), m(v)) \in E'$. As a consequence, $\forall v \in V \cup V', |m(v)| = 1$ (because m is a bijective matching) and $\forall (u, v) \in E \cup E', |m(u, v)| = 1$ (because $\forall (u, v) \in V \times V, (u, v) \in E \Leftrightarrow (m(u), m(v)) \in E'$). So, $\delta^{iso}_m(G, G') = 0$ and therefore $\delta^{iso}(G, G') = 0$.

(2) $\Rightarrow$ (1). If $\delta^{iso}(G, G') = 0$, there exists a matching m such that $\delta^{iso}_m(G, G') = 0$. Given the definition of δ^{iso}_{vertex}, m is such that $\forall v \in V \cup V', |m(v)| = 1$. As a consequence, the matching m is a bijective matching. Furthermore, if $\delta^{iso}_m(G, G') = 0$, then, $\forall (u, v) \in E \cup E', |m(u, v)| = 1$. As a consequence, each edge of both graphs is matched with exactly one edge of the other graph, so $(u, v) \in E \Leftrightarrow (m(u), m(v)) \in E'$. So, m defines an isomorphic matching between the two graphs and G and G' are isomorphic.

Partial Subgraph Isomorphism (or Monomorphism)

Problem Definition

Given two graphs $G = (V, E)$ and $G' = (V', E')$ such that $|V| \leq |V'|$, the partial subgraph isomorphism problem (or monomorphism problem) consists in deciding

---
[1] Let us recall that for univalent matchings, when the set of vertices matched with a vertex v is a singleton, i.e., $|m(v)| = 1$, we note $m(v)$ to denote the single vertex, which is an element of $m(v)$.

if the graph G is isomorphic to a partial subgraph of the graph G', i.e., in finding an injective matching $m \subseteq V \times V'$ such that $\forall (u,v) \in V \times V, (u,v) \in E \Rightarrow (m(u), m(v)) \in E'$. The partial subgraph isomorphism problem is used to decide if a graph is included into another graph.

Measure Definition

To solve the partial subgraph isomorphism problem using our distance measure, we have to define vertex and edge distance functions such that these functions return 0 if an element of G is matched with one element (in order to preserve vertices and edges of G) and $+\infty$ otherwise (in order to avoid noninjective matching). Distance functions for vertices and edges of G' just forbid nonunivalent matchings. More formally:

$$G \begin{cases} \forall v \in V, \forall s_v \subseteq V', \delta^{psub}_{vertex}(v, s_v) = 0 & \text{if } |s_v| = 1 \\ \qquad\qquad\qquad\qquad\qquad\qquad\quad = +\infty & \text{otherwise} \\ \forall (u,v) \in E, \forall s_e \subseteq E', \delta^{psub}_{edge}(u,v,s_e) = 0 & \text{if } |s_e| = 1 \\ \qquad\qquad\qquad\qquad\qquad\qquad\quad = +\infty & \text{otherwise} \end{cases}$$

$$G' \begin{cases} \forall v \in V', \forall s_v \subseteq V, \delta^{psub}_{vertex}(v, s_v) = 0 & \text{if } |s_v| \leq 1 \\ \qquad\qquad\qquad\qquad\qquad\qquad\quad = +\infty & \text{otherwise} \\ \forall (u,v) \in E', \forall s_e \subseteq E, \delta^{psub}_{edge}(u,v,s_e) = 0 & \text{if } |s_e| \leq 1 \\ \qquad\qquad\qquad\qquad\qquad\qquad\quad = +\infty & \text{otherwise} \end{cases}$$

$$\delta^{psub} = <\delta^{psub}_{vertex}, \delta^{psub}_{edge}, \otimes_\Sigma>$$

Theorem 2. *Given two graphs $G = (V, E)$ and $G' = (V', E')$, the two following properties are equivalent:*

1. *The graph G is a partial subgraph of G'*
2. *$\delta^{psub}(G, G') = 0$*

Proof. (1) $\Rightarrow$ (2). By definition, if G is a partial subgraph of G', there exists an injective matching $m \subseteq V \times V'$ such that $\forall (u,v) \in V \times V, (u,v) \in E \Rightarrow (m(u), m(v)) \in E'$. As a consequence, $\forall v \in V, |m(v)| = 1, \forall v \in V', |m(v)| \leq 1$, and $\forall (u,v) \in E', |m(u,v)| \leq 1$ (because m is an injective matching). Furthermore, $\forall (u,v) \in E, |m(u,v)| = 1$ (because $(u,v) \in E \Rightarrow (m(u), m(v)) \in E'$). So, given the definition of δ^{psub}_{vertex} and δ^{psub}_{edge}, $\delta^{psub}_m(G, G') = 0$ and therefore $\delta^{psub}(G, G') = 0$.

(2) $\Rightarrow$ (1). If $\delta^{psub}(G, G') = 0$, then, there exists a matching m such that $\delta^{psub}_m(G, G') = 0$. Given the definition of δ^{psub}_{vertex}, $\forall v \in V, |m(v)| = 1$ and $\forall v \in V', |m(v)| \leq 1$. As a consequence, m is an injective matching. Furthermore, $\forall (u,v) \in E, |m(u,v)| = 1$. As a consequence, each edge of G is matched with exactly one edge of G' and $(u,v) \in E \Rightarrow (m(u), m(v)) \in E'$. So, there exists an injective matching $m \subseteq V \times V'$ that preserves all the edges of G and, by definition, G is a partial subgraph of G'.

Induced Subgraph Isomorphism

Problem Definition

Given two graphs $G = (V, E)$ and $G' = (V', E')$ such that $|V| \leq |V'|$, the induced subgraph isomorphism problem consists in deciding if the graph G is isomorphic to an induced subgraph of G', i.e., in finding an injective matching $m \subseteq V \times V'$ such that $\forall (u, v) \in V \times V, (u, v) \in E \Leftrightarrow (m(u), m(v)) \in E'$. The induced subgraph isomorphism problem is a special case of partial subgraph isomorphism: it adds the constraint that for each couple $(u, v) \in V^2$, if (u, v) is not an edge of G, then, the corresponding vertices in m must neither be an edge of G'.

Measure Definition

The induced subgraph problem between $G = (V, E)$ and $G' = (V', E')$ adds a constraint on each couple of vertices of V (to be or not matched with an edge of G'). To check these constraints, the edge distance function δ_{edge} has to be defined for each couple $(u, v) \in V \times V$ of vertices of G and each subset $s_e \subseteq E'$ of edges of G'. As a consequence, one has to compare the complete graph $G'' = (V, V \times V)$ to the graph $G' = (V', E')$. The vertex distance function must return $+\infty$ if the matching is not injective (rules a, d, and e) and 0 otherwise. The edge distance function must return $+\infty$ if an edge of G is not matched (rule b) or if a couple (u, v) of vertices of G which is not an edge is matched with an edge of G' (rule c) and 0 otherwise. More formally, given a graph $G = (V, E)$ and a graph $G' = (V', E')$, we have to compare the graphs $G'' = (V, V \times V)$ and G' with the two following distance functions:

$$G'' \begin{cases} a & \forall v \in V, \forall s_v \subseteq V', \delta^{sub}_{vertex}(v, s_v) = 0 \quad \text{if } |s_v| = 1 \\ & \qquad\qquad\qquad\qquad\qquad\qquad\qquad = +\infty \quad \text{otherwise} \\ b & \forall (u, v) \in V^2, \forall s_e \subseteq E', \delta^{sub}_{edge,G}(u, v, s_e) = 0 \quad \text{if } (u, v) \in E \wedge |s_e| = 1 \\ c & \qquad\qquad\qquad\qquad\qquad\qquad\qquad\qquad = 0 \quad \text{if } (u, v) \notin E \wedge s_e = \emptyset \\ & \qquad\qquad\qquad\qquad\qquad\qquad\qquad\qquad = +\infty \quad \text{otherwise} \end{cases}$$

$$G' \begin{cases} d & \forall v \in V', \forall s_v \subseteq V, \delta^{sub}_{vertex}(v, s_v) = 0 \quad \text{if } |s_v| \leq 1 \\ & \qquad\qquad\qquad\qquad\qquad\qquad\qquad = +\infty \quad \text{otherwise} \\ e & \forall (u, v) \in E', \forall s_e \subseteq E, \delta^{sub}_{edge,G}(u, v, s_e) = 0 \quad \text{if } |s_e| \leq 1 \\ & \qquad\qquad\qquad\qquad\qquad\qquad\qquad\qquad = +\infty \quad \text{otherwise} \end{cases}$$

$$\delta^{sub}_G = \ < \delta^{sub}_{vertex}, \delta^{sub}_{edge,G}, \otimes_\Sigma >$$

Theorem 3. *Given two graphs $G = (V, E)$ and $G' = (V', E')$, the two following properties are equivalent:*

1. *The graph G is an induced subgraph of G'*
2. *$\delta^{sub}_G(G'', G') = 0$, where $G'' = (V, V \times V)$*

Proof. (1) $\Rightarrow$ (2). By definition, if G is a subgraph of G', there exists an injective matching $m \subseteq V \times V'$ such that $\forall (u, v) \in V \times V, (u, v) \in E \Leftrightarrow (m(u), m(v)) \in E'$.

As a consequence, $\forall v \in V, |m(v)| = 1$, $\forall v \in V', |m(v)| \leq 1$, and $\forall (u,v) \in E', |m(u,v)| \leq 1$ (because m is an injective matching). Furthermore, $\forall (u,v) \in E, |m(u,v)| = 1$ (because $\forall (u,v) \in V \times V, (u,v) \in E \Rightarrow (m(u), m(v)) \in E'$) and $\forall (u,v) \in (V \times V) - E, m(u,v) = \emptyset$ (because $\forall (u,v) \in V \times V, (u,v) \notin E \Rightarrow (m(u), m(v)) \notin E'$). So, given the definition of δ_{vertex}^{sub} and $\delta_{edge,G}^{sub}$, $\delta_{mG}^{sub}(G'', G') = 0$ and $\delta_G^{sub}(G'', G') = 0$.

(2) $\Rightarrow$ (1). If $\delta_G^{sub}(G'', G') = 0$, there exists a matching m such that $\delta_{mG}^{sub}(G'', G') = 0$. Given the definition of δ_{vertex}^{sub}, $\forall v \in V, |m(v)| = 1$ and $\forall v \in V, |m(v)| \leq 1$. As a consequence, m is an injective matching. Furthermore, if m involves a distance equal to 0, then, $\forall (u,v) \in E, |m(u,v)| = 1$. As a consequence, each edge of G is matched with exactly one edge of G', so $\forall (u,v) \in V \times V, (u,v) \in E \Rightarrow (m(u), m(v)) \in E'$. Finally, $\forall (u,v) \in (V \times V) - E, m(u,v) = \emptyset$, and each couple of vertices of G that is not an edge of G is linked to a couple of vertices of G' that is neither an edge of G'. As a consequence, m is an injective matching such that $\forall (u,v) \in V \times V, (u,v) \in E \Leftrightarrow (m(u), m(v)) \in E'$ and G is an induced subgraph of G'.

Generalization of the Subgraph Isomorphism Problem

Problem Definition

Zampelli et al. propose [16] a generalization of the subgraph isomorphism problem. This problem is called "approximate subgraph matching" and consists in looking for a pattern graph into a target graph. It is used for the analysis of biochemical networks. The specificity of this problem is that the pattern graph is composed of mandatory vertices and edges (i.e., vertices and edges that must be preserved by the matching), optional vertices (i.e., vertices that may not be matched), and forbidden edges (i.e., edges that must not be preserved by the matching). Note that an edge having an optional endpoint is optional until its endpoints are matched.[2] More formally, an approximate pattern graph is defined by a tuple $G_p = (V_p, O_p, E_p, F_p)$ where (V_p, E_p) is a graph, $O_p \subseteq V_p$ is the set of optional nodes, and $F_p \subseteq (V_p \times V_p) - E_p$ is the set of forbidden edges. An approximate subgraph matching m between an approximate pattern graph $G_p = (V_p, O_p, E_p, F_p)$ and a target graph $G_t = (V_t, E_t)$ is an univalent matching $m \subseteq V_p \times V_t$ such that:

1. $\forall v \in V_p - O_p, |m(v)| = 1$
2. $\forall (u,v) \in V_p \times V_p, |m(u)| = 1 \wedge |m(v)| = 1$
   $\wedge (u,v) \in E_p \Rightarrow (m(u), m(v)) \in E_t$
3. $\forall (u,v) \in V_p \times V_p, |m(u)| = 1 \wedge |m(v)| = 1$
   $\wedge (u,v) \in F_p \Rightarrow (m(u), m(v)) \notin E_t$
4. $\forall v' \in V_T, |m(v')| \leq 1$

---
[2] This notion of optional vertices is only useful when we look for a matching satisfying some other constraints. Otherwise, we just have to remove optional vertices and their edges from the pattern graph.

Measure Definition

Solving an approximate subgraph matching problem consists in finding an univalent matching m between G_p and the graph $G' = (V_t, V_t \times V_t)$ such that each mandatory vertex is matched with exactly one vertex (rule a), each optional vertex is matched with at most one vertex (rule b), each edge (u, v) is either matched with a couple of vertices (u', v') of G' which is an edge of G_t (rule d) or is not matched at all if one of its optional endpoints is not matched (rule c), each forbidden edge is not matched (rule e). Finally, the matching must be univalent (rules f and g). More formally, one has to compute the distance between $G = (V_p, E_p \cup F_p)$ and $G' = (V_t, E' = V_t \times V_t)$ with the following vertex and edge distance functions:

$$G \begin{cases} a & \forall v \in V_p - O_p, \forall s_v \subseteq V_t, \delta_{vertex}^{agm}(v, s_v) = 0 \quad \text{if } |s_v| = 1 \\ & \qquad\qquad\qquad\qquad\qquad\qquad\qquad\qquad = +\infty \quad \text{otherwise} \\ b & \forall v \in O_p, \forall s_v \subseteq V_t, \delta_{vertex}^{agm}(v, s_v) = 0 \quad \text{if } |s_v| \leq 1 \\ & \qquad\qquad\qquad\qquad\qquad\qquad\qquad\qquad = +\infty \quad \text{otherwise} \\ & \forall (u, v) \in E_p, \forall s_e \subseteq V_t \times V_t, \\ c & \qquad \delta_{edge, G_t}^{agm}(u, v, s_e) = 0 \quad \text{if } s_e = \emptyset \\ d & \qquad\qquad\qquad\qquad\qquad\qquad = 0 \quad \text{if } s_e = \{(u', v')\} \\ & \qquad\qquad\qquad\qquad\qquad\qquad\qquad \wedge (u', v') \in E_t \\ & \qquad\qquad\qquad\qquad\qquad\qquad = +\infty \quad \text{otherwise} \\ e & \forall (u, v) \in F_p, \forall s_e \subseteq E', \delta_{edge, G_t}^{agm}(u, v, s_e) = 0 \quad \text{if } s_e = \{(u', v')\} \\ & \qquad\qquad\qquad\qquad\qquad\qquad\qquad\qquad \wedge (u', v') \notin E_t \\ & \qquad\qquad\qquad\qquad\qquad\qquad\qquad\qquad = +\infty \quad \text{otherwise} \end{cases}$$

$$G' \begin{cases} f & \forall v \in V_t, \forall s_v \subseteq V_p, \delta_{vertex}^{agm}(v, s_v) = 0 \quad \text{if } |s_v| \leq 1 \\ & \qquad\qquad\qquad\qquad\qquad\qquad\qquad\qquad = +\infty \quad \text{otherwise} \\ g & \forall (u, v) \in E', \forall s_e \subseteq E_p \cup F_p, \delta_{edge, G_t}^{agm}(u, v, s_e) = 0 \quad \text{if } |s_e| \leq 1 \\ & \qquad\qquad\qquad\qquad\qquad\qquad\qquad\qquad = +\infty \quad \text{otherwise} \end{cases}$$

$$\delta_{G_t}^{agm} = \langle \delta_{vertex}^{agm}, \delta_{edge, G_t}^{agm}, \otimes_{\sum} \rangle$$

Theorem 4. *Given an approximate pattern graph $G_p = (V_p, O_p, E_p, F_p)$, a target graph $G_t = (V_t, E_t)$ and a mapping $m \subseteq V \times V'$, the two following properties are equivalent:*

1. *m is a solution of the approximate subgraph matching problem between the approximate pattern graph $G_p = (V_p, O_p, E_p, F_p)$ and the target graph $G_t = (V_t, E_t)$*
2. *$\delta_{m, G_t}^{agm}(G, G') = 0$ where $G = (V_p, E_p \cup F_p)$ and $G' = (V_t, V_t \times V_t)$*

Proof. (1) $\Rightarrow$ (2). If m is a solution of the approximate subgraph matching problem then $\forall v \in V_p - O_p, |m(v)| = 1$ (condition 1), $\forall v \in V_t, |m(v)| \leq 1$ and $\forall (u, v) \in V_t \times V_t, |m(u, v)| \leq 1$ (condition 4), $\forall (u, v) \in E_p, m(u, v) = \{(u', v')\} \wedge (u', v') \in E_t$ (condition 2), and $\forall (u, v) \in F_p, m(u, v) = \{(u', v')\} \wedge (u', v') \notin E_t$ (condition

3). As a consequence, given the definition of the vertex and edge distance functions, $\delta_{m,G_t}^{agm}(G,G') = 0$.

(2) ⇒ (1). If the distance $\delta_{m,G_t}^{agm}(G,G') = 0$, then the matching m is univalent because, given the vertex and edge distance functions, all nonunivalent matchings give rise to an infinite distance. Furthermore, if $\delta_{m,G_t}^{agm}(G,G') = 0$, then $\forall v \in V_p - O_p, |m(v)| = 1$ so that m respects condition 1. Furthermore, $\forall (u,v) \in E_p, (m(u) \neq \emptyset \wedge m(v) \neq \emptyset) \Rightarrow (m(u,v) = \{(u',v')\} \wedge (u',v') \in E_t)$ and as a consequence, m respects condition 2. Finally, $\forall (u,v) \in F_p, (m(u) \neq \emptyset \wedge m(v) \neq \emptyset) \Rightarrow (m(u,v) = \{(u',v')\} \wedge (u',v') \notin E_t)$ and as a consequence, m respects condition 3 and m is a solution of the approximate subgraph matching problem.

4.2 Error Tolerant Graph Matchings

In this section we show how to model error tolerant graph matching problems as graph distance measures. For all these problems, we are looking for an univalent matching between the vertices of two graphs. As a consequence, the vertex and edge distance functions are chosen in such a way that a nonunivalent matching always gives an infinite positive distance. Furthermore, as these problems are optimization problems, the objective is always to find the matching that gives the lower distance.

Maximum Common Partial Subgraph

Problem Definition

Given two graphs G and G' the maximum common partial subgraph problem consists in finding the size of the largest partial subgraph G'' of G that is isomorphic to a partial subgraph of G'. For this problem, the size of a graph $G = (V, E)$ is defined by the number of its vertices and edges, i.e., $|G| = |V| + |E|$. The maximum common partial subgraph problem is used to quantify the intersection of two graphs and therefore, it can be used to define a graph similarity measure. Indeed, the similarity of two objects a and b is usually defined as $size(a \cap b)/size(a \cup b)$ [17, 18].

Measure Definition

We have to use vertex and edge distance functions that forbid multivalent matchings while encouraging vertices and edges of G and G' to be matched. As a consequence, the vertex and edge distance functions must return $+\infty$ if the element is matched with more than one element, 1 if it is not matched and 0 if the element is matched with exactly one element, i.e.:

$$\forall v \in V \cup V', \forall s_v \subseteq V \cup V', \delta_{vertex}^{mcps}(v, s_v) = 1 \quad \text{if } s_v = \emptyset$$
$$= 0 \quad \text{if } |s_v| = 1$$
$$= +\infty \quad \text{otherwise}$$

$$\forall (u,v) \in E \cup E', \forall s_e \subseteq E \cup E', \delta_{edge}^{mcps}(u,v,s_e) = 1 \quad \text{if } s_e = \emptyset$$
$$= 0 \quad \text{if } |s_e| = 1$$
$$= +\infty \quad \text{otherwise}$$

$$\delta^{mcps} = \, <\delta_{vertex}^{mcps}, \delta_{edge}^{mcps}, \otimes_{\sum}>$$

Theorem 5. *Given two graphs $G = (V, E)$ and $G' = (V', E')$, and a mapping $m \subseteq V \times V'$, the two following properties are equivalent:*

1. *m is a mapping that minimizes the distance δ_m^{mcps}*
2. *The subgraph G_m of G induced by the matching m is a maximum common partial subgraph of G and G'*

Proof. The proof is decomposed into two steps, we first show that for every matching $m \subseteq V \times V'$ such that $\delta_m^{mcps}(G, G') = d \neq +\infty$, the induced subgraph G_m of G is a common partial subgraph of G and G' and $|G_m| = (|G| + |G'| - d)/2$. In a second step, we show that, if there exists a subgraph G''' of G isomorphic to a partial subgraph of G', then, we can find a matching m having a distance d equal to $|G| + |G'| - 2 * |G'''|$ and such that $G''' = G_m$, the subgraph induced by the mapping m. Then, as we prove that each common partial subgraph G''' corresponds to a mapping inducing a noninfinite distance inverse to the size of G''' (and conversely), the property holds.

$\delta_m^{mcps}(G, G') = d \neq +\infty \Rightarrow G_m$ is a common subgraph of G and G' such that $|G_m| = (|G| + |G'| - d)/2$. Given the vertex and edge distance functions, if $\delta_m^{mcps}(G, G') \neq +\infty$ then m is an univalent matching (because all nonunivalent matchings give an infinite distance). By definition, the subgraph $G_m = (V_m, E_m)$ of G induced by m is a partial subgraph of G and the subgraph $G'_m = (V'_m, E'_m)$ of G' induced by m is a partial subgraph of G'. Given the definition of an induced subgraph and knowing that the mapping is univalent, the matching m is a bijective matching between the vertices of G_m and G'_m such that $(u, v) \in E_m \Leftrightarrow (m(u), m(v)) \in E'_m$. As a consequence, G_m and G'_m are isomorphic and G_m is a common partial subgraph of both G and G'. Given the vertex and edge distance functions, if $\delta_m^{mcps}(G, G') = d \neq +\infty$ then $d = |G| + |G'| - |G_m| - |G'_m|$. As G_m and G'_m are isomorphic, then $|G_m| = |G'_m|$. As a consequence, $|G_m| = (|G| + |G'| - d)/2$ and the property holds.

G''' is a common subgraph of G and $G' \Rightarrow \exists m$ such that $\delta_m^{mcps}(G, G') = |G| + |G'| - 2 * |G'''|$ and $G''' = G_m$. If there exists a common subgraph $G''' = (V''', E''')$ of $G = (V, E)$ and $G' = (V', E')$, then, by definition of a common subgraph, there exists at least one graph $G'''' = (V'''' \subseteq V', E'''' \subseteq E')$ and a bijective matching $m \subseteq V''' \times V''''$ such that $(u, v) \in E''' \Leftrightarrow (m(u), m(v)) \in E''''$. As a consequence, the matching m is such that $\forall v \in V'' \cup V''', |m(v)| = 1$ (because m is a bijective matching), $\forall (u, v) \in E'' \cup E''', |m(u, v)| = 1$ (because m is such that $(u, v) \in E'' \Leftrightarrow (m(u), m(v)) \in E'''$. Furthermore, by definition, m is such that $\forall v \in V - V'', m(v) = \emptyset$, $\forall v \in V' - V''', m(v) = \emptyset$, $\forall (u, v) \in E - E'', m(u, v) = \emptyset$, and $\forall (u, v) \in E' - E''', m(u, v) = \emptyset$. As a consequence, $\delta_m^{mcps}(G, G') = |G| + |G'| -$

$|G'''| - |G''''|$. G''' and G'''' are isomorphic, so, $|G'''| = |G''''|$ and $\delta_m^{mcps}(G, G') = |G| + |G'| - 2 * |G'''|$. The property holds.

Maximum Common Induced Subgraph

Problem Definition

Given two graphs G and G' the maximum common induced subgraph problem consists in finding the largest induced subgraph G''' of G that is isomorphic to an induced subgraph of G'. For this problem, the size of a graph $G = (V, E)$ is defined by the number of its vertices, i.e., $|G| = |V|$. As the maximum common partial subgraph, the maximum common induced subgraph problem is used to define an intersection between two graphs and a corresponding graph similarity measure [10].

Measure Definition

To solve the maximum common subgraph problem using our distance measure, we have to use vertex and edge distance functions encouraging vertices of G to be matched while forbidding matchings that do not correspond to common induced subgraph. So, similarly to the induced subgraph isomorphism problem, the edge distance function must check a constraint (and so be defined) for each couple of vertices of both the graphs. As a consequence, complete graphs must be compared. The vertex distance function encourages the vertices of G to be matched (rule a) and the edge distance function returns $+\infty$ when a couple of vertices (u, v) of G (resp. (u', v') of G') is linked to a couple of vertices (u', v') of G' (resp. (u, v) of G) such that $(u, v) \in E \not\Leftrightarrow (u', v') \in E'$ (rule b) (resp. rule d). Finally, the matching must be univalent (rule c). More formally, we have to compute the distance of the graph $G_2 = (V, V \times V)$ with the graph $G'_2 = (V', V' \times V')$ by using the following vertex and edge distance functions:

$$G_2 \begin{cases} a\ \forall v \in V, \forall s_v \subseteq V', \delta_{vertex}^{mcs}(v, s_v) = 1 & \text{if } s_v = \emptyset \\ \qquad\qquad\qquad\qquad\qquad\qquad\quad = 0 & \text{if } |s_v| = 1 \\ \qquad\qquad\qquad\qquad\qquad\qquad\quad = +\infty & \text{otherwise} \\ \quad\ \forall (u, v) \in V^2, \forall s_e \subseteq V'^2, \\ b\ \ \delta_{edge,GG'}^{mcs}(u, v, s_e) = 0 & \text{if } s_e = \emptyset \\ \qquad\qquad\qquad\qquad\quad\ = 0 & \text{if } s_e = \{(u', v')\} \\ \qquad\qquad\qquad\qquad\qquad\ \wedge ((u, v) \in E \Leftrightarrow (u', v') \in E') \\ \qquad\qquad\qquad\qquad\quad\ = +\infty & \text{otherwise} \end{cases}$$

$$G'_2 \begin{cases} c\ \forall v \in V', \forall s_v \subseteq V, \delta_{vertex}^{mcs}(v, s_v) = 0 & \text{if } |s_v| \leq 1 \\ \qquad\qquad\qquad\qquad\qquad\qquad\quad = +\infty & \text{otherwise} \\ \quad\ \forall (u, v) \in V'^2, \forall s_e \subseteq V^2, \\ d\ \ \delta_{edge,GG'}^{mcs}(u, v, s_e) = 0 & \text{if } s_e = \emptyset \\ \qquad\qquad\qquad\qquad\quad\ = 0 & \text{if } s_e = \{(u', v')\} \\ \qquad\qquad\qquad\qquad\qquad\ \wedge ((u, v) \in E' \Leftrightarrow (u', v') \in E) \\ \qquad\qquad\qquad\qquad\quad\ = +\infty & \text{otherwise} \end{cases}$$

$$\delta_{GG'}^{mcs} = <\delta_{vertex}^{mcs}, \delta_{edge,GG'}^{mcs}, \otimes_\sum >$$

Theorem 6. *Given two graphs $G = (V, E)$ and $G' = (V', E')$, and a mapping $m \subseteq V \times V'$, the two following properties are equivalent:*

1. *m is a mapping that minimizes the distance $\delta_{m,GG'}^{mcs}$*
2. *The subgraph G_m of G induced by the matching m is a maximum common induced subgraph of G and G'*

Proof. The proof is decomposed into two steps. We first show that, for every matching $m \subseteq V \times V'$ such that $\delta_{GG'm}^{mcs}(G, G') = d \neq +\infty$, the subgraph G_m of G induced by the mapping m is an induced common subgraph of G and G' such that $|G_m| = |G| - d$. In a second step, we show that, if there exists an induced subgraph G'' of G isomorphic to an induced subgraph of G', then, we can find a matching m having a distance d equal to $|G| - |G''|$ and such that $G'' = G_m$, the subgraph of G induced by the matching m. Then, as we prove that each common induced subgraph G'' corresponds to a mapping inducing a noninfinite distance inverse to the size of G'' (and reversely), the property holds.

$\delta_{mGG'}^{mcs}(G_2, G'_2) = d \neq +\infty \Rightarrow G_m$ is a common induced subgraph of G and G' such that $|G_m| = |G| - d$. Given the vertex and edge distance functions, if $\delta_{mGG'}(G_2, G'_2) \neq +\infty$ then m is an univalent matching (because all nonunivalent matchings give a distance equal to $+\infty$). By definition, the subgraph $G_{2m} = (V_{2m}, E_{2m})$ of G_2 induced by m is a partial subgraph of G_2 and of G. Furthermore, given the definition of the edge distance function, $(u, v) \in E_{2m} \Rightarrow (u, v) \in E$ and $(u, v) \notin E_{2m} \Rightarrow (u, v) \notin E$. As a consequence, G_{2m} is an induced (i.e., a nonpartial) subgraph of G and $G_{2m} = G_m$. In the same way, we can also prove that the subgraph $G'_{2m} = (V'_{2m}, E'_{2m})$ of G'_2 induced by m is an induced subgraph of G' and that $G'_{2m} = G'_m$. Finally, m is a univalent matching and, given the definitions of the vertex and edge distance functions, m is such that $(u, v) \in E_m \Leftrightarrow (m(u), m(v)) \in E'_m$ so, m defines an isomorphism matching between G_m and G'_m. As a consequence G_m is a common induced subgraph of G and G'. Finally, as only the number of nonrecovered vertices of G influences (positively) the distance, $|G_m| = |G| - d$.

G'' is a common induced subgraph of G and $G' \Rightarrow \exists m$ such that $\delta_{mGG'}^{mcs}(G_2, G'_2) = |G| - |G''|$ and such that $G_m = G''$. If there exists a common induced subgraph $G'' = (V'', E'')$ of $G = (V, E)$ and $G' = (V', E')$, then, by definition of an induced common subgraph, there exists at least one induced subgraph $G''' = (V''', E''')$ of G' and one bijective matching $m \subseteq V'' \times V'''$ such that $(u, v) \in E'' \Leftrightarrow (m(u), m(v)) \in E'''$. Given the vertex and edge distance functions, we can see that the distance $\delta_{mGG'}^{mcs}(G_2, G'_2)$ is equal to $|G| - |G''|$ and that $G_m = G''$.

Graph Edit Distance (ged)

Problem Definition

Given two labeled graphs G and G' (i.e., graphs where a label is associated with each vertex and each edge), the graph edit distance of G and G' is the minimum cost set of weighted operations needed to transform G into G'. Considered operations are insertions, substitutions (i.e., relabeling), and deletions of vertices and edges. Bunke shows in [10] that, when considering appropriate weight definitions, ged is closely related to the maximum common subgraph, and therefore it is also closely related to the similarity measure based on it.

Bunke and Jiang define formally the graph edit distance in [19]. A *labeled graph* is defined by a tuple $G = (V, E, L, \alpha, \beta)$ where V is a set of vertices, E is a set of edges, L is a set of labels, $\alpha : V \to L$ is a total function labeling the vertices of G and $\beta : E \to L$ is a total function labeling the edges of G. Given two labeled graphs $G = (V, E, L, \alpha, \beta)$ and $G' = (V', E', L', \alpha', \beta')$, an *error tolerant graph matching* is an univalent matching $m \subseteq V \times V'$. The vertex $u \in V$ is *substituted* by the vertex v if $m(u) = v$. If $\alpha(u) = \alpha'(m(u))$, the substitution is called an *identical substitution*, otherwise, it is a *nonidentical substitution*. Each vertex $v \in V$ such that $m(v) = \emptyset$ is *deleted* by m and each vertex $v' \in V'$ such that $m(v') = \emptyset$ is *inserted* by m. The same terms are used for the substituted, deleted, and inserted edges of the graphs. A cost c_{vs} (resp. c_{vi} and c_{vd}) is associated with the nonidentical vertex substitutions (resp. insertions and deletions) and a cost c_{es} (resp. c_{ei} and c_{ed}) is associated with the nonidentical edge substitutions (resp. insertions and deletions). Once the six operation costs are set, the *cost of an error tolerant graph matching* m is defined as the sum of the costs of each operation induced by m. Finally, the *graph edit distance* between two graphs is defined as the minimum cost error-tolerant graph matching.

Measure Definition

Each univalent graph matching of our model corresponds to an error-tolerant graph matching of Bunke and Jiang [19]. As a consequence, if the vertex and edge distance functions are defined in such a way that they reproduce the cost of each operation while forbidding nonunivalent matchings, the distance between G_1 and G_2 with respect to an univalent mapping m corresponds to the cost of the error-tolerant graph matching defined by m. More formally, to compute the graph edit distance between two labeled graphs $G = (V, E, L, \alpha, \beta)$ and $G' = (V', E', L', \alpha', \beta')$, we have to compare the graphs $G_1 = (V, E)$ and $G_2 = (V', E')$ with the following vertex and edge distance functions:

$$G_1 \begin{cases} \forall v \in V, \forall s_v \subseteq V', \\ \delta^{ged}_{vertex,GG'}(v, s_v) = c_{vd} \quad \text{if } s_v = \emptyset \\ \qquad\qquad\qquad\quad = 0 \quad \text{if } s_v = \{v'\} \wedge \alpha(v) = \alpha'(v') \\ \qquad\qquad\qquad\quad = c_{vs} \quad \text{if } s_v = \{v'\} \wedge \alpha(v) \neq \alpha'(v') \\ \qquad\qquad\qquad\quad = +\infty \quad \text{if } |s_v| > 1 \\ \forall (u,v) \in E, \forall s_e \subseteq E', \\ \delta^{ged}_{edge,GG'}(u, v, s_e) = c_{ed} \quad \text{if } s_e = \emptyset \\ \qquad\qquad\qquad\quad = 0 \quad \text{if } s_e = \{(u', v')\} \wedge \beta((u,v)) = \beta'((u',v')) \\ \qquad\qquad\qquad\quad = c_{es} \quad \text{if } s_v = \{(u', v')\} \wedge \beta((u,v)) \neq \beta'((u',v')) \\ \qquad\qquad\qquad\quad = +\infty \quad \text{if } |s_e| > 1 \end{cases}$$

$$G_2 \begin{cases} \forall v \in V', \forall s_v \subseteq V, \delta^{ged}_{vertex,GG'}(v, s_v) = c_{vi} & \text{if } s_v = \emptyset \\ \phantom{\forall v \in V', \forall s_v \subseteq V, \delta^{ged}_{vertex,GG'}(v, s_v)} = 0 & \text{if } |s_v| = 1 \\ \phantom{\forall v \in V', \forall s_v \subseteq V, \delta^{ged}_{vertex,GG'}(v, s_v)} = +\infty & \text{if } |s_v| > 1 \\ \forall (u, v) \in E', \forall s_e \subseteq E, \delta^{ged}_{edge,GG'}(u, v, s_e) = c_{ei} & \text{if } s_e = \emptyset \\ \phantom{\forall (u, v) \in E', \forall s_e \subseteq E, \delta^{ged}_{edge,GG'}(u, v, s_e)} = 0 & \text{if } |s_e| = 1 \\ \phantom{\forall (u, v) \in E', \forall s_e \subseteq E, \delta^{ged}_{edge,GG'}(u, v, s_e)} = +\infty & \text{if } |s_e| > 1 \end{cases}$$

$$delta^{ged}_{GG'} = \;<\delta^{ged}_{vertex,GG'}, \delta^{ged}_{edge,GG'}, \otimes_{\sum}>$$

Theorem 7. *Given two labeled graphs G and G' ($G = (V, E, L, \alpha, \beta)$ and $G' = (V', E', L', \alpha', \beta')$), the graph edit distance of Bunke and Jiang [19] is equal to the distance $\delta^{ged}_{GG'}(G_1, G_2)$, where $G_1 = (V, E)$ and $G_2 = (V', E')$.*

Proof. The proof of correctness is trivially done first by proving the equivalence between the set of error-tolerant graph matchings and the set of univalent graph matchings and second, by proving that, given an univalent matching m, the computed distance with respect to m is equal to the cost of the error-tolerant graph matching m.

4.3 Multivalent Graph Matchings

In this section we show how to model different multivalent graph matching problems as graph distance measures. As these problems are optimization problems, the objective is always to find the matching that gives the lowest distance.

Extended Graph Edit Distance

Problem Definition

Ambauen et al. [2] propose to extend the graph edit distance with two new operations: vertex splitting – to split one vertex of G into several vertices of G' – and vertex merging – to merge several vertices of G into one single vertex of G'. These two new operations are added in order to merge over-segmented regions and to split under-segmented regions. Each of these new operations is weighted by a cost c_{split} and c_{merge} (but, in [2], these costs are set to 0). Finally, nonoverlapping constraints are added on the two kinds of "multivalent matching" operations (vertex merging and splitting): if one wants to link two vertices u and v of one graph to another vertex u', one has to merge u and v. As a consequence, it will not be possible anymore to link u with a vertex v' without linking v to v'.

Measure Definition

We cannot model the extended graph edit distance in the same way as that for (nonextended) graph edit distance: the nonoverlapping constraint could not be checked. To take into account this constraint, the matching m must represent the operations that are done. We introduce an "operation graph" $G_O = (V_O, E_O = V_O \times V_O)$. This graph is a complete graph that has as many vertices as the two graphs to compare, i.e., $|V_O| = |V| + |V'|$. Its vertices must be matched with the vertices of the two graphs to compare, i.e., we are looking for a matching $m \subseteq V_O \times (V \cup V')$. Depending on the way the vertices of G_O are matched with the vertices of G and G', the matching m represents a set of edit operations between G and G'. When a vertex of G_O is only matched with a vertex v of G, the vertex v is deleted. When a vertex of G_O is only matched with a vertex v' of G', the vertex v' is inserted. When a vertex of G_O is matched with a vertex v of G and a vertex v' of G', the vertex v is substituted by the vertex v'. In the same way, the edges of G_O model the edge deletions, insertions, and substitutions. When a vertex of G_O is matched with some vertices of G (resp. G'), these vertices are merged (resp. splitted). If the vertices of G and G' must be matched with exactly one vertex of G_O, every matching satisfying this constraint corresponds to a set of edition operations of the extended graph edit distance satisfying the nonoverlapping constraint.

More formally, to model the extended graph edit distance between $G = (V, E, L, \alpha, \beta)$ and $G' = (V', E', L', \alpha', \beta')$, with our generic graph distance measure, one have to compare the graph $G'' = (V'' = V \cup V', E'' = E \cup E')$ (let us recall that $V \cap V' = \emptyset$) and the complete graph $G_O = (V_O, E_O = V_O \times V_O)$ such that $|V_O| = |V| + |V'|$ (because there is at most one edition operation for each vertex of G and G'). The distance functions δ_{vertex}^{eged} and δ_{edge}^{eged} must constrain the vertices of G and G' to be matched with exactly one vertex of G_O. The cost of the edition operations must be computed on the vertices of the graph G_O:

$$\begin{cases} \forall v \in V'', \forall s_v \subseteq V_O, \delta_{vertex}^{eged}(v, s_v) = 0 & \text{if } |s_v| = 1 \\ \qquad\qquad\qquad\qquad\qquad\qquad = +\infty & \text{otherwise} \\ \forall (u,v) \in E'', \forall s_e \subseteq E_O, \delta_{edge}^{eged}((u,v), s_e) = 0 \\ \forall v_o \in V_O, \forall s_v \subseteq V'', \\ \qquad \delta_{vertex}^{eged}(v_o, s_v) = 0 & \text{if } s_v = \emptyset \\ \qquad\qquad\qquad\qquad = match_v(s_v \cap V_1, s_v \cap V_2) & \text{otherwise} \\ \forall (u_o, v_o) \in E_O, \forall s_e \subseteq E'', \\ \qquad \delta_{edge}^{eged}((u_o, v_o), s_e) = 0 & \text{if } s_e = \emptyset \\ \qquad\qquad\qquad\qquad = match_e(s_e \cap E_1, s_e \cap E_2) & \text{otherwise} \end{cases}$$

$$\delta^{eged} = <\delta_{vertex}^{eged}, \delta_{edge}^{eged}, \otimes_{\sum}>$$

where $match_v(s_v, s'_v)$ (resp. $match_e(s_e, s'_e)$) is the cost needed to match the (possibly empty) set of vertices s_v (resp. edges s_e) of G_1 to the (possibly empty) set of vertices s'_v (resp. edges s'_e) of G_2. More formally, the functions $match_v : \wp(V_1) \times \wp(V_2) \to [0, +\infty[$ et $match_e : \wp(E_1) \times \wp(E_2) \to [0, +\infty[$ are defined by:

a $\forall s_v \subseteq V, \forall s'_v \subseteq V',$
$\quad match_v(s_v, s'_v) = merge(s_v) + merge(s'_v)$
$\qquad\qquad\qquad +subst_v(s_v, s'_v)$ if $s_v \neq \emptyset \wedge s'_v \neq \emptyset$
b $\qquad\qquad\qquad = merge(s_v) + del_v(s_v)$ if $s_v \neq \emptyset \wedge s'_v = \emptyset$
c $\qquad\qquad\qquad = merge(s'_v) + ins_v(s'_v)$ if $s_v = \emptyset \wedge s'_v \neq \emptyset$

$$d \:\forall s_e \subseteq V, \forall s'_e \subseteq V',$$
$$match_e(s_e, s'_e) = subst(s_e, s'_e) \quad \text{if } s_e \neq \emptyset \wedge s'_e \neq \emptyset$$
$$e \qquad\qquad\qquad = del_e(s_e) \quad\;\; \text{if } s_e \neq \emptyset \wedge s'_e = \emptyset$$
$$f \qquad\qquad\qquad = ins_e(s'_e) \quad\;\; \text{if } s_e = \emptyset \wedge s'_e \neq \emptyset$$

The function $merge(s_v)$ is the cost needed to merge the vertices of the set s_v, the function $subst_v(s_v, s'_v)$ (resp. $subst_e(s_e, s'_e)$) is the cost needed to substitute the vertices (resp. the edges) of the set s_v (resp. s_e) by the vertices (resp. the edges) of the set s'_v (resp. s'_e). $ins_v(s_v)$ (resp. $ins_e(s_e)$) is the cost need to insert the vertices (resp. edges) of the set s_v (resp. s_e) and $del_v(s_v)$ (resp. $del_e(s_e)$) is the cost of their deletion.

Theorem 8. *Given two (mono)-labeled graphs $G = (V, E, L, \alpha, \beta)$ and $G' = (V', E', L', \alpha', \beta')$, the extended graph edit distance is equal to $\delta^{eged}(G_O, G'')$ where $G'' = (V_1 \cup V_2, E_1 \cup E_2)$ and $G_O = (V_O, V_O \times V_O)$ such that $|V_O| = |V_1| + |V_2|$.*

Proof. The proof of correctness is easy: each matching m giving rise to a noninfinite distance correspond to a sequence of edition operations of the extended graph edit distance (and reversely). Furthermore, the vertex and edge distance functions are defined in such a way that the cost of this sequence is equal to the distance induced by m.

Nonbijective Graph Matching Problem

Definition

Boeres et al. [4] propose a nonbijective graph similarity measure to compare medical images of brains to an image model of a brain. The model has a schematic aspect easy to segment whereas the real image is noised and generally over-segmented. As a consequence, when comparing the image graph to the model graph, one has to use a nonbijective graph matching where the vertices of the model graph may be linked to a set of vertices of the image graph in order to merge over-segmented regions of the image graph. The similarity between an image graph and its model is computed with respect to vertex and edge similarity matrices and the problem consists in finding the best matching (the one with the highest similarity) that satisfies application-dependent constraints. More formally, two graphs are used to represent the problem: the model graph $G = (V, E)$ and the image graph $G' = (V', E')$ (with $|V| \leq |V'|$). A solution is a matching $m \subseteq V \times V'$ between G and G' such that each vertex of G is linked to a nonempty set of connected vertices of G' (i.e., $\forall v \in V, |m(v)| \geq 1$ and the subgraph induced by $m(v)$ is a connected graph), and each vertex of G' is linked to exactly one vertex of G (i.e., $\forall v \in V', |m(v)| = 1$). Finally, some couples of vertices cannot be matched together. Given any matching that respects these constraints, a similarity measure $sim[Boeres]_m$ is computed with respect to a vertex and an edge similarity function $sm_v : V \times V' \to [0, 1]$ and $sm_e : E \times E' \to [0, 1]$ as follows:

$$sim[Boeres]_m = \frac{\sum_{(u,v)\in m} sm_v(u,v)}{|V|.|V'|} + \frac{\sum_{(u,v)\in (V\times V')-m} 1 - sm_v(u,v)}{|V|.|V'|} +$$

$$\frac{\sum_{((u,u'),(v,v'))\in E\times E', \{(u,v),(u',v')\}\in m} sm_e((u,u'),(v,v'))}{|E|.|E'|} +$$

$$\frac{\sum_{((u,u'),(v,v'))\in E\times E', \{(u,v),(u',v')\}\notin m} 1 - sm_e((u,u'),(v,v'))}{|E|.|E'|}$$

Measure Definition

By properly choosing vertex and edge distance functions δ_{vertex} and δ_{edge}, we can model the similarity of Boeres et al. as a graph distance measure. The vertex distance function returns $+\infty$ when the matching violates a constraint and both the vertex and edge distance functions reproduce the similarity matrices sm_v and sm_e. More formally:

$$G \begin{cases} \forall v \in V, \forall s_v \subseteq V', \delta_{vertex}^{nbgm}(v, s_v) = \sum_{v' \in s_v} 1 - sm_v(v, v') \\ \qquad\qquad\qquad\qquad\qquad\quad + \sum_{v' \in V' - s_v} sm_v(v, v') \\ \qquad\qquad\qquad\qquad\qquad\quad \text{if } connected(s_v) \\ \qquad\qquad\qquad\qquad\qquad = +\infty \text{ otherwise} \\ \forall (u,v) \in E, \forall s_e \subseteq E', \\ \delta_{edge}^{nbgm}((u,v), s_e) = \sum_{(u',v')\in s_e} 1 - sm_e((u,v),(u',v')) \\ \qquad\qquad\qquad\qquad\quad + \sum_{(u',v')\in E'-s_e} sm_e((u,v),(u',v')) \end{cases}$$

$$G' \begin{cases} \forall v \in V', \forall s_v \subseteq V, \delta_{vertex}^{nbgm}(v, s_v) = 0 \text{ if } allowed(v, s_v) \\ \qquad\qquad\qquad\qquad\qquad\qquad\quad = +\infty \text{ otherwise} \\ \forall (u',v') \in E', \forall s_e \subseteq E, \delta_{edge}^{nbgm}((u',v'), s_e) = 0 \end{cases}$$

$$\delta^{nbgm} = <\delta_{vertex}^{nbgm}, \delta_{edge}^{nbgm}, \otimes_\sum>$$

where *connected* and *allowed* are two predicates introduced to check the constraints. *connected* is false when a vertex of the model is not matched or when it is matched with a nonconnected set of vertices and true otherwise. *allowed* is false when a vertex of the image is not matched with only one allowed vertex of the model and true otherwise:

$$\forall v \in V, \forall s_v \subseteq V', connected(s_v) = \text{true if } s_v \text{ is a nonempty set of}$$
$$\qquad\qquad\qquad\qquad\qquad\qquad\qquad \text{connected vertices}$$
$$\qquad\qquad\qquad\qquad\qquad\qquad\quad \text{false otherwise}$$

$$\forall v \in V', \forall s_v \subseteq V, allowed(v, s_v) = \text{true if } s_v = \{v'\} \wedge (v, v') \text{ is allowed}$$
$$\qquad\qquad\qquad\qquad\qquad\qquad\qquad \text{false otherwise}$$

Theorem 9. *If the matching m minimizing the distance $\delta_m^{nbgm}(G, G')$ gives rise to a noninfinite distance, then m is the matching that maximizes the similarity of Boeres et al. otherwise, there does not exist a mapping that satisfies the hard constraints of the similarity of Boeres et al.*

Proof. We can easily prove that, thanks to the predicates *connected* and *allowed*, the distance between G and G' with respect to a matching m is equal to $+\infty$ if and only if m is a matching that violates at least one hard constraint. Finally, by decomposing the vertex and edge distance functions, we can prove that the distance δ^{nbgm} is inverse to the similarity of [4] and as a consequence, the matching minimizing the distance δ^{nbgm} is the matching that maximizes the similarity of Boeres et al.

5 Comparison with the Graph Similarity Measure of Champin and Solnon

In [15], we show that the similarity of Champin and Solnon [6] is generic in the sense that, by properly instantiating parameters of this measure, it can be used to solve all the graph matching problems listed earlier. In this section, we briefly present the graph similarity measure of Champin and Solnon and we show that this measure and our graph distance measure are equivalent.

5.1 Definition of the Graph Similarity of Champin and Solnon

The measure of Champin and Solnon is defined for multilabeled graphs, i.e., graphs where a nonempty set of labels is associated with each vertex and each edge of the graphs. More formally, given a set L_V of vertex labels and a set L_E of edge labels, a multilabeled graph G is defined by a tuple $G = \langle V, r_V, r_E \rangle$ such that:

- V is a finite set of vertices
- $r_V \subseteq V \times L_V$ is a relation associating labels to vertices, i.e., r_V is the set of couples (v_i, l) such that vertex v_i is labeled by l
- $r_E \subseteq V \times V \times L_E$ is a relation associating labels to edges, i.e., r_E is the set of triples (v_i, v_j, l) such that edge (v_i, v_j) is labeled by l. Note that the set E of edges of the graph can be defined by $E = \{(v_i, v_j) | \exists l, (v_i, v_j, l) \in r_E\}$

The first step for measuring graph similarity of two graphs $G = \langle V, r_V, r_E \rangle$ and $G' = \langle V', r_{V'}, r_{E'} \rangle$ defined over the same set L_V and L_E of vertex and edge labels is to match their vertices. The matching m considered here is multivalent, i.e., $m \subseteq V \times V'$.

Once a multivalent mapping is defined, the next step is to identify the set of features that are common to the two graphs with respect to this matching. This set contains all the features from both G and G' whose vertices (resp. edges) are matched by m to at least one vertex (resp. edge) that has the same feature. More formally, the set of common features $G \sqcap_m G'$, with respect to a matching m, is defined as follows:

$$G \sqcap_m G' \doteq \{(v,l) \in r_V | \exists v' \in m(v), (v',l) \in r_{V'}\}$$
$$\cup \{(v',l) \in r_{V'} | \exists v \in m(v'), (v,l) \in r_V\}$$
$$\cup \{(v_i, v_j, l) \in r_E | \exists (v'_i, v'_j) \in m(v_i, v_j), (v'_i, v'_j, l) \in r_{E'}\}$$
$$\cup \{(v'_i, v'_j, l) \in r_{E'} | \exists (v_i, v_j) \in m(v'_i, v'_j), (v_i, v_j, l) \in r_E\}$$

Given a multivalent matching m, we also have to identify the set of split vertices, i.e., the set of vertices that are matched with more than one vertex, each split vertex v being associated with the set s_v of its mapped vertices:

$$splits(m) = \{(v, m(v)) | v \in V \cup V', |m(v)| \geq 2\}$$

The *similarity* of G and G' with respect to a mapping m is then defined by:

$$sim_m(G, G') = \frac{f(G \sqcap_m G') - g(splits(m))}{f(r_V \cup r_E \cup r_{V'} \cup r_{E'})} \quad (3)$$

where f and g are two functions that are introduced to weight features and splits, depending on the considered application.

Finally, the *absolute similarity* $sim(G, G')$ of two graphs G and G' is the highest similarity with respect to all possible mappings:

$$sim(G, G') = \max_{m \subseteq V \times V'} \frac{f(G \sqcap_m G') - g(splits(m))}{f(r_V \cup r_E \cup r_{V'} \cup r_{E'})} \quad (4)$$

5.2 Our Graph Distance Measure and the Graph Similarity of Champin and Solnon

Both our graph distance measure and the graph similarity of Champin and Solnon have been shown to be generic in the sense that they can be used to model many other graph distance/similarity measures from the literature. We show here that these two measures have the same ability to represent graph matching problems.

Theorem 10. *Given two sets of vertex and edge labels L_V and L_E and two functions f and g that define a graph similarity measure, there exists a distance measure $\delta = <\delta_{vertex}, \delta_{edge}, \otimes >$ such that for any pair of labeled graphs $G_1 = \langle V_1, r_{V1}, r_{E1} \rangle$ and $G_2 = \langle V_2, r_{V2}, r_{E2} \rangle$ defined over L_V and L_E, the matching $m \subseteq V_1 \times V_2$ that maximizes $sim_m(G_1, G_2)$ also minimizes $\delta_m(G'_1, G'_2)$ where G'_1 and G'_2 are the nonlabeled graphs corresponding to G_1 and G_2, i.e., $G'_1 = (V_1, E_1)$ and $G'_2 = (V_2, E_2)$ with $E_1 = \{(u,v)/\exists (u,v,l) \in r_{E1}\}$ and $E_2 = \{(u,v)/\exists (u,v,l) \in r_{E2}\}$.*

Proof. In order to make the proof, we show that it is possible to define the distance functions δ_{vertex} and δ_{edge} in such a way that the arguments of the function $\otimes$ contains all the information required to reconstitute the matching done. As a consequence, the function $\otimes$ can be defined with the functions f and g.

Let us define a bijective function $num : \wp(V_2) \to N$ that associates an unique integer value with every different subset of vertices of G'_2. The function num is used by the vertex distance function δ_{vertex} to return the set of vertices of G'_2 matched with each vertex of G'_1:

$$\forall v \in V_1, \forall s_v \subseteq V_2, \delta_{vertex}(v, s_v) = num(s_v)$$
$$\forall v \in V_2, \forall s_v \subseteq V_1, \delta_{vertex}(v, s_v) = 0$$
$$\forall (u,v) \in E_1, \forall s_e \subseteq E_2, \delta_{edge}((u,v), s_e) = 0$$
$$\forall (u',v') \in E_2, \forall s_e \subseteq E_1, \delta_{edge}((u',v'), s_e) = 0$$

With such vertex and edge distance functions, the function $\otimes_{sim}$ can be defined with the functions f and g of the similarity measure:[3]

$$\otimes_{sim}(S) = g(split(m_s)) - f(G_1 \sqcap_{m_S} G_2)$$

where m_S is defined as follows:

$$m_S = \{(u, u')/\exists (u,d) \in S \wedge u \in V_1 \wedge u' \in num^{-1}(d)\}$$

Theorem 11. *Given a distance definition $\delta =< \delta_{vertex}, \delta_{edge}, \otimes >$, there exists a graph similarity measure sim of Champin and Solnon (defined by the two functions f and g) such that for any pair of graphs G_1 and G_2, the matching $m \subseteq V_1 \times V_2$ that minimizes the distance $\delta_m(G_1, G_2)$ also maximize $sim_m(G'_1, G'_2)$, where G'_1 and G'_2 are two labeled graphs corresponding to G_1 and G_2.*

Proof. In order to make the proof, we show that, by properly choosing the multi-labeled graphs G_1 and G_2 to compare, the set $G_1 \sqcap_m G_2$ can contain all the information required to know the matching m done. As a consequence, the function f that takes this set as parameter can be defined with the functions $\delta_{vertex}, \delta_{edge}$, and $\otimes$ of the graph distance measure.

Given two graphs $G_1 = (V_1, E_1)$ and $G_2 = (V_2, E_2)$, we define the multilabeled graphs $G'_1 = \langle V_1, r_{V1}, r_{V2}\rangle$ and $G'_2 = \langle V_2, r_{V2}, r_{E2}\rangle$ and the sets L_V and L_E of vertex and edge labels such that:

$$L_V = \{(u,v), u \in V_1, v \in V_2\}, L_E = \{l_e\}$$
$$r_{V1} = \{(u, (u,v)), u \in V_1, v \in V_2\}, r_{E1} = \{(u, v, l_e), (u,v) \in E_1\}$$
$$r_{V2} = \{(v, (u,v)), u \in V_1, v \in V_2\}, r_{E2} = \{(u, v, l_e), (u,v) \in E_2\}$$

With such labeled graphs, the function f can be defined with the functions $\delta_{vertex}, \delta_{edge}$ and $\otimes$:

$$f(S) = - \otimes (\{(v, \delta_{vertex}(v, m_S(v)))/v \in V_1 \cup V_2\}$$
$$\cup \{((u,v), \delta_{edge}((u,v), m_S(u,v)))/(u,v) \in E_1 \cup E_2\})$$

where the matching m_S is defined by:

$$m_S = \{(u,v)/\exists (u, (u,v)) \in S\}$$

---

[3] Note that in one case the problem is to minimize the distance and in the other case, the problem is to maximize the similarity. So, the function $\otimes$ must be defined in such a way that $\forall m \subseteq V_1 \times V_2, \delta_m(G_1, G_2) = -sim_m(G_1, G_2)$.

6 Computing the Distance Between two Graphs

All matching problems described in Sect. 4 are NP-complete or NP-hard problems, except for the graph isomorphism problem, the complexity of which is not exactly stated.[4] As a consequence, computing the distance between two graphs is also a NP-hard problem in the general case.

Complete algorithms have been proposed for computing the matching which maximizes the similarity of Champin and Solnon [6] and for computing the extended graph edit distance of Ambauen et al. [2]. This kind of algorithms based on an exhaustive exploration of the search space combined with pruning techniques, guarantees solution optimality. However, these algorithms are limited to very small graphs. Therefore, incomplete algorithms, that do not guarantee optimality but have a polynomial time complexity, appear to be good alternatives. We propose in [6,15,20,21] three incomplete algorithms for computing the similarity of Champin and Solnon. These algorithms may be adapted to our graph distance in a very straightforward way.

Greedy Algorithm

We propose in [6] a greedy algorithm. The algorithm starts from an empty matching $m = \emptyset$, and iteratively adds to m couples of vertices chosen within the set of candidate couples $cand = V \times V' - m$. This greedy addition of couples to m is iterated until m is locally optimal, i.e., until no more couple addition can increase the similarity. At each step, the couple to be added is randomly chosen within the set of couples that most increase the similarity. This greedy algorithm has a polynomial time complexity of $\mathcal{O}((|V| \times |V'|)^2)$, provided that the computation of the f and g functions have linear time complexities with respect to the size of the matching.

Reactive Tabu Search

The greedy algorithm of [6] returns a "locally optimal" matching in the sense that adding or removing one couple of vertices to this matching cannot improve it. However, it may be possible to improve it by adding and/or removing more than one couple to this matching. In order to improve the matching returned by the greedy algorithm, we propose in [6,15] a reactive tabu local search.

A local search [25,26] tries to improve a solution by locally exploring its neighborhood: the neighbors of a matching m are the matchings which can be obtained by adding or removing one couple of vertices to m.

From an initial matching, computed by the greedy algorithm, the search space is explored from neighbor to neighbor until the optimal solution is found (when the optimal value is known) or until a maximum number of moves have been performed.

---

[4] For particular graphs (such as trees or planar graphs) the graph isomorphism problem is polynomial [22–24]; in general case, the graph isomorphism problem clearly belongs to NP but has neither be proven to belong in P nor to be NP-complete.

The tabu metaheuristic [25,27] is used to choose the next neighbor to move on. At each step, the best neighbor, i.e., the one that most increase the similarity, is chosen. To avoid staying around locally optimal matchings by always performing the same moves, a tabu list is used. This list has a length k and memorizes the last k moves (i.e., the last k added/removed couples of vertices) in order to forbid backward moves (i.e., to remove/add a couple recently added/removed).

The length k of the tabu list is a critical parameter that is hard to set: if the list is too long, search diversification is too strong so that the algorithm converges too slowly; if the list is too short, intensification is too strong so that the algorithm may be stuck around local maxima and fail in improving the current solution. To solve this parameter tuning problem, Battiti and Protasi [28] introduced *Reactive Search* where the length of the tabu list is dynamically adapted during the search. We have used the same idea to build a reactive tabu search algorithm to compute our generic graph distance measure.

Ant Colony Optimization

We also proposed in [20,21] to use the Ant Colony Optimization (ACO) metaheuristic approach to compute the similarity of Champin and Solnon. The ACO metaheuristic is a bioinspired approach [29,30] that has been used to solve many hard combinatorial optimization problems. The main idea is to model the problem to solve as a search for an optimal path in a graph – called the construction graph – and to use artificial ants to search for "good" paths.

The behavior of artificial ants mimics the behavior of real ones: (1) ants lay pheromone trails on the components of the construction graph to keep track of the most promising components, (2) ants construct solutions by moving through the construction graph and choose their path with respect to probabilities which depend on the pheromone trails previously laid, and (3) pheromone trails decrease at each cycle simulating in this way the evaporation phenomena observed in the real world.

Given two graphs $G = (V, E)$ and $G' = (V', E')$ to match, the construction graph is the complete nondirected graph that associates a vertex $< (u, u') >$ with each couple $(u, u') \in V \times V'$. Each elementary path into this graph represents a matching $m \subseteq V \times V'$.

At each cycle, each ant of a colony constructs a matching in a randomized greedy way: starting from an empty matching $m = \emptyset$, the ant iteratively adds couples of vertices that are chosen within the set $cand = \{(u, u') \in V \times V' - m\}$. As usually in ACO algorithm, the choice of the next couple to be added to m is done with respect to a probability that depends on pheromone and heuristic factors (i.e., the similarity improvement when adding the couple). A simple local search procedure may be applied on built matchings to improve their quality.

Once each ant of the colony has built a matching, pheromone trails are updated according to the best matching found. Pheromone is laid on each vertex $< (u, u') >$ of the best found matching in a quantity proportional to the similarity induced by the matching. As a consequence, the amount of pheromone on a vertex $< (u, u') >$ represents the learnt desirability to match u with u'. This process stops iterating either when an ant has found an optimal matching, or when a maximum number of cycles has been performed.

Experimental Results

These three algorithms have been experimentally compared on three different test suites: graph and subgraph isomorphism problems, randomly generated multivalent problems, and the nonbijective graph matching problems of Boeres et al. [4]. Each of these problems has been transformed into our generic graph similarity measure computing problem and, as a consequence, we always use exactly the same code whatever the problem to solve is.

Experimental results showed us that on graph and subgraph isomorphism problems, our algorithms are not competitive with dedicated algorithms: our reactive tabu search and ACO algorithms are able to solve these problems but are clearly longer than dedicated algorithms such as Nauty [31] or VFLIB [32, 33]. These results can be explained by the fact that our algorithms do not use any kind of filtering techniques and potentially explore all kinds of mappings, even multivalent ones. On the seven instances of the nonbijective graph matching problem, our algorithms obtain better results than $LS+$, the reference algorithm of [4] (six instances over seven are better solved by reactive tabu search and seven instances over seven are better solved by ACO algorithm). On all these instances, ACO obtains better results than reactive tabu search but reactive tabu search finds the solutions in shorter times than ACO. On multivalent graph matching problems, reactive tabu search and ACO obtain similar results. However, reactive tabu search finds the solutions in shorter times than ACO.

As a consequence, ACO usually obtains better results but is slower than reactive tabu search. These two algorithms are complementary: if we have to quickly compute a "good" solution of hard instances or if instances are easy, we can use reactive tabu search but if we have more time to spend on computation or if we want to solve very hard instances, we can use ACO.

7 Conclusion

In this chapter, we propose a graph distance measure. This distance is generic: it is based on multivalent matchings of the graph vertices and it is parameterized by two distance functions δ_{vertex} and δ_{edge} used to introduce the application-dependent distance knowledge on vertices and edges and a function $\otimes$ used to aggregate these local preferences. We have shown that we can use our graph distance measure to solve many graph matching problems including the problem of computing the generic graph similarity of Champin and Solnon. We quickly describe three algorithms to compute this generic distance measure: a greedy algorithm, a reactive tabu local search, and an Ant Colony Optimization algorithm. These algorithms are generic so that they can be used to solve any kind of graph matching problem.

References

1. H. Bunke. Graph matching: Theoretical foundations, algorithms, and applications. In *Proceedings on Vision Interface 2000*, Montreal, pages 82–88, 2000
2. R. Ambauen, S. Fischer, and H. Bunke. Graph edit distance with node splitting and merging, and its application to diatom identification. In E. Hancock and M. Vento, editors, *IAPR-TC15 Wksp on Graph-based Representation in Pattern Recognition*, volume 2726 of *LNCS*, Springer, Berlin Heidelberg New York, pages 95–106, 2003
3. R. Baeza-Yates and G. Valiente. An image similarity measure based on graph matching. In *Proceedings of 7th International Symposium on String Processing and Information Retrieval*, pages 28–38. IEEE Computer Science Press, 2000
4. M. Boeres, C. Ribeiro, and I. Bloch. A randomized heuristic for scene recognition by graph matching. In *WEA 2004*, pages 100–113, 2004
5. A. Hlaoui and S. Wang. A new algorithm for graph matching with application to content-based image retrieval. *LNCS*, volume 2396, 2002
6. P.-A. Champin and C. Solnon. Measuring the similarity of labeled graphs. In *5th International Conference on Case-Based Reasoning (ICCBR 2003)*, volume Lecture Notes in Artificial Intelligence 2689-Springer, Berlin Heidelberg New York, pages 80–95, 2003
7. T. Akutsu. Protein structure alignment using a graph matching technique, citeseer.nj.nec.com/akutsu95protein.html, 1995
8. L. Holm and C. Sander. Mapping the protein universe. *Science*, 273:595–602, 1996
9. A. Schenker, M. Last, H. Bunke, and A. Kandel. Classification of web documents using graph matching. *International Journal of Pattern Recognition and Artificial Intelligence*, 18(3): 475–496, 2004
10. H. Bunke. On a relation between graph edit distance and maximum common subgraph. *Pattern Recognition Letters*, 18: 689–694, 1997
11. D. Conte, P. Foggia, C. Sansone, and M. Vento. Thirty years of graph matching in pattern recognition. *International Journal of Pattern Recognition and Artificial Intelligence*, 18(3): 265–298, 2004
12. A. Deruyver, Y. Hod, E. Leammer, and J.-M. Jolion. Adaptive pyramid and semantic graph: Knowledge driven segmentation. In L. Brun and M. Vento, editors, *Graph-Based Representations in Pattern Recognition: 5th IAPR International Workshop, GbRPR 2005, Poitiers, France, April 11–13, 2005. Proceedings*, volume 3434 of *LNCS*, page 213. Springer, Berlin Heidelberg New York, 2005
13. H. Bunke. Recent developments in graph matching. In *ICPR 2000*, pages 2117–2124, 2000
14. J.M. Jolion. Graph matching: what are we really talking about? In *3rd IAPR-TC15 workshop on Graph-based Representations in Pattern Recognition*, pages 170–175, 2001
15. S. Sorlin and C. Solnon. Reactive tabu search for measuring graph similarity. In L. Brun and M. Vento, editors, *5th IAPR-TC-15 workshop on Graph-based Representation in Pattern Recognition*, Springer, Berlin Heidelberg New York, pages 172–182, 2005
16. S. Zampelli, Y. Deville, and P. Dupont. Approximate constrained subgraph matching. In *11th International Conference on Principles and Practice of Constraint Programming*, number 3709 in LNCS, Springer, Berlin Heidelberg new York, pages 832–836, 2005
17. D. Lin. An Information-theoretic definition of similarity. In *Proceedings of ICML 1998, 15th International Conference on Machine Learning*, Morgan Kaufmann, Los Altos, CA, pages 296–304, 1998
18. A. Tversky. Features of Similarity. In *Psychological Review*, volume 84, American Psychological Association Inc., pages 327–352, 1977

19. H. Bunke and X. Jiang. *Graph matching and similarity*, In H.-N. Teodorescu, D. Mlynek, A. Kandel, and H.-J. Zimmermann, editors, *Intelligent Systems and Interfaces*, Chapter 1. Kluwer, Dordrecht, 2000
20. O. Sammoud, C. Solnon, and K. Ghdira. Ant algorithm for the graph matching problem. In *5th European Conference on Evolutionary Computation in Combinatorial Optimization (EvoCOP 2005)*, volume 3448 of *LNCS*, Springer, Berlin Heidelberg New York, pages 213–223, April 2005
21. O. Sammoud, S. Sorlin, C. Solnon, and K. Ghdira. A comparative study of ant colony optimization and reactive search for graph matching problems. In *6th European Conference on Evolutionary Computation in Combinatorial Optimization (EvoCOP 2006)*, volume to appear of *LNCS*. Springer, Berlin Heidelberg New York, April 2006
22. A.V. Aho, J.E. Hopcroft, and J.D. Ullman. *The Design and Analysis of Computer Algorithms*. Addison Wesley, Reading, MA, USA, 1974
23. J.E. Hopcroft and J.-K. Wong. Linear time algorithm for isomorphism of planar graphs. 6th *Annual ACM Symposium on Theory of Computing*, pages 172–184, 1974
24. E.M. Luks. Isomorphism of graphs of bounded valence can be tested in polynomial time. *Journal of Computer System Science*, pages 42–65, 1982
25. F. Glover. Tabu search – part I. *Journal on Computing*, pages 190–260, 1989
26. S. Kirkpatrick, S. Gelatt, and M. Vecchi. Optimisation by simulated annealing. In *Science*, volume 220, pages 671–680, 1983
27. S. Petrovic, G. Kendall, and Y. Yang. A Tabu Search Approach for Graph-Structured Case Retrieval. In *STAIRS 2002*, pages 55–64, 2002
28. R. Battiti and M. Protasi. Reactive local search for the maximum clique problem. In Springer, editor, *Algorithmica*, volume 29, pages 610–637, 2001
29. M. Dorigo and G. Di Caro. The ant colony optimization meta-heuristic. In D. Corne, M. Dorigo, and F. Glover, editors, *New Ideas in Optimization*. McGraw Hill, London, UK, pages 11–32, 1999
30. M. Dorigo and T. Stützle. *Ant Colony Optimization*. MIT, Cambridge, MA, 2004
31. B.D. McKay. Practical graph isomorphism. *Congressus Numerantium*, Nauty, 1981
32. L.P. Cordella, P. Foggia, and M. Vento C. Sansone. An improved algorithm for matching large graphs. *3rd IAPR-TC15 Workshop on Graph-based Representations in Pattern Recognition*, 2001
33. L.P. Cordella, P. Foggia, C. Sansone, and M. Vento. Performance evaluation of the vf graph matching algorithm. In *Proceedings of the 10th International Conference on Image Analysis and Processing (ICIAP'99)*, IEEE, New York, page 1172, 1999

Learning from Supervised Graphs

Joseph Potts, Diane J. Cook and Lawrence B. Holder

Summary. We describe an approach to learning patterns in relational data represented as a graph. The approach, implemented in the Subdue system, searches for patterns that maximally compress the input graph. Subdue can be used for supervised learning, as well as unsupervised pattern discovery and clustering.

Mining graph-based data raises challenges not found in linear attribute-value data. However, additional requirements can further complicate the problem. In particular, we describe how concepts can be learned from training examples which are embedded into a single connected graph, or *supervised graph*. We demonstrate the technique using data from a NASA SST domain as well as a homeland security domain.

1 Introduction

Much of current data mining research focuses on algorithms to discover sets of attributes that can discriminate data entities into classes, such as shopping or banking trends for a particular demographic group. In contrast, we are developing data mining techniques to discover patterns consisting of complex relationships between entities. The field of relational data mining, of which graph-based relational learning is a part, is a new area investigating approaches to mining relational information by finding associations involving multiple tables in a relational database.

Two main approaches have been developed for mining relational information: logic-based approaches and graph-based approaches. Logic-based approaches fall under the area of inductive logic programming (ILP) [1]. ILP embodies a number of techniques for inducing a logical theory to describe the data, and many techniques have been adapted to relational data mining [2]. Graph-based approaches differ from logic-based approaches to relational mining in several ways, the most obvious of which is the underlying representation. Furthermore, logic-based approaches rely on the prior identification of the predicate or predicates to be mined, while graph-based approaches are more data-driven, identifying any portion of the graph that has high support. However, logic-based approaches allow the expression of more complicated patterns involving, e.g., recursion, variables, and constraints among variables. These representational limitations of graphs can be overcome, but at a computational cost.

Our research is particularly applicable to domains in which the data is event driven, such as counter-terrorism intelligence analysis, and domains where distinguishing characteristics can be object attributes or relational attributes. This ability has also become a crucial challenge in many security-related domains. For example, the US House and Senate Intelligence Committees' report on their inquiry into the activities of the intelligence community before and after the September 11, 2001 terrorist attacks revealed the necessity for "connecting the dots" [3]; that is, focusing on the relationships between entities in the data, rather than merely on an entity's attributes. A natural representation for this information is a graph, and the ability to discover previously unknown patterns in such information could lead to significant improvement in our ability to identify potential threats. Similarly, identifying characteristic patterns in spatial or temporal data can be a critical component in acquiring a foundational understanding of important research in many of the basic sciences.

Learning systems capable of utilizing graph-based data typically require training examples to be represented using disjoint graphs, one for each example. In a highly relational domain, however, the positive and negative examples of a concept are not easily separated. We call such a graph a *supervised graph*, because the graph as a whole contains embedded class information which may not be easily separated into individual labeled components. In this chapter we describe a method of learning concepts from examples in supervised graphs that builds upon the capabilities of the Subdue graph-based data mining system.

2 Related Work

Graph-based data mining (GDM) is the task of finding novel, useful, and understandable graph-theoretic patterns in a graph representation of data. Several approaches to GDM exist based on the task of identifying frequently occurring subgraphs in graph transactions, i.e., those subgraphs meeting a minimum level of support. Kuramochi and Karypis [4] developed the FSG system for finding all frequent subgraphs in large graph databases. FSG starts by finding all frequent single and double edge subgraphs. Then, in each iteration, it generates candidate subgraphs by expanding the subgraphs found in the previous iteration by one edge. In each iteration the algorithm checks how many times the candidate subgraph occurs within an entire graph. The candidates, whose frequency is below a user-defined level, are pruned. The algorithm returns all subgraphs occurring more frequently than the given level.

Yan and Han [5] introduced gSpan, which combines depth-first search and lexicographic ordering to find frequent subgraphs. Their algorithm starts from all frequent one-edge graphs. The labels on these edges together with labels on incident vertices define a code for every such graph. Expansion of these one-edge graphs maps them to longer codes. The codes are stored in a tree structure such that if $\alpha = (a_0, a_1, ..., a_m)$ and $\beta = (a_0, a_1, ..., a_m, b)$, the β code is a child of the α code. Since every graph can map to many codes, the codes in the tree structure are not unique. If there are two codes in the code tree that map to the same graph and one is smaller then the other, the branch with the smaller code is pruned during the depth-first search traversal of the code tree. Only the minimum code uniquely defines the graph. Code ordering and pruning reduces the cost of matching frequent subgraphs in gSpan.

Inokuchi et al. [6] developed the apriori-based graph mining (AGM) system, which uses an approach similar to Agrawal and Srikant's [7] apriori algorithm for discovering frequent itemsets. AGM searches the space of frequent subgraphs in a bottom-up fashion, beginning with a single vertex, and then continually expanding by a single vertex and one or more edges. AGM also employs a canonical coding of graphs in order to support fast subgraph matching. AGM returns association rules satisfying user-specified levels of support and confidence.

We distinguish graph-based relational learning (GBRL) from graph-based data mining in that GBRL focuses on identifying novel, but not necessarily most frequent, patterns in a graph representation of data [8]. Only a few GBRL approaches have been developed to date. Two specific approaches, Subdue [9] and GBI [10], take a greedy approach to finding subgraphs maximizing an information theoretic measure. Subdue searches the space of subgraphs by extending candidate subgraphs by one edge. Each candidate is evaluated using a minimum description length metric [11], which measures how well the subgraph compresses the input graph if each instance of the subgraph were replaced by a single vertex. GBI continually compresses the input graph by identifying frequent triples of vertices, some of which may represent previously compressed portions of the input graph. Candidate triples are evaluated using a measure similar to information gain. Kernel-based methods have also been used for supervised GBRL [12].

3 Graph-based Relational Learning in Subdue

The Subdue graph-based relational learning system[1] [9,13] encompasses several approaches to graph-based learning, including discovery, clustering, and supervised learning, which will be described in this section. Subdue uses a labeled graph $G = (V, E, L)$ as both input and output, where $V = \{v_1, v_2, \ldots, v_n\}$ is a set of vertices, $E = \{(v_i, v_j) | v_i, v_j \in V\}$ is a set of edges, and L is a set of labels that can appear on vertices and edges. The graph G can contain directed edges, undirected edges, self-edges (i.e., $(v_i, v_i) \in E$), and multi-edges (i.e., more than one edge between vertices v_i and v_j). The input graph need not be connected, but the learned patterns must be connected subgraphs (called substructures) of the input graph. The input to Subdue can consist of one large graph or several individual graph transactions, and in the case of supervised learning, the individual graphs are classified as positive or negative examples.

---

[1] Subdue source code, sample datasets, and publications are available at http://ailab.uta.edu/subdue.

3.1 Substructure Discovery

SUBDUE's discovery algorithm is shown in Fig. 1 and is given as the input graph, the beam length, and a limit on the total number of substructures considered by the algorithm.

Subdue searches for a substructure that best compresses the input graph. Subdue uses a variant of beam search for its main search algorithm. A substructure in Subdue consists of a subgraph definition and all its occurrences throughout the graph. The initial state of the search is the set of substructures consisting of all uniquely labeled vertices. The only operator of the search is the *ExtendSubstructure* operator. As its name suggests, it extends a substructure in all possible ways by a single edge and a vertex, or by only a single edge if both vertices are already in the subgraph.

The search progresses by applying the *ExtendSubstructure* operator to each substructure in the current state. The resulting state, however, does not contain all the substructures generated by the *ExtendSubstructure* operator. The substructures are kept on a queue and are ordered based on their description length (or sometimes referred to as value) as calculated using the MDL principle described later.

The search terminates upon reaching a user-specified limit on the number of substructures extended, or upon exhaustion of the search space. Once the search

Subdue (Graph, Beam, Limit)

    queue Q = {v | v is a vertex in Graph having a unique label}

    bestSub = first substructure in Q

    repeat

        newQ = { }

        for each substructure $S \in Q$

            newSubs = Extend-Substructure (S, Graph)

                in all possible ways

        Evaluate (newSubs)

        newQ = newQ $\cup$ newSubs mod Beam

        Limit = Limit − 1

        if best substructure in newQ better than bestSub

        then bestSub = best substructure in Q

        Q = newQ

    until Q is empty **or** Limit ≤ 0

    return bestSub

Fig. 1. SUBDUE's discovery algorithm

terminates and Subdue returns the list of best substructures found, the graph can be compressed using the best substructure. The compression procedure replaces all instances of the substructure in the input graph by single vertices, which represent the substructure definition. Incoming and outgoing edges to and from the replaced instances will point to, or originate in the new vertex that represents the instance. The Subdue algorithm can be invoked again on this compressed graph. This procedure can be repeated a user-specified number of times, and is referred to as an iteration.

Subdue's search is guided by the minimum description length (MDL) [11] principle, which seeks to minimize the description length of the entire data set. The evaluation heuristic based on the MDL principle assumes that the best substructure is the one that minimizes the description length of the input graph when compressed by the substructure [9]. The description length of the substructure S given the input graph G is calculated as $DL(S) + DL(G|S)$, where $DL(S)$ is the description length of the substructure and $DL(G|S)$ is the description length of the input graph compressed by the substructure. Description length $DL()$ is calculated as the number of bits in a minimal encoding of the graph. Subdue seeks a substructure S that maximizes compression as calculated in (1).

$$Compression = \frac{DL(G)}{DL(S) + DL(G|S)} \quad (1)$$

As an example, Fig. 2a shows a collection of geometric objects described by their shapes and their "ontop" relationship to one another. Figure 2b shows the graph representation of a portion ("triangle on square") of the input graph for this example and also represents the substructure minimizing the description length of the compressed graph. Figure 2c shows the input example after being compressed by the substructure.

3.2 Graph-based Clustering

Given the ability to find a prevalent subgraph pattern in a larger graph and then compress the graph with this pattern, iterating over this process until the graph can no longer be compressed will produce a hierarchical, conceptual clustering of the input

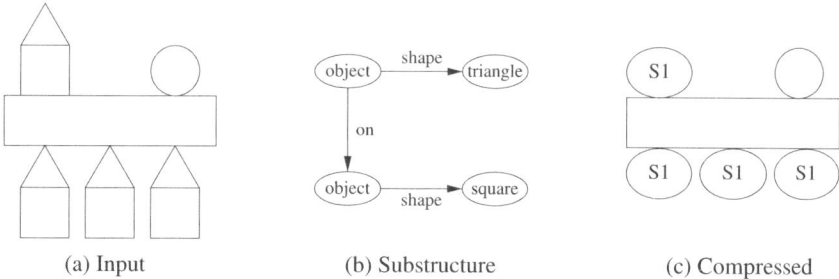

(a) Input   (b) Substructure   (c) Compressed

Fig. 2. Example of Subdue's substructure discovery capability

data. On the ith iteration, the best subgraph S_i is used to compress the input graph, introducing new vertices labeled Si in the graph input to the next iteration. Therefore, any subsequently discovered subgraph S_j can be defined in terms of one or more S_i, where $i < j$. The result is a lattice, where each cluster can be defined in terms of more than one parent subgraph. For example, Fig. 3 shows such a clustering done on a DNA molecule. See [14] for more information on graph-based clustering. The idea of clustering graphs has been explored by others such as Günter and Bunke [15, 16] who also determine the optimal number of clusters automatically, and by Giugno and Shasha [17], who provide graph querying and clustering tools for a wide variety of graph types. Our approach is unique in employing a discovery algorithm to perform the clustering, and in yielding a hierarchical lattice of graph clusters from the original graph data.

3.3 Supervised Learning from Graphs

Extending a graph-based discovery approach to perform supervised learning involves, of course, the need to handle negative examples (focusing on the two-class scenario). In the case of a graph the negative information can come in two forms. First, the data may be in the form of numerous small graphs, or graph transactions, each labeled either positive or negative. Second, data may be composed of two large graphs: one positive and one negative.

The first scenario is closest to the standard supervised learning problem in that we have a set of clearly defined examples. Figure 4a depicts a simple set of positive and negative examples. Let G^+ represent the set of positive graphs, and G^- represent the set of negative graphs. Then, one approach to supervised learning is to find a subgraph that appears often in the positive graphs, but not in the negative graphs. This amounts to replacing the information-theoretic measure with simply an error-based measure. For example, we would find a subgraph S that minimizes

$$\frac{|\{g \in G^+ | S \not\subseteq g\}| + |g \in G^- | S \subseteq g\}|}{|G^+| + |G^-|},$$

where $S \subseteq g$ means S is isomorphic to a subgraph of g. The first term of the numerator is the number of false negatives, and the second term is the number of false positives.

This approach will lead the search toward a small subgraph that discriminates well, e.g., the subgraph in Fig. 4b. However, such a subgraph does not necessarily compress well, nor represent a characteristic description of the target concept. We can bias the search toward a more characteristic description by using the information-theoretic measure to look for a subgraph that compresses the positive examples, but not the negative examples. If $I(G)$ represents the description length (in bits) of the graph G, and $I(G|S)$ represents the description length of graph G compressed by subgraph S, then we can look for an S that minimizes $I(G^+|S) + I(S) + I(G^-) - I(G^-|S)$, where the last two terms represent the portion of the negative graph incorrectly compressed by the subgraph. This approach

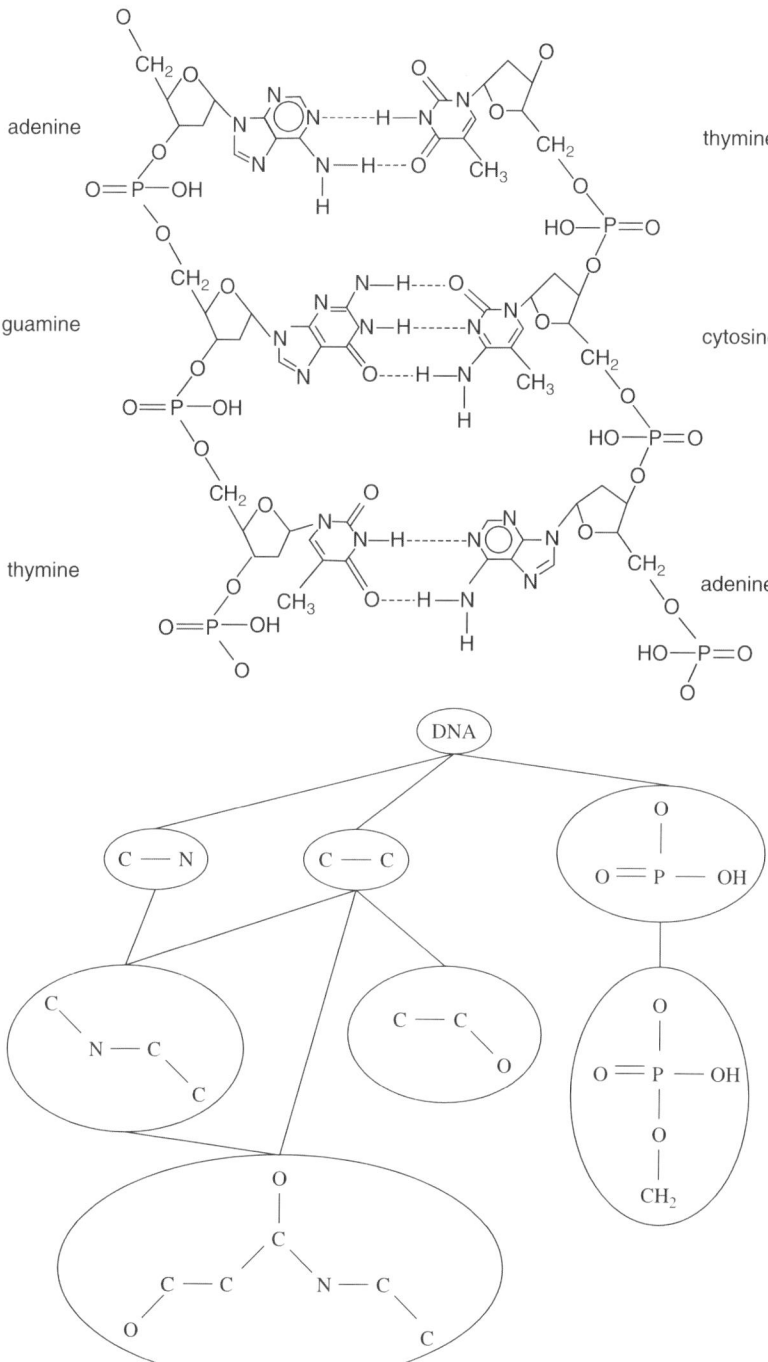

Fig. 3. Example of Subdue's clustering (bottom) on a portion of DNA (top)

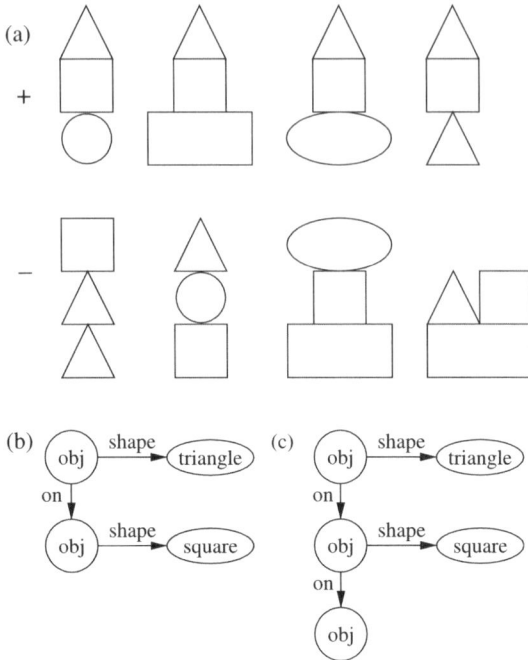

Fig. 4. Graph-based supervised learning example with (**a**) four positive and four negative examples, (**b**) one possible graph concept, and (**c**) another possible graph concept

will lead the search toward a larger subgraph that characterizes the positive examples, but not the negative examples, e.g., the subgraph in Fig. 4c.

Finally, this process can be iterated in a set-covering approach to learn a disjunctive hypothesis. If using the error measure, then any positive example containing the learned subgraph would be removed from subsequent iterations. If using the information-theoretic measure, then instances of the learned subgraph in both the positive and negative examples (even multiple instances per example) are compressed to a single vertex. See [18] for more information on graph-based supervised learning.

4 Learning from Supervised Graphs

Learning systems capable of utilizing graph-based input have typically required the training examples to be represented as disjoint graphs. Input for these systems consists of training examples represented as individual graphs, each of which is an example of one of n classes. The goal is to learn a concept which can be used to determine to which class a previously unseen graph belongs.

In a domain where training examples are naturally embedded (and possibly overlap) in a single graph, efficiently transforming the data for input to systems such as these can be quite difficult. If a system requires individual graphs for each example, then it is necessary to excise each example along with some amount of surrounding graph structure to create a disconnected graph containing that example. If the examples are close enough to each other in the original graph, then this surrounding data may overlap with the surrounding data of another example. In fact, the training example graph may even have to include all or part of another example. This overlap will result in some data appearing in more than one example graph.

Determining just how much structure to include in an example is tricky. Taking too large a region around the example causes extra data to be handled. Taking too small a region may result in the loss of potentially vital information. Since processing graph-based data is very resource intensive, any redundant information can have a drastic effect on performance. Failure to include enough data may result in the inability of the system to learn.

We hypothesize that a compression-based graph mining algorithm can be used to learn class information embedded in a single, connected graph. We develop a learner that allows the input graph, containing all the training examples for all classes to be input with a minimum of preprocessing and a minimum of added or redundant information. In a highly complex relational domain, positive and negative examples of a concept are not easily separated into nonoverlapping graphs. We call such a graph with embedded, possibly overlapping examples a *supervised graph*, or a graph that contains embedded class information which may not be easily separated into individual labeled components.

For example, consider a social network in which we look for patterns distinguishing various income levels. Individuals of a particular income level can appear anywhere in the graph and may be interact with or be related to individuals at other income levels, so we cannot easily partition the graph into separate training cases without potentially severing the target relationships.

To validate our hypothesis, we propose a representation requiring the addition to the input graph of one vertex for each example. We also propose an enhancement of the Subdue algorithm which will construct substructures capable of identifying the examples of each class guided by a new performance metric called *classification compression*. Finally, we propose a representation for these learned substructures called a *classification sequence* which facilitates the determination of class membership for new observations.

4.1 Problem Statement

Our approach to learning concepts from supervised graphs is embodied in the Subdue-EC algorithm. In addition to the labeled graph defined earlier as $G = (V, E, L)$, Subdue-EC also expects as input a set of examples, X, where each $x \in X$ is a set of one or more vertices in V and a set C of class designations, one for each example from the set of n classes. Subdue-EC the learns a concept which is expressed as a set of subgraphs, S, which can be used to assign a class to designated sets of vertices in graphs.

For simplicity in the following discussion, we will consider the two-class learning problem. However, the algorithm, the performance measures, and the classification concepts are applicable to problems with any number of classes. Furthermore, the examples are represented as sets of vertices. Again, for simplicity, we will use single-vertex examples, but any number of vertices may be part of a training example.

4.2 Evaluating Concepts

To be able to perform inductive learning on a single graph with both positive and negative examples, compression of the input graph becomes a less desirable evaluation metric because the graph contains examples of all classes. To allow the MDL principle to guide us in classification, we have to look not at the graph, but at the classification itself. That is, we assume that our receiving agent already has the graph and all of the examples it contains. What we need to send is the classification of those examples. The straightforward way to do that is simply send the class number for each example. Since the examples are in the same order in the receiver's copy of the graph as they are in the sender's, we can just send the class number for examples $1 \ldots n$ and the agent will be able to classify each example. The description length of this naive classification, C_{naive}, is simply the number of bits required to provide a class number for each of the examples. Thus $DL(C_{naive})$ is $nlog_2k$, where n is the number of classes and k is the number of examples.

An alternative to just telling the receiver the class of each example, is to provide the concept as a sequence of substructures $s_1, s_2, \ldots, s_j$, each with an associated class. If an instance of one of the substructures is found in the new graph, then the class associated with that substructure is assigned to all vertices in the substructure instance. The description length of this encoding is thus the description length of the substructure sequence with classes, or classification sequence CS. This is computed as the sum of the description length of the substructures in the sequence, $DL(CS) = \Sigma_i DL(s_i)$.

Of course, this approach may misclassify or fail to classify some examples. In this case, we must inform the agent of the correct classification for those examples. Thus the descriptive length of our alternative message is the sum of the descriptive lengths of each substructure, the class number for each substructure, and encoded exceptions for each class. The description length of this exception list, $DL(EL)$, will require $(k+m+u)log_2(n+1)$ bits, where m is the number of misclassified examples and u is the number of examples left unclassified by the substructure sequence.

If the description length of CS together with ES is smaller than the description length of the naive classification, $DL(CS) + DL(EL) < DL(C_{naive})$, then we will have reduced the message size required to convey the classification to the receiver. We will thus have compressed the classification using our concept, or classification sequence CS. In the same way that Subdue searches for a subgraph that best compresses the input graph, Subdue-EC searches for a classification sequence which provides the best compression of the naive classification. We can now calculate compression as

$$Compression = \frac{DL(CS) + DL(EL)}{DL(C_{naive})} = \frac{\Sigma_i DL(s_i) + (k + m + u)log_2(n+1)}{nlog_2 k}.$$

As before, we take the reciprocal of the compression and use the resulting value as the evaluation measure for potential concepts (classification sequences). The classification sequence that yields the largest value is selected by Subdue-EC as the best concept.

4.3 Identifying Examples

Now that we have a metric for evaluating potential concepts, the remaining issue is how to identify the embedded examples and their associated classes. This is accomplished by the addition of a vertex to the graph for each training example. The vertex is labeled with the class name of the example and is connected by an edge to each vertex in the graph that is part of the training example. We do not need to mark the edges of the example since Subdue-EC will include them in the classifying substructure if they are needed for classification purposes. This vertex is relabeled by Subdue-EC to "EXAMPLE."

Observe that with this representation, vertices and edges in the original graph can be members of more than one training example, possibly from different classes. This is the type of representational freedom that we desire. An individual may interact with one group that is represents a terrorist threat and at the same time do business with other groups that are not under suspicion. In fact, these types of overlaps are sometimes critical to finding the desired concept.

In addition, note that now we can make the initial state of the search algorithm much smaller by starting with only one substructure. All we need are the instances of the single vertex subgraph "EXAMPLE," since all classifying substructures must start there. This "focuses" the search immediately on the right place. Subdue-EC is constrained to never add an "EXAMPLE" node during substructure extension since no classifying substructure can have more than one such vertex.

When the example in Fig. 5 is processed by Subdue-EC, the following five classifying substructures are discovered:

- <u>D</u> (negative)
- <u>B</u>→A→<u>C</u> (positive)
- <u>C</u> (negative)
- <u>B</u>→A→B (positive)
- <u>B</u> (negative)

In this description, the underlined vertices are the ones being classified (the vertices to which an "EXAMPLE" vertex is connected). Thus the first vertex labeled "B" in the B→A→B subgraph is being classified as positive, not the second one. The second "B" vertex is later classified as negative.

Two points should be addressed here. The order that the classifying substructures are applied must be the same as the order in which they were discovered. This facilitates the discover of smaller substructures. For example, the substructure B→A→B

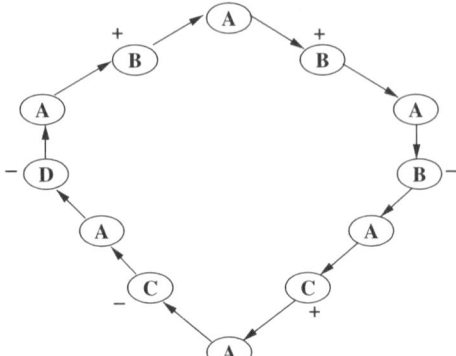

Fig. 5. Embedded examples. Positive-labeled vertices are connected to vertices in the figure with a "+," and negative-labeled vertices are connected to vertices the figures with a "−"

compresses away both positive B vertices. Any remaining B vertices are thus part of negative examples.

5 Experimental Results Using NASA Data

To validate the effectiveness of Subdue-EC to discover concepts from supervised graphs, we chose a simple classification task on a large set of data. We obtained sea surface temperature (SST) data from NASA [19]. This data is averaged over a five-day period and placed on a one degree global grid. The data contains a fill value for grid points for which the SST is not available such as land or due to missing information. We first determined for each grid point whether the temperature *increased*, *decreased*, or stayed the *same* from January 8, 1990 to February 7, 1990. We then placed the nonfill temperature values into one of 9 equal width bins.

We represent this data as a graph, as shown in Fig. 6. Vertices are used to represent each month, discretized latitude and longitude values, hemisphere, and change in temperature from one month to the next. Vertices labeled with "increase" represent regions with increasing temperatures and "decrease" vertices represent regions with decreasing temperatures. Each vertex is connected to its northern and western neighbors, continued in a circle around the globe. This results in a graph that looks like a mesh cylinder, containing 259,200 vertices and 323,640 edges. Each grid point also was connected to a unique vertex containing its temperature binand to another unique vertex labeled N or S, indicating the position's hemisphere.

Note that this is an example dataset where there may exist overlap between instances of the same or different classes. While instances could be extracted from the graph and presented as separate subgraphs for training, the amount of information surrounding each region node that is critical for learning the concept is not known. As a result, the instances cannot be effectively extracted without jeopardizing the accuracy of the result and greatly affecting the runtime of the system due to redundancy in the instances.

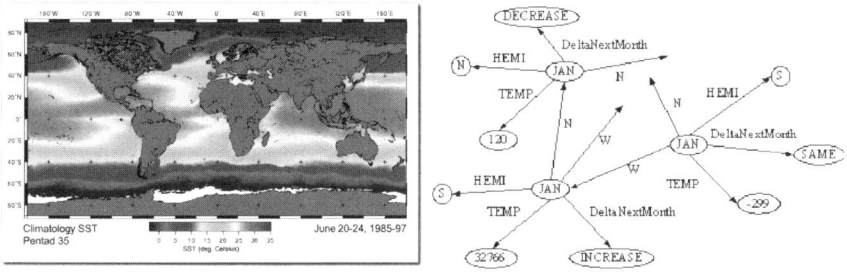

Fig. 6. NASA's SST data (*left*) and Subdue graph representation (*right*)

Table 1. Accuracy results on NASA SST data

run	#substructures generated	time (seconds)	accuracy on training data	accuracy on test data
0	106	52822	86.07%	85.31%
1	104	49669	85.81%	85.26%
2	109	76336	85.57%	85.25%
3	100	71679	85.81%	85.32%
4	104	78874	85.66%	85.69%
5	111	80388	85.84%	86.22%
6	108	73174	85.84%	85.00%
7	112	77236	85.81%	85.39%
8	99	75392	85.64%	84.57%
9	108	80497	85.76%	84.10%
Min	99	49669	85.57%	84.10%
Max	112	80497	86.07%	86.22%
Avg	106.1	71607	85.78%	85.21%

Table 1 shows the results of tenfold cross-validation testing applied to the NASA SST data. For each fold 90% of the grid nodes were randomly selected as training data and the remaining 10% were assigned to a second copy of the graph and was used for testing. The accuracy was good, and accuracy for the test data was fairly consistent with the accuracy on the training data.

We also conducted some tests varying Subdue parameters such as beam size and limit (the number of substructures extended and evaluated). These tests were conducted on 100% of the examples. That is, class vertices were attached to all 64,800 grid points. Surprisingly, the accuracy did not change a lot even when the number of substructures decreased substantially. This is due to the tradeoff in the numerator of classification compression between substructure size and misclassifications. For the NASA data, adding one more vertex adds about 16 bits to the size of the substructure. Since there are about 64,000 examples, the penalty for a misclassification is about 16 bits (that is how many bits it takes to tell the sender the example number of the misclassified example). Thus eliminating two misclassifications more than pays for making the substructure one vertex bigger. This tends to drive substructure growth larger and larger until terminated by the limit parameter. On the other hand,

Table 2. Accuracy with increasing limit

limit	substructures generated	time	accuracy
6	16	17345	83.70%
7	11	10603	84.41%
8	28	28099	85.17%
9	28	26768	85.17%
10	26	26077	85.12%
20	71	53370	85.46%
40	113	77107	85.87%
60	135	82789	85.90%
80	137	92012	86.09%
100	146	101324	86.23%

the unclassified examples are then classified on a subsequent iteration. Thus there are more substructures and larger substructures as the limit is increased, but the accuracy does not significantly improve (see Table 2).

Our final test on the NASA data is to train Subdue-EC with all of the 1990 data and use the learned concept to classify 1991 data. We created a graph using the same representation for data from January 8, 1991 to February 7, 1991, and calculated the accuracy of the learned substructures on this new data. Using a limit of ten substructures, Subdue-EC achieved 81.98% accuracy, showing that the learned substructures have classification value even for subsequent years.

The learned substructures are what one might intuitively expect. The first in the sequence addresses the large number of *same* examples. These are primarily land areas which are still land masses 30 days later and therefore still have fill values for the temperature and receive the same classification. Otherwise, the concepts represent the ideas that the northern hemisphere gets colder in winter and the southern hemisphere gets warmer. Interestingly, temperature bin 0 classifies as *same*. This may be because the coldest areas do not change temperature much throughout the year. In addition, southern hemisphere grid points north of temperature bin 6 decrease. This is consistent with the fact that these areas are on the equator and therefore start to cool off as winter drags on and they get less sun. Finally, it should be noted that none of the tests ever leave any data unclassified. On these data there always seems to be benefit to including a catchall classification substructure at the end that has enough correct classifications to pay for its misclassifications.

6 Experimental Results Using Security Data

As part of a government-sponsored program, a domain has been built to simulate the evidence available about terrorist groups and their plans prior to execution. This domain is motivated from an understanding of the real problem of intelligence data analysis. The domain consists of a number of concepts, including threat and nonthreat actors, threat, and nonthreat groups, targets, exploitation modes (vulnerability modes are exploited by threat groups, productivity modes are exploited by threat and nonthreat groups), capabilities, resources, communications, visits to targets, and transfer of resources between actors, groups, and targets.

The domain follows a general plan of starting a group, recruiting members with needed capabilities, acquiring needed resources, visiting a target, and then exploiting the target. The data we use for our experiments represents the activities of terrorist organizations as they attempt to exploit vulnerable targets, represented by the execution of five different event types. They are:

- *Two-way-communication.* Involves one initiating person and one responding person.
- *N-way-communication.* Involves one initiating person and multiple respondents.
- *Generalized-transfer.* One person transfers a resource.
- *Applying-capability.* One person applies a capability to a target.
- *Applying-resource.* One person applies a resource to a target.

All data is generalized so that no specific names are used. The simulator generates evidence related to all of these events, and this evidence is passed through filters varying the degree of observability and noise in the final evidence.

For our experiments, a graph was creating in which vertices are used to represent member agents from threat and nonthreat groups. Anyone which whom these agents communicates is also added to the graph and connected to the agent with an undirected "association" edge. Communication events between associates are similarly represented with "association" edges.

In addition, each person may be described using attribute and capability vertices. In the simulated data, every individual is assigned at least two strong "trust-link" attributes (e.g., school, place of worship, former military unit, extended family) and at least two weaker "culture-link" attributes (e.g., nationality, language, religion) that are commonly applied in social network development. Capabilities refer to unique abilities exhibited by the individual. Figure 7 shows a portion of the graph generated for this dataset.

Our experiments were conducted on a large graph (graph1) consisting of 435,429 vertices and 763,504 edges representing 61,105 people as well as a smaller graph (graph2) consisting of 217,901 vertices and 314,793 edges representing 30,715 people. Class vertices labeled *threat* were attached to members of known threat groups, and *nonthreat* vertices were attached to members of nonthreat groups.

Our goal for the experiments was to see how well Subdue-EC could classify threat and nonthreat individuals, given training examples embedded in a single connected graph. In the original graphs there is a large predominance of nonthreat individuals (58,373, in contrast to the 1,732 threat individuals). To provide a stronger sample to the learning algorithm, we randomly sampled an equal number of threat and nonthreat individuals.

Table 3 summarizes the results for graph1. For the individuals that included one or more of the classifying substructures, Subdue's classification accuracy was 72.98%. However, the computational limitations of the discovery algorithm prevented further substructures from being discovered in a reasonable amount of time,

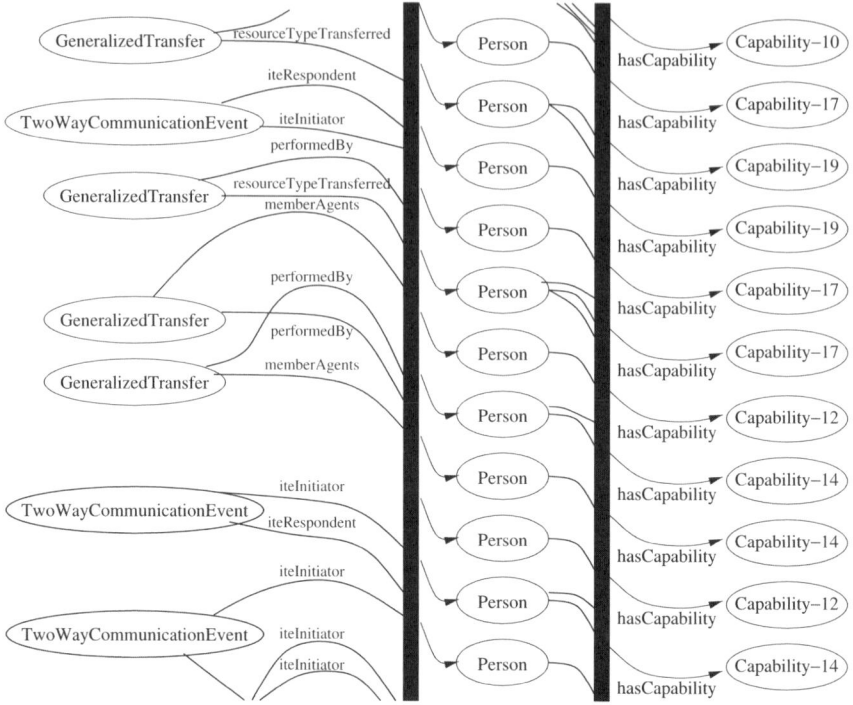

Fig. 7. A section of the graph representation for the counter-terrorism data

Table 3. Classification results on graph1

	total	correct	incorrect	unclassified
threats	1,732	765	35	932
nonthreats	1,732	70	290	1,372

so 2,304 individuals remained unclassified. The greatest number of misclassifications were false positives (classified as threats when the true classification is nonthreat), which is a preferred type of mistake for this problem.

Of the substructures that were discovered, many consisted of an individual exhibiting a particular capability. However, a few of the substructures, such as the one shown in Fig. 8, highlight an association between two individuals in addition to attributes and capabilities of the individuals.

The fact that Subdue discovered useful substructures that highlight relationships between the individuals to be classified highlights the strength of Subdue-EC. If the individuals have been separated into disjoint examples, this relationship could not have been found. If we tried to extract individuals with a large enough neighborhood of information around them to find these discoveries, several difficulties would arise. First, how much information do we retain? The user cannot always know a priori how much of a neighborhood must be extracted in order to retain all potentially

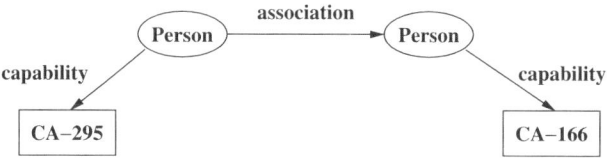

Fig. 8. A sample discovered substructures. This substructure highlights an association between two individuals, each with certain capabilities. The individual on the left is a known threat

Table 4. Results of graph1 testing on all individuals

	total	correct	incorrect	unclassified
threats	1,732	765	35	932
nonthreats	59,373	1,830	12,840	44,703

Table 5. Classification results on graph2

		total	correct	incorrect	unclassified
set 1	threats	1,225	463	28	734
	nonthreats	1,225	38	221	966
set 2	Threats	1,225	463	28	734
	nonthreats	29,490	876	5,596	23,018

useful information. Second, when the neighborhood of information is extracted, it is in essence reproduced for each example object that requires the information. This results in substantial cost increase both in memory and in processing time for the discovery algorithm.

To determine the effect of the sample size on Subdue's classification accuracy, we performed another classification experiment on graph1 in which training and testing were performed on the entire set of input threat and nonthreat individuals. The results are summarized in Table 4. As can be seen, the results did not change for threat individuals. While the number of correctly classified nonthreat individuals did increase, so did the number of misclassifications, resulting in a poorer performance than the earlier experiment.

In a separate experiment, we evaluated the generalizability of Subdue's results by using the substructures discovered in the first two experiments to classify individuals from a separate dataset, graph2. Table 5 shows the results of this experiment. As can be seen, while the percentage of accurate classifications does drop for the new dataset, Subdue still is able to perform fairly well on previously unseen data.

7 Conclusions

The handling of supervised graphs is an important direction for mining structural data. To extend our current work, we would like to handle embedded instances without a single representative instance node (the "increase" and "decrease" nodes in our NASA example) and instances that may possibly overlap. In addition, improved scalability of graph operations is necessary to learn patterns, evaluate their accuracy on test cases, and ultimately to use the patterns to find matches is future intelligence data. The graph and subgraph isomorphism operations are a significant bottleneck to these capabilities. We are currently designing faster approximate versions of these operations to improve the scalability of graph-based relational learning.

8 Acknowledgments

This research is sponsored by the Air Force Research Laboratory (AFRL) under contract F30602-01-2-0570. The views and conclusions contained in this document are those of the authors and should not be interpreted as necessarily representing the official policies, either expressed or implied of AFRL or the United States Government.

References

1. S. Muggleton, editor. *Inductive Logic Programming*. Academic, New York, 1992
2. S. Dzeroski and N. Lavrac, editors. *Relational Data Mining*. Springer, Berlin Heidelberg New York, 2001
3. U. Senate and H. C. on Intelligence. Joint inquiry into intelligence community activities before and after the terrorist attacks of september 11, 2001, December 2002
4. M. Kuramochi and G. Karypis. Frequent subgraph discovery. In *Proceedings of the First IEEE Conference on Data Mining*, 2001
5. X. Yan and J. Han. gSpan: Graph-based substructure pattern mining. In *Proceedings of the International Conference on Data Mining*, 2002
6. A. Inokuchi, T. Washio, and H. Motoda. Complete mining of frequent patterns from graphs: Mining graph data. *Machine Learning*, 50(3): 321–354, 2003
7. R. Agrawal and R. Srikant. Fast algorithms for mining association rules. In *Proceedings of the Twentieth Conference on Very Large Databases*, pages 487–499, 1994
8. L. Holder and D. Cook. Graph-based relational learning: Current and future directions. *ACM SIGKDD Explorations*, 5(1): 90–93, 2003
9. D. J. Cook and L. B. Holder. Substructure discovery using minimum description length and background knowledge. *Journal of Artificial Intelligence Research*, 1: 231–255, 1994
10. K. Yoshida, H. Motoda, and N. Indurkhya. Graph-based induction as a unified learning framework. *Journal of Applied Intelligence*, 4: 297–328, 1994
11. J. Rissanen. *Stochastic Complexity in Statistical Inquiry*. World Scientific, Singapore, 1989
12. H. Kashima and A. Inokuchi. Kernels for graph classification. In *Proceedings of the International Workshop on Active Mining*, 2002
13. D. Cook and L. Holder. Graph-based data mining. *IEEE Intelligent Systems*, 15(2): 32–41, 2000
14. I. Jonyer, D. Cook, and L. Holder. Graph-based hierarchical conceptual clustering. *Journal of Machine Learning Research*, 2: 19–43, 2001
15. S. Gunter and H. Bunke. Self-organizing map for clustering in the graph domain. *Pattern Recognition Letters*, 23: 401–417, 2002

16. S. Gunter and H. Bunke. Validation indices for graph clustering. *Pattern Recognition Letters*, 24(8): 1107–1113, 2003
17. R. Giugno and D. Shasha. Graphgrep: A fast and universal method for querying graphs. In *Proceedings of the IEEE International Conference on Pattern Recognition*, pages 112–115, 2002
18. J. Gonzalez, L. Holder, and D. Cook. Graph-based relational concept learning. In *Proceedings of the Nineteenth International Conference on Machine Learning*, 2002
19. JPL. Physical oceanograpy DACC, WOCE global data, v2.0, satellite data, sea surface temperature, 2000

Part III

Special Applications

Graph-Based and Structural Methods for Fingerprint Classification

Gian Luca Marcialis, Fabio Roli and Alessandra Serrau

Summary. Automatic Fingerprint Identification Systems (AFISs) are widely used for criminal investigations for matching the latent fingerprints found at the crime scene with those registered in the police database. As databases usually contain an enormous number of fingerprints, the time required to identify potential suspects can be extremely long. Therefore, a classification phase is performed to whittle down and thus speed up the search. Latent fingerprints are classified into five classes known as Henry classes. In this way each fingerprint only need to be matched against records of the corresponding class contained in the database. Many fingerprint classification methods have been proposed to date, but only a few of these exploit graph-based, or structural, representations of fingerprints. The results reported in the literature indicate that classical statistical methods outperform structural methods for benchmarking fingerprint databases. However, recent works have shown that graph-based methods can offer some advantages for fingerprint classification which warrant further investigation, especially when combined with statistical methods. This chapter opens with a critical review of the main graph-based and structural fingerprint classification methods. Then, these methods are compared with the statistical methods currently used for fingerprint classification. Experimental comparisons using a benchmarking fingerprint database are described, and the benefits of fusing graph-based and statistical methods are investigated. The chapter closes with some considerations on the present utility and future potential of graph-based methods for fingerprint classification.

1 Introduction

Over the last few years, personal recognition by Automatic Fingerprint Identification Systems (AFISs) has been gaining increasing importance in many applications (e.g., for criminal identification through crime scene fingerprint recognition) [1, 2].

The "identification time" strictly depends on the number of fingerprints contained in the database, as identification is performed by matching the fingerprint against each one stored in the database. As the number of fingerprints in real databases can be very large, this can be a very lengthy process. As an example, if the database contains 200 million fingerprints (like the FBI repository) and comparing two fingerprints takes 0.2 s (a realistic value for a modern one-to-one fingerprint matcher running on a standard PC [3]), average identification time will be 246 days. To reduce this time, it is customary to classify the fingerprint into one of the five classes proposed by Sir Edward Henry, thus limiting the search to the set of fingerprints corresponding to that class [4]. This strategy can significantly reduce identification time. For the above example and assuming the five classes to be uniformly distributed, identification time can be reduced to 49 days and even further using specialized algorithms and computers.

Henry found that many differently oriented ridge lines converge around the so-called "singularity" points, named "core" and "delta" [4]. Figure 1 shows the five Henry classes with relative singularities. Core and delta points are indicated for each class by squares and triangles, respectively. The A class has no singularities, the L, R, T classes have two singularities (one core and one delta point), and the W class has four singularities (two cores and two deltas).

Unfortunately, several factors complicate fingerprint classification. These include poor fingerprint image quality which can mislead singularity detection, and the existence of ambiguous fingerprints which cannot be reliably classified even by human experts. In particular, the key issue of ambiguous fingerprints arises from the large within-class variability and small between-class separation [1]. In some cases, fingerprints which cannot be reliably assigned to a single class even by human experts are labeled with two classes, and are known as "cross-referenced" fingerprints [1]. Several approaches to fingerprint classification have been proposed to address the above issues and to achieve the performances required by many AFIS applications, and much research continues to be done in this area [5–22].

For the purposes of this chapter, the proposed approaches to fingerprint classification can be divided into two main categories: statistical and structural approaches. Statistical methods are characterized by the use of the decision-theoretic approach to pattern classification, namely, a set of characteristic measurements, called feature vector, is extracted from fingerprint images and used for classification [7–11]. Structural approaches basically adopt the syntactic or structural pattern recognition methods [12–18]. Fingerprints are described by production rules or relational graphs and parsing processes or graph matching algorithms are used for classification. Recently, the use of multiple fingerprint classifiers combining different fingerprint classification approaches has been proposed. In particular, some experimental results recently

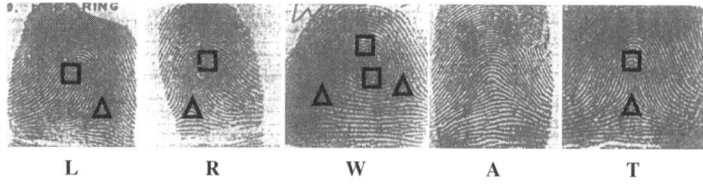

Fig. 1. The five fingerprint classes in the Henry system: (**L**) left loop (**R**) right loop (**W**) whorl (**A**) arch (**T**) tented arch. The "core" and "delta" points are indicated for each class by squares and triangles

reported by the authors and other researchers have demonstrated the potential of integrating structural with statistical methods [19–22].

In this chapter, we review the main approaches to fingerprint classification and perform an experimental comparison of some graph-based and structural methods. The comparison is then extended to a state-of-the-art statistical approach and to its combination with the investigated structural methods.

The chapter is organized as follows. Section 2 reviews the main approaches to fingerprint classification by focusing on the structural methods and their integration with the statistical ones. Section 3 describes the structural methods we used for our experiments. Section 4 deals with the fusion algorithms adopted. Section 5 describes our experiments. The chapter closes with some considerations on the present utility and future potential of graph-based and structural methods for fingerprint classification.

2 State-of-the-Art of Fingerprint Classification: An Overview

Automatic fingerprint classification has been widely researched but none of the proposed approaches provides a satisfactory answer to this intrinsically complex problem. The aim of this Section is to provide an overview of the main approaches to fingerprint classification. Further details about the state of the art of fingerprint classification can be found in [1, 6].

First of all, it should be pointed that the Henry classes are determined according to the location and number of the "singularity" points named "core" and "delta" (Fig. 1). Therefore, it might seem natural to classify fingerprints once these points have been detected [5]. Unfortunately, the singularity points are very difficult to detect because of the noise affecting fingerprint images, which can deceive automatic singularity detection algorithms. Moreover, the small between-class separation may not permit reliable fingerprint classification, because of irregular ridge flow patterns (e.g., T class fingerprints, which exhibit singularities, can appear as A class fingerprints). Finally, singularity detection is often not possible when some parts of the fingerprint image are lost (e.g., smudged fingerprint parts). Because of these limitations, most of the proposed approaches to automatic fingerprint classification are only partially based on singularities detection (e.g., on detection of the core point).

In this chapter, we categorize automatic fingerprint classification approaches according to the representation they use. We have defined three categories namely "statistical," "structural," and "structural–statistical." The first type of approach uses a fixed length feature vector, i.e., a set of measures extracted from the fingerprint image. The second uses structured data types such as graphs and trees. The third approach combines structural and statistical representations. Standard classifiers are often used to classify patterns characterized by the given representation. Thus, we briefly describe the classifier only if it is a special purpose classifier designed around the used representation.

The remainder of this section provides a review of fingerprint classification methods according to these categories, pointing out the common features, the differences and the pros and cons of the various methods. This review is by no means exhaustive, as a plethora of literature exists on fingerprint classification. A more comprehensive review can be found in [6].

2.1 Statistical Approaches for Fingerprint Classification

The core of statistical approaches is the feature extraction step. Depending on the feature extraction mechanism, they can be divided into three categories: (1) those which extract features from the fingerprint image directly [2]; (2) those which extract features from the orientation field of the fingerprint [7, 9]; and (3) those which concatenate, in a single feature vector, features extracted in the above two ways with other features, named "structural" features, which attempt to describe concisely the structure of the fingerprint classes [10, 11].

In the following, we briefly describe the rationale behind each of the above types of approach.

Features Extracted from the Fingerprint Image: The Fingercode

It is very difficult to extract reliable features from the fingerprint image, which may be corrupted by noise. However, this may be quicker than subjecting the fingerprint images to other preprocessing steps.

Jain et al. proposed characterizing each fingerprint by a numerical feature vector named FingerCode [8]. We describe this method in greater detail as we used it for comparison and combination with other structural methods investigated in this chapter. FingerCode computation commences by identifying the "core" point in the fingerprint input image and by defining a spatial tessellation of the image region around this point. This spatial tessellation is a circle decomposed into 48 sectors. Then, four orientation-selective bandpass Gabor filters (0, $\pi/4$, $\pi/2$, and $3\pi/4$) are applied to the tessellated image, generating four orientation-filtered images. Each filtered image accentuates ridge structures along one direction. Finally, the standard deviation of grey levels is computed for each filtered image and for each sector and a 192-dimensional FingerCode feature vector produced. Jain et al. used this feature vector as input for a two-stage classification architecture using a K-nearest neighbor classifier to find the two most probable classes of fingerprints and ten two-class neural networks for decision making.

As can be observed, with this approach it is possible to extract features in a very fast and simple way. Problems may arise if the fingerprint core cannot be detected for one reason or another. In this case, another reference point should be found (as for the A class fingerprints), or another form of tessellation defined [3]. Another problem is that FingerCode representation is not rotation invariant. One way of solving this problem is to generate a number of FingerCodes for different fingerprint image rotations, but this slows the procedure down [3]. Finally, there are some parameters which have to be tuned, like for example, the number of Gabor filter orientations and the number of tessellation sectors. Careful tuning of these parameters will depend on image resolution and on the degree of precision required for an accurate description of the fingerprint texture.

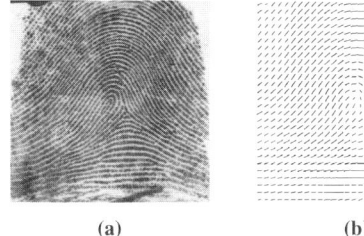

Fig. 2. (**a**) Example of fingerprint image and (**b**) corresponding orientation field represented by a graph. Note that each pixel in the orientation field is the average ridge and valley orientation for a given block of the original image. In this example, the orientation field is a 28 × 30-dimensional real-valued matrix from a 480 × 512-dimensional fingerprint image

Features Extracted from the Orientation Field

The orientation field is a map of the average ridge-flow orientations of the fingerprint image. Figure 2 shows an example extracted from a fingerprint image. The orientation field can be obtained after appropriate preprocessing for increasing the contrast between ridges and valleys [1].

In this category of statistical approaches, we briefly review two methods which illustrate the different uses of the orientation field for feature extraction.

The first method was proposed by Candela et al. [7]. They do not extract features from the orientation field, but use the orientation field itself as feature vector with 1,680 elements. To reduce the dimensionality and to avoid redundant information in the orientation field, the KL-transform is applied [23]. A probabilistic neural network is used for classifying the KL-transformed feature vector.

The second method was proposed by Cappelli et al. [9]. They introduce a generalized KL-transform of the orientation field, named multiple KL-transform (MKL). The main idea is to represent each fingerprint class in multiple KL-subspaces, and to construct a feature vector consisting of a set of distances between the input fingerprint and these subspaces. This feature vector is classified according to the nearest neighbor criterion.

The papers by Candela et al. [7] and Cappelli et al. [9] well represent the current use made of the orientation field for fingerprint classification. The main problem that may arise adopting the approach proposed by Cappelli et al. concerns the number of samples available for each class. Moreover, these approaches are not image rotation invariant. Accordingly, a fingerprint "registration" phase is necessary. Fingerprint registration depends on the location of singularities. Thus, singularity detection plays a key role in the performance of these methods.

Features Extraction by Concatenating Structural and Statistical Features

These methods couple the features extracted from the fingerprint image or its orientation field with other features that attempt to represent fingerprint structure. This can be done by identifying particular configurations in the "neighborhood" of a certain pixel of the fingerprint orientation field, or using additional structural data types such as graphs. In the following, we briefly review two approaches which exploit this idea.

The first was proposed by Nagaty [10]. Nagaty uses a 186-dimensional feature vector made up of statistical and "structural" features such as the input to a neural network. The structural features are represented by a 180-dimensional binary feature vector that constitutes the orientation field code. The statistical features are represented by a 6-dimensional feature vector. Each component of this feature vector is a sort of distance of the given fingerprint texture from a predefined "textural model" of each class (Nagaty divided the W class into two subclasses, thus obtaining a 6-class problem).

The second method was proposed by Yao et al. [11]. They use a 212-dimensional feature vector made up of statistical and "structural" features for characterizing fingerprints. The first 192 features constitute the FingerCode [8]. The remaining features are obtained by compressing a graph-based representation obtained from the orientation field into a 20-dimensional real-valued feature vector. This compression is obtained through a machine learning architecture explicitly aimed at learning complex data structures [24, 25]. The above 212-dimensional feature vector is the input to a series of support vector machines (SVMs) [23]. In order to cover as far as possible ambiguous fingerprints due to the large within-class variability and the small between-class separation, a set of 25 SVMs is first trained on five one-vs-all tasks (e.g. A class vs. LRTW classes), 10 two-vs-three tasks (e.g. AT vs. LRW), 10 pairwise tasks (e.g. A vs. T). The SVMs decisions are codified into a 25-dimensional binary feature vector. An appropriate "distance" of this feature vector from the rows of a 5×25 Error Correcting-Code matrix is defined and computed [26]. As each matrix row is associated to a given fingerprint class, the smallest distance indicates the final classification.

Clearly the second approach codifies the fingerprint with a structural representation (graphs) and then reduces data complexity through a machine learning architecture. The rationale behind this approach is that the machine learning architecture is able to "filter" the noise generated by the orientation field in a vector containing the distributed representation of the fingerprint. However, this produces coding noise when few samples are available. Coding depends on the architecture parameters, which are adjusted according to the data and to the labels attached to each graph node.

Nagaty's approach is simpler but the main issue here is to define an appropriate "textural model" for each class. This problem is very similar to finding an appropriate graph-prototype for each fingerprint class in the structural methods.

Finally, neither Nagaty's nor Yao et al.'s approach are image rotation invariant.

2.2 Structural Approaches for Fingerprint Classification

The structural approaches describe each fingerprint class with data structures such as grammars or graphs. These approaches can be further divided into syntactical approaches and template-matching-based approaches. Syntactical approaches associate a grammar to each fingerprint class. A parsing process is required for computing the grammar to which a test fingerprint belongs. The template-matching-based approaches associate a given template to each fingerprint class, usually through a graph-based representation. Fingerprints are compared by inexact graph-matching with each template, in order to find the model nearest to the given fingerprint.

Syntactical Approaches

Moayer and Fu and Rao and Balck provide a syntactical description of the fingerprint [12, 13]. They define a set of terminal symbols, based on the directional coding of the orientation field and a set of production rules in order to create a grammar representing each class. A parsing algorithm is applied to perform the final classification. Each terminal symbol has to take into account the wide range of distortions due to the noise added to the fingerprint orientation field images. Moreover, a large number of primitives is required to deal with the large within-class variations and the small between-class separation. As an example, in [12] the final number of terminal symbols considered is 69. This number increases with the increasing number of possible fingerprint image rotations.

Because of their high computational complexity, as well as the difficulty in finding an appropriate number of terminal symbols and production rules for each grammar, these methods are no longer investigated. The noise in fingerprint images is another key issue which contributes to complicating the design of effective syntactical approaches. To the best of our knowledge, no significant improvements have been made recently in syntactical fingerprint classification.

Graph-Based Approaches

These approaches are based on the observation that relational graphs are independent of the fingerprint image rotations. Therefore, the relational graph appears to be suitable for overcoming the rotation issue in fingerprint representations. However, each graph node needs to be enriched with a set of features which are usually dependent on fingerprint rotation, thus the problem is only partially solved. The first ideas for obtaining graph-based representations of fingerprints are based on the segmentation of the orientation field into regions having homogeneous directions. A node of the graph is associated to each region and an edge joins two nodes associated to adjacent regions.

This idea has been taken up by Lumini et al., who perform inexact graph matching with a template graph for each class to determine the best match for the final classification [14]. Due to the high variability of the segmentations obtained, it is very difficult to find a set of graph prototypes. A large number of prototypes are needed to represent as many variations as possible, but the high computational complexity of graph matching algorithms only increases classification time.

Following the same idea as Lumini et al., Cappelli et al. found a set of prototype – segmentations for each class, and designed an algorithm to "guide" the orientation field segmentation in order to produce a class-dependent segmentation [15]. A cost is associated to each segmentation. The orientation field of an input fingerprint is segmented according to five dynamic masks. The least cost segmentation corresponds to the "structure" that best represents a given orientation field, and the relative class is associated to the fingerprint. However, detected prototypes are not optimal for effectively discriminating between fingerprint classes, because of the small between-class separation.

In [16], the graph-prototype search is completely avoided by using machine learning architectures, called recursive neural networks, explicitly trained to learn complex data structures [24, 25]. These networks take as input a structured representation of fingerprints in terms of directed positional acyclic graphs (DPAGs) and perform learning in a manner similar to multiple layer perceptrons (backpropagation-based algorithms). Similarly to Lumini et al. and Cappelli et al., the orientation field is segmented and an algorithm designed and implemented for deriving the DPAG from this segmentation. The main difficulty of this approach is tuning the parameters both for the structured representation (e.g., the maximum number of nodes joined by another one must be fixed) and the classifier (e.g., the number of neural connections). Moreover, the use of DPAG in the place of relational graphs can affect the expressive power of the representations. The advantage is that there is no need for graph-prototypes because the possible DPAG "configurations" are learned by examples.

The problem of finding an effective set of graph-prototypes has recently been addressed by Neuhaus and Bunke in [17, 18]. In [17], each graph prototype is based on the location of the cores and deltas of the relative fingerprint class, and the number of ridges along the line joining each core and each delta. In fact, a node is associated to the core, the delta, and to each ridge along this line. Similarly, a graph is derived from an input fingerprint. The class associated to the graph prototype nearest the input graph is selected by inexact graph matching [27]. In [18], the problem of finding an appropriate set of graph-prototypes is solved from another point of view. The main issue of the current graph-based approaches is that fingerprints of the same class can generate very different graphs. Neuhaus and Bunke propose combining different graph-based representations of the same fingerprint, thus obtaining a more robust and reliable graph-based representation. The solution proposed by these authors concerns pattern recognition problems in general, but it can also be applied to fingerprint images with promising results.

2.3 Structural–Statistical Approaches for Fingerprint Classification

It is well-known that combining data from multiple sources can affect classification system performance [28]. The main idea behind the structural–statistical approaches is that the fusion of multiple classifiers based on structural and statistical fingerprint representations can improve performance. These methods "fuse" multiple fingerprint classifiers by combining their outputs. These outputs can be the class associated to the input fingerprint by the classifiers or the posterior probabilities of each class for a given input fingerprint. In the first case the fusion takes place at the classifier decision level (decision-level fusion), in the second at the classifiers output level (measurement-level fusion). The combination functions are generally called "fusion rules." These rules can be fixed (e.g., majority voting at the decision-level, or sum of the outputs at the measurement-level), trained (e.g., weighted majority on the basis of the reliability of each classifier at the decision-level, or weighted sum at the measurement-level) or "stacked," whereby additional classifiers are used for implementing the fusion rule. In the following, we review some of these approaches with fusion rules at different levels and with different structural and statistical representations.

Cappelli et al. combine the structural method presented in [15] with the multiple KL-transform presented in [9, 20]. The feature vector, consisting of the distance set between the input fingerprint and the multiple KL-subspaces is first classified according to the nearest neighbor rule and the k-nearest neighbor rule. Combining these two decisions with the dynamic mask method is also examined using the majority voting rule.

Senior proposes a fingerprint classification system based on integrating hidden Markov models (HMM) with decision trees (DT) [19]. The HMM-based classifier is trained by a set of novel features extracted from the skin ridge flow. This feature extraction step is performed as follows. A set of horizontal and vertical "fiducial" lines intersects the skeletonized fingerprint image at different points. At each "fiducial line" to "ridge line" intersection, a set of measures is computed. A multilayered HMM is designed considering the so-computed set of features at each fiducial line as the input to each layer. The decision tree classifier is trained on another set of features extracted from the skin ridge pattern. These features are used to encode the ridge line shape. The output from the DT and HMM classifiers is used as input to a feed-forward neural network for the final classification (Senior fuses the A and T class fingerprints into a single class, obtaining a four-class classification problem).

In [11,21], the authors presented the results of measurement-level fusion of structural and statistical approaches. The statistical approach consists of a multilayer perceptron trained with the FingerCode [8]. The structural approach is that investigated in [16]. The output of the statistical and structural classifiers, in terms of class posterior probabilities, are combined using the k-nn classifier.

In [22], the authors presented the preliminary results of the comparison of three structural approaches [14, 15, 21]. This chapter can be considered as a follow up to that preliminary work. Accordingly, in the following we describe the state-of-the-art of three structural approaches to fingerprint classification and perform an experimental comparison. The investigation is then extended to the measurement-level fusion of these approaches also including the statistical approach proposed in [8].

3 Investigated Graph-Based Fingerprint Classification Algorithms

Previous works have shown that fingerprints have an intrinsic "structure" that can be extracted by segmenting fingerprint images into regions containing ridges having homogeneous orientations. Figure 3 shows fingerprint structure for the five Henry

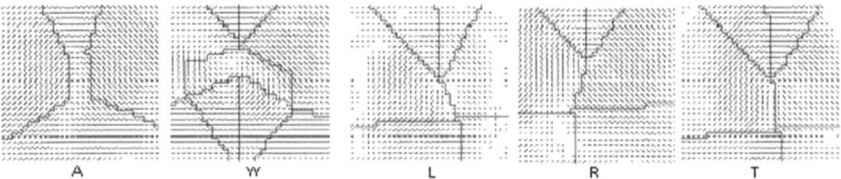

Fig. 3. Segmentation of the orientation fields of the fingerprint images for the five Henry classes

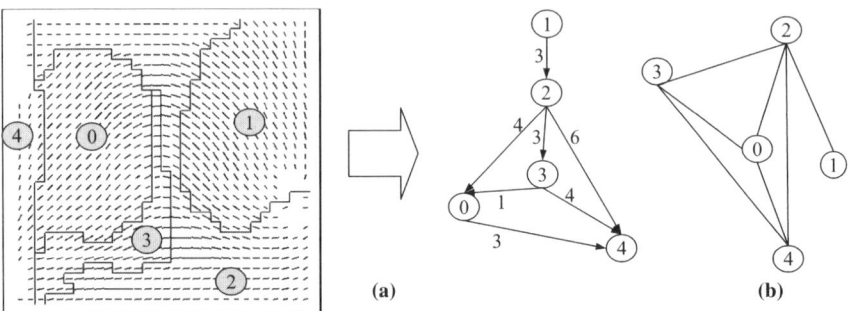

Fig. 4. Graph-based representation obtained from segmented orientation field of a fingerprint image: (**a**) DPAG representation obtained from the algorithm described in [29]; (**b**) relational graph representation. Node labels represent the number associated to each region as shown in the orientation field and the edge labels represent the position of each child-node with respect to the father-node

classes. According to this definition, one can argue that structural information can prove very useful for identifying class A fingerprints, as this class exhibits a structure (i.e., topology of regions containing ridges with homogeneous orientations) very different from the other classes. Conversely, structural information is not so useful for identifying fingerprints belonging to the other classes, especially classes R, L and T, as they have similar structures, with only minor differences due to the different positions of the two singularity points.

An important issue when using structural methods is how to describe fingerprint structure through an appropriate data type. The approaches described below use different structural representations: from simple relational graphs for the inexact graph matching to DPAGs for recursive neural networks. Figure 4 shows an example of relational graph and DPAG for the same fingerprint orientation field segmentation.

The dynamic mask method is based on orientation field segmentation; however it does not explicitly use any kind of graph but simply serves to guide the segmentation. Nevertheless, this approach is considered graph-based because the masks represent the structure extracted from the orientation field segmentation. Hence, the masks could be represented with graphs.

3.1 Fingerprint Classification by Inexact Graph Matching

One of the most natural structural representations of fingerprint orientation field segmentation (i.e., of the segmentation of the fingerprint image in regions with homogeneous orientation of ridge and valley, as shown in Fig. 3) is the relational graph. Relational graphs appear to be appropriate, as nodes might naturally correspond to the regions extracted by the segmentation algorithm [14]. Each graph node can be associated to a segmentation region and the edges join two nodes according to the adjacency relationship of the respective regions. The representation is completed by associating to each node and edge a feature vector containing characteristics of the regions and geometrical differences between contiguous regions, respectively. In this chapter the orientation field segmentations are obtained using the algorithm proposed in [29].

In order to compare relational graphs, the *error-correcting graph matching* is used [27, 30]. This method computes a measure of the "dissimilarity" between the graph representing the input pattern to be classified and a certain graph-prototype. This dissimilarity measure is called "edit-distance." It is based on a deformation model that exploits edit operations such as substitution, deletion and insertion of nodes and/or edges.

Hence, let G_I and G_P be the input and prototype graphs, respectively. Let $T(G_I, G_P)$ be the set of all possible sequences of edit operations which transform G_I into G_P. Let $S = (o_1, \ldots, o_n) \in T(G_I, G_P)$ be a sequence of edit operations $o_1, \ldots, o_n$. Let $C(o_i) \in \Re^+ \cup \{0\}$ be a nonnegative real value, called "cost associated to o_i" ($i \in \{1, \ldots, n\}$). A nonnegative real value, called "cost of S," is associated to an instance of $T(G_I, G_P)$, namely, S, by the cost function $F_c : T(G_I, G_P) \to \Re^+ \cup \{0\}$. Cost function $F_c(S)$ is defined as the sum of all costs $C(o_i)$, for $i \in \{1, \ldots, n\}$. The sequence $S^* = (o_1^*, \ldots, o_n^*)$ that provides the minimum cost is ultimately the edit distance $D_e(G_I, G_P)$ between G_I and G_P graphs:

$$D_e(G_I, G_P) = \min_{S \in T(G_i, G_P)} F_c(S), \text{ where } F_c(S) = \sum_{i=1}^{n} C(o_i) \quad (1)$$

The cost of each edit operation $C(o_i)$ depends on the values of each feature related to the node/edge v of the input graph and to the node/edge w of the prototype graph according to the following equation:

$$C(o_i) = \sqrt{\sum_{h=1}^{N_f} \left(W_h \left(f_h(v) - f_h(w) \right)^2 \right)} \quad (2)$$

where f_h is the h-th feature associated to a node/edge and the W_h are weights for assigning different weights to each feature. If the deletion or insertion operation is applied, the feature value f_h related to the input or prototype graph node/edge is set to zero. Accordingly:

$$C(o_i) = \sqrt{\sum_{h=1}^{N_f} \left(W_h f_h^2(v) \right)} \quad (3)$$

A good cost function definition should take into account the fact that a higher cost should be associated to a significant distortion of the input graph with respect to another. Moreover, an edit operation that occurs frequently should be assigned a lower cost than one that seldom occurs. Weights W_h should therefore be selected accordingly.

Note that, in order to obtain a distance, $D_e(G_I, G_P) = D_e(G_P, G_I)$ must hold. Hence, the same weight W_h should be associated to both delete and insert operations.

So we have to find the sequence S^* of edit operations that yields the least cost according to (1). For this purpose, a search tree is constructed containing all possible edit operation sequences $T(G_I, G_P)$. Each path from the root to the leaves corresponds to a sequence S of edit operations. The search for the best sequence S^*, i.e. the least cost path, is performed using an algorithm similar to A* [27].

Once edit distances between input graph and all prototypes have been computed, the class corresponding to the nearest neighbor is associated to the input graph.

To obtain the conditional class probabilities for a given input pattern G_I to be classified, the least edit distance d_I^k, for the comparison between this graph and the most similar prototypes belonging to the k-th class, are converted using the *softmax* formula:

$$P(C = k|G_I) = \frac{e^{-\alpha d_I^k}}{\sum_{c=1}^{5} e^{-\alpha d_I^c}} \quad (4)$$

where α is a "smoothing" parameter. We empirically found that $\alpha = 9$ is the best value for converting edit distances into posterior probabilities.

3.2 Fingerprint Classification by Recursive Neural Networks

As mentioned earlier, a key issue in structural fingerprint classification is finding the graph that best represents each fingerprint class, so as to apply a template-matching algorithm. In particular, it is very difficult to discriminate between L, R, and T class fingerprint structures using a simple relational graph-based representation. Another problem is making this fingerprint representation robust against the large within-class variability and the small between-class separation, which is accentuated in real applications by noisy sensitive data. As each region is derived from the segmentation algorithm, the robustness degree depends essentially on this algorithm. However, to the best of our knowledge, none of the proposed segmentation algorithms is sufficiently robust. As a result, very different segmentations for fingerprints of the same class and similar segmentations for fingerprints of different classes (L, R, T, and W especially) are often obtained.

In order to overcome these problems, recursive neural networks (RNNs) explicitly designed to handle structured data have been proposed [11,16,21]. As RNNs are trained to classify complex data structures on examples, the problem of designing a set of templates for each class is avoided. The main limitation of this approach is that RNNs can learn to classify only data structures in terms of DPAGs. A DPAG is a directed acyclic graph in which the "child-nodes" (nodes linked by another node, also called "parent-node") are arranged according to a certain rule. For example, a DPAG node can have the first, second and fourth child, while the third is missing. The maximum number of child-nodes is called "outdegree." Moreover, the "super-source" node, i.e. the node dominating all other nodes, is defined within a DPAG.

In this chapter, we refer to the algorithm proposed in [21] for converting an orientation field segmentation obtained with the algorithm proposed in [29] into a DPAG. Similarly to the relational graph-based representation, a feature vector containing the local characteristics of the regions and the geometrical and spectral differences between neighboring regions is associated to each node. Figure 4a gives an example of DPAG generation from an orientation field segmentation. Note that the edges of the DPAG are labeled with the position of each child-node with respect to its parent-node. This position is computed using the DPAG generation algorithm.

A RNN performs a "recursive transduction" that maps a graph V into another graph Z_V having the same topology. Hence, the node n_{Z_V} of Z_V corresponds to the node n_V of V. As shown in Fig. 5, a "state" vector $X(n_{Z_V}) \in \Re^N$ is associated to each node n_{Z_V}, computed on the basis of the n_V's feature vector, denoted with $U(n_V)$, using the equation:

$$X(n_{Z_V}) = f(X(u_1), \ldots, X(u_k), U(n_V)), \text{ where } u_j \in Child(n_{Z_V}) \quad (5)$$

f is called "state transition function." It combines a vector encoding the feature vector of n_V with the state vectors of $\{u_1, \ldots, u_k\}$, which is the *ordered* set of n_{Z_V}'s children. Computation proceeds recursively from the childless nodes to the super-source node (the node dominating all other nodes). The base step for (5) is $X(u) = 0$ if u is a missing child.

Figure 5 shows that the vector $X(n_{Z_V})$ contains the distributed representation of the sub-graph dominated by n_{Z_V} (i.e., all nodes that can be reached via n_{Z_V}).

In our case, the transition function f is computed by a multilayer perceptron (MLP), which is replicated at each node in the DPAG, sharing weights among replicas

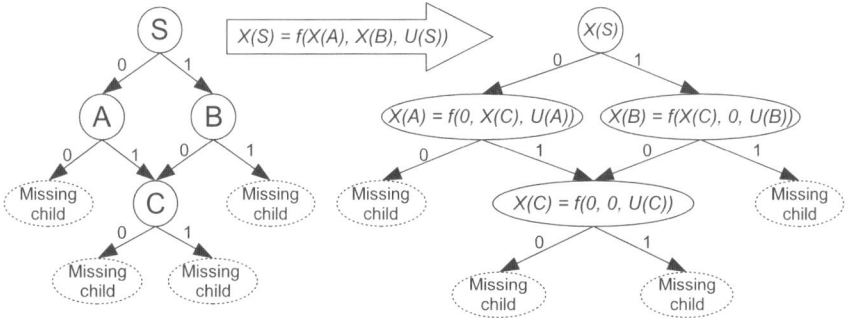

Fig. 5. Recursive transduction from a simple input DPAG ($outdegree = 2$) to the state-DPAG labeled with state vectors. The node labeled with the state vector $X(S)$ contains the distributed representation of the entire input DPAG. Each child can be univocally identified through its position with respect to its parent: e.g., A is the first child of S, B the second. Note that C is childless. In this case, the base-step is applied for each missing child

[31]. Classification with recurrent neural networks is performed by adding an output function g that takes as input the hidden state vector $X(S)$ associated with the super-source S: $Y = g(X(S))$. Note that $X(S)$ can be extracted and independently used as a "structural" feature vector, as it contains the distributed representation of the entire DPAG [11]. Function g is also implemented by a multilayer perceptron.

The output layer in this case uses the *softmax* functions (normalized exponentials), so that Y can be interpreted as a vector of conditional class probabilities for a given input graph, i.e. $Y_i = P(C = i|V)$, where C is a multinomial class variable [31]. Training relies on maximum likelihood and uses a gradient-descent approach to weight optimization of functions f and g called "back-propagation through structure" [32]. Further details can be found in [24, 25, 33].

In order to take into account the cross-referenced fingerprints, which have two classes instead of one, a "soft" target vector was introduced into the training phase. The two cross-referenced classes were regarded as having the same probability for a given pattern. Thus, the target-vector takes a value of 0.5 for the two target classes of the cross-referenced fingerprints. For the standard fingerprints, the target-vector takes a value of 1.0 for the target class.

3.3 Fingerprint Classification by Dynamic Masks

This method was introduced by Cappelli et al. to overcome the large variability in segmentations of similar fingerprints, when the segmentation algorithm described in [29] is applied. The basic idea of this approach is to perform a "guided" segmentation of the orientation field of the fingerprint image so as to reduce variability during the segmentation process [15]. To this end, five filters, called "dynamic masks," one for each class, "guide" the orientation field segmentation, producing a class-dependent segmentation. These dynamic masks can be regarded as "prototypes" of images segmented by the orientation field. Using these filters the number of segmentation regions and their rough shape are fixed. Each dynamic mask is obtained following the four steps:

1. For each class, select a set of representative fingerprints.
2. Compute the respective orientation fields.
3. Apply a genetic algorithm to segment the orientation field.
4. Identify an "average" ensemble of fixed and moving vertices and segments that define the mask. These vertices are located around the singularity points ("core" and "delta"), according to the fingerprint structure shown in Fig. 3.

To classify fingerprints the orientation field of an input fingerprint is segmented according to the five dynamic masks (one for each class). The "cost" for each mask provides a measure of the difficulty of the guided segmentation process. Accordingly, the least cost means that the segmentation process can easily produce a segmented image very similar to the used mask. The cost vector is then converted into a posterior probabilities vector by (4) (we empirically found $\alpha = 1$ to be the best value). The class associated with the maximum posterior probability is associated to the fingerprint. Further details about mask design can be found in [15].

4 Measurement-Level Fusion of Statistical and Structural Approaches

In this section, we describe our approach for measurement-level fusion of structural and statistical fingerprint classifiers. The general scheme is quite similar to the commonly used multiple classifiers fusion scheme at the measurement-level [28]. The system is characterized by different representations. One classifier is designed for each representation.

We used the FingerCode to statistically represent the fingerprint [8]. This vector is the input of a multilayer perceptron trained according to the maximum likelihood cost function. The outputs of the network have the same meaning as the recursive neural network output, i.e., estimation of the conditional probability of each class for a given fingerprint pattern. Hence, the output of each network is a 5-dimensional probability vector. Structural representations of fingerprints are those described in Sect. 3.

We used two types of fusion algorithms (or "fusion rules"). The first was based on "fixed" transformation of the classifier outcomes [28]. We shall call these "fixed" fusion rules as they require no parameter estimation. In particular, we used the mean and the product rules.

The second fusion algorithm follows the so-called "meta-classification" (or "stacked") approach which uses an additional classifier for combination [28]. In particular, a k-nearest neighbor classifier was used. The input of this classifier is a "feature-vector" consisting of the outcome of each classifier to be combined. Clearly this fusion strategy is more complex than that based on fixed rules.

5 Experimental Results

5.1 The Data Set

The NIST-4 database, created by the National Institute of Standards and Technology, is a reference dataset widely used for assessing and comparing fingerprint classification algorithms [34]. It contains 4,000 ink-on-paper fingerprint images, equally divided into five classes (A, L, R, W and T). Each fingerprint was acquired twice. The first acquisition denotes fingerprints from f0001 to f2000, the second fingerprints from s0001 to s2000. We followed the experimental protocol generally used for this data set (see for e.g., [8,11,19]). The first 1,800 fingerprints (f0001–f900 and s0001–s900) were used for classifier training. The next 200 fingerprints were used as validation set, necessary for early stopping of neural classifiers (RNN and MLP) training, and the last 2,000 fingerprints as test set.

Seven hundred fingerprint images in the NIST4 dataset were assigned to two classes instead of just one ("cross-referenced" fingerprints), as they could not be reliably categorized in a single class even by human experts.

To use the FingerCode statistical representation, we had to discard sixty three fingerprint images as the poor quality made it impossible to detect the "core" point.

Note that Jain et al. also disregarded the same fingerprint images in their experiments [8].

5.2 Comparative Analysis of Structural Approaches

In this section, we describe the performance results of the structural classifiers described in Sect. 3. First we compared the performance of structural classifiers so as to analyze the main pros and cons for fingerprint classification. Then, we investigated their measurement-level fusion as described in to Sect. 4.

Table 1 shows classification accuracy rate (second to sixth columns) and overall classification accuracy (seventh column). The second row refers to the dynamic masks method ("Masks"), the third to the recursive neural networks-based approach ("RNN"), and the fourth to the graph matching approach ("GM").

The best performance is exhibited by the RNN classifier (Table 1, seventh column). This can probably be explained by the fact that RNNs do not require class prototypes, as class representations are automatically learnt by examples. Therefore, RNNs are better able to handle the intrinsically small class-separation of fingerprints, which makes it difficult to find a representative set of class prototypes to use with GM. On the other hand, the RNN classifier performs worst for class T. One likely explanation is the enormous number of cross-referenced fingerprints contained in the NIST4 data set. Another reason for this result is the introduction of the soft-target, which made it possible to reduce the "noise labeling" effect, be it at the expense of class T training effectiveness, as there are fewer class T patterns assigned to a single class than cross-referenced ones (70.0% of class T fingerprints are cross-referenced).

Though, on average, the GM classifier did not perform as well as the RNN, they exhibit similar behavior. Both GM and RNN performed well for class A, and worst for class T fingerprints. The satisfactory results obtained for class A confirm the utility of structural features for discriminating among strongly structured classes. Accordingly, it can be hypothesized that the performance of the GM classifier could be significantly improved were a more robust orientation field segmentation algorithm, or more effective methods for class prototypes selection, available. In fact, although we used the sophisticated fingerprint segmentation algorithm described in [29], segmentations of class L, R, and T fingerprints often contained errors that made their graphs very similar. Clearly future efforts should be directed towards graph representation and matching techniques that can handle these segmentation errors.

Table 1. Class accuracy rates and overall accuracy of dynamic masks method ("Masks"), recursive neural networks ("RNN"), and graph matching approach ("GM")

	A	L	R	T	W	overall
Masks	48.1	84.5	82.1	66.0	78.4	71.5
RNN	90.7	79.1	83.3	36.2	81.4	76.8
GM	71.9	62.3	69.4	52.7	66.3	65.2

Table 2. Accuracy rate of measurement-level fusion of structural classifiers by the mean rule, the product rule, and K-nearest neighbor (KNN).

fusion of	mean	product	KNN
Masks–RNN (76.8)	80.6	79.2	81.7
Masks–GM (71.5)	79.1	77.8	79.8
RNN–GM (76.8)	76.1	76.7	76.6
Masks–RNN–GM (76.8)	82.4	82.2	83.6

Overall accuracy of the best single classifier is shown in brackets in the first column

The dynamic masks classifier performed quite differently from the others. In particular, it yielded poor results for class A. In our opinion, this poor performance can be explained by the absence of singularity points in class A, which make it rather difficult to design an appropriate dynamic mask for that class. In this regard, note that for the other classes, which exhibit at least two singularities, this classifier performed far better.

Table 2 shows the performance of the single classifiers and their measurement-level fusion for different combination rules.

When all three structural classifiers are combined the best performance is achieved for the KNN-based fusion rules. But the simple mean rule also yields good results (Table 2, fifth row). Improvement in classification performance is around 7% compared to the best single structural classifier. This result shows that each representation contributes significantly to enhancing performance.

5.3 Comparison with the Statistical Approach

Table 3 shows the accuracy for the statistical classifier test set mentioned in Sect. 2, i.e., the multilayer perceptron using FingerCodes. Above all, Table 3 shows that the overall accuracy of the statistical classifier is greater than any of the structural classifiers or a combination thereof. However, as the Tables 1 and 3 clearly show, the structural classifier far outperform the statistical classifiers for class A (except for the dynamic masks method).

To investigate the advantages of structural approaches for discriminating among classes with a clear structure, for which the standard statistical classifier often does not perform satisfactorily, we examined in detail the degree of misclassification between classes A and T. Table 4 shows the degree of misclassification between classes A and T (i.e., the percentage of class A fingerprints misclassified as class T fingerprints) for each structural classifier, their best fusion, and the statistical classifier. Note that interclass confusion is a well-known problem in state-of-the-art statistical classifiers. Table 4 shows that structural approaches significantly reduce the effect of misclassification on ultimate fingerprint classification accuracy. Even for the dynamic masks and the GM classifiers, which by themselves do not outperform the statistical classifier, Table 4 shows their combination considerably reduces this effect.

Table 3. Classification accuracy rate and overall accuracy of multilayer perceptron trained with FingerCodes

	class A	class L	class R	class T	class W	overall
statistical classifier	80.5	91.8	89.5	79.1	89.4	86.0

Table 4. Percent of class A–T misclassification for the single classifiers (Masks, RNN, GM), their best fusion, and the statistical classifier. The best combination yielded the best overall accuracy shown in Table 2

method	accuracy	method	accuracy
Masks	19.8	Masks–RNN	4.5
RNN	2.7	Masks–GM	5.0
GM	19.4	RNN–GM	2.9
statistical	16.7	Masks–RNN–GM	5.2

Table 5. Overall accuracy rate of investigated fusion rules

fusion of	mean	product	KNN
statistical-Masks	84.4	83.5	86.2
statistical-RNN	87.6	87.8	88.5
statistical-GM	86.7	87.0	88.6
statistical-RNN–GM	86.0	87.5	89.0
statistical-Masks–GM	86.8	85.8	88.0
statistical-Masks–RNN	88.2	85.8	88.8
statistical-Masks–RNN–GM	88.6	87.0	89.6

5.4 Fusion of Statistical and Structural Approaches

Table 5 gives the overall accuracy obtained using the product, mean, and KNN fusion rules (second, third, fourth columns, respectively).

The sharp improvement in overall accuracy, very close to other state-of-the-art results (90.0% in [8], 92.2% in [20]), clearly shows the advantage of combining structural and statistical classifiers. In particular, the degree of misclassification of A–T classes is about 5.0% for the four-classifier combination. The improvement with respect to the statistical classifier alone is more than 10.0%. Therefore, by combining structural and statistical approaches it is possible to deal effectively with the well-known problems inherent in fingerprint classification algorithms, and also to improve their overall performance.

Further experimental evidence is provided in Fig. 6 which shows the accuracy–rejection curves of the best classifiers and fusion approaches investigated. The rejection option makes sense when a fingerprint cannot be classified without a large margin of uncertainty, thus increasing the probability of misclassification. Obviously the use of the rejection option ultimately increases identification time (Sect. 1). Therefore, a good trade-off is needed between the percentage of rejected fingerprints and the required classification accuracy. As the FBI requirements for the NIST

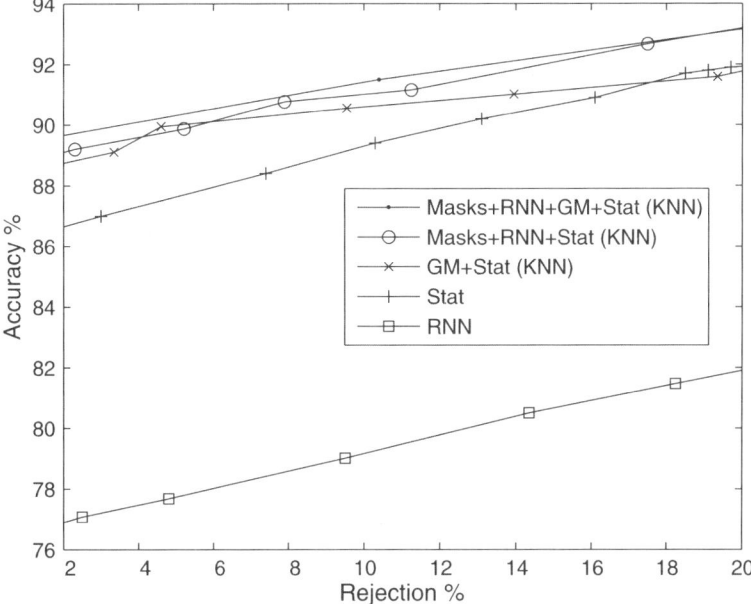

Fig. 6. Accuracy–rejection curves of the best classifiers and their best fusion algorithm

databases are 99% classification accuracy with 20% rejection rate, we investigated accuracy for rejection rates ranging from 2% to 20% in our experiments [1, 5, 19].

We followed Chow's rule for pattern rejection, that is, its maximum posterior probability should exceed a certain reject threshold otherwise it is considered "rejected" or not classified [35]. From Fig. 6, it clearly emerges that by increasing the number of structural classifiers, it is possible to gradually enhance performance and also to improve classification accuracy as rejection rate increases. This confirms that each classifier contributes to significantly improving performance, and also positively impacts the reliability of the classification system.

6 Conclusions

This chapter aimed to: (a) provide an overview of the main approaches to structural fingerprint classification, (b) investigate the use of some measurement-level fusion rules for exploiting the potential of their fusion with statistical approaches.

We believe that our investigation has contributed to demonstrating the present utility of structural approaches in fingerprint classification. Our review has clearly shown that structural approaches receive less attention than do statistical approaches. The main reason for this is the lack of effective learning mechanisms and the high computational complexity of structural and graph-based methods. The large within-class variability and the small between-class separation, the noise corrupting sensitive data, and the image processing errors (e.g., image segmentation errors), make learning of prototypes or class descriptions for syntactic and structural methods a challenging task.

However, structural and graph-based approaches exhibit some interesting properties (e.g., fingerprint rotation invariance) and effectively address some of the problems inherent in statistical approaches (e.g., degree of class A–T confusion). Finally, the fusion of structural methods with a statistical method using simple rules significantly enhanced performance with respect to state-of-the-art statistical approaches. It can be argued that the design of special purpose fusion rules could well develop into a promising research direction for exploiting the potential of structural and statistical methods.

Acknowledgments

The authors wish to thank Anil K. Jain for providing them with the FingerCode representation of the NIST-4 dataset, and Raffaele Cappelli and Davide Maltoni for the results of their image segmentation algorithm. Thanks also go to Paolo Frasconi and Horst Bunke who introduced the authors to the use of recursive neural networks and to the graph matching theory.

References

1. D. Maltoni, D. Maio, A.K. Jain, and S. Prabhakar, *Handbook of Fingerprint Recognition*, Springer, Berlin Heidelberg New York, 2003
2. A.K. Jain, R. Bolle, and S. Pankanti (Eds.), *BIOMETRICS – Personal Identification in Networked Society*, Kluwer, Boston Dordrecht London, 1999
3. A. Ross, A.K. Jain, and J. Reisman, A Hybrid Fingerprint Matcher, *Patt. Recog.*, 36 (7), 1661–1673, 2003
4. E.R. Henry, *Classification and Uses of Fingerprints*, Routledge, London, 1900
5. K. Karu and A.K. Jain, Fingerprint Classification, *Patt. Recog.*, 29 (3), 389–404, 1996
6. N. Yager and A. Amin, Fingerprint Classification: A Review, *Patt. Anal. Appl.*, 7 (1), 77–93, 2004
7. G.T. Candela, P.J. Grother, C.I. Watson, R.A. Wilkinson, and C.L. Wilson, *PCASYS – A Pattern-Level Classification Automation System for Fingerprints*, NIST Technical Report NISTIR 5647, 1995
8. A.K. Jain, S. Prabhakar, and L. Hong, A Multichannel Approach to Fingerprint Classification, *IEEE Trans. PAMI*, 21 (4), 348–358, 1999
9. R. Cappelli, D. Maio, and D. Maltoni, Fingerprint classification based on Multi-space KL, *Proc. AutoID'99*, Summit, NI, pp. 117–120, 1999
10. K.A. Nagaty, Fingerprint Classification Using Artificial Neural Networks: A Combined Structural and Statistical Approach, *Neural Networks*, 14 (9), 1293–1305, 2001
11. Y. Yao, G.L. Marcialis, M. Pontil, P. Frasconi, and F. Roli, Combining Flat and Structural Representations with Recursive Neural Networks and Support Vector Machines, *Patt. Recog.*, 36 (2), 397–406, 2003
12. B. Moayer and K.S. Fu, A Syntactic Approach to Fingerprint Pattern Recognition, *Patt. Recog.*, 7, 1–23, 1975

13. K. Rao and K. Balck, Type Classification of Fingerprints: A Syntactic Approach, *IEEE Trans. PAMI*, 2 (3), 223–231, 1980
14. A. Lumini, D. Maio, and D. Maltoni, Inexact Graph Matching for Fingerprint Classification, *Machine Graphics and Vision*, 8 (2), 241–248, 1999
15. R. Cappelli, A. Lumini, D. Maio, and D. Maltoni, Fingerprint Classification by Directional Image Partitioning, *IEEE Trans. PAMI*, 21 (5), 402–421, 1999
16. G.L. Marcialis, F. Roli, and P. Frasconi, Fingerprint Classification by Combination of Flat and Structural Approaches, *Proc. of 3rd Int. Conf. on Audio- and Video- Based Biometric Person Authentication AVBPA'01*, in: J. Bigun and F. Smeraldi (Eds.), Springer, Berlin Heidelberg New York, LNCS 2091, pp. 241–246, 2001
17. M. Neuhaus and H. Bunke, A Graph Matching Based Approach to Fingerprint Classification Using Directional Variance, *Proc. of 5th Int. Conf. on Audio- and Video-Based Biometric Person Authentication AVBPA'05*, in: T. Kanade, A.K. Jain, N. Ratha (Eds.), Springer, Berlin Heidelberg New York, LNCS 3546, pp. 191–200, 2005
18. M. Neuhaus and H. Bunke, Graph-Based Multiple Classifier System – A Data Levels Fusion Approach, *Proc. of 13th Int. Conf. on Image Analysis and Processing ICIAP'05*, in: F. Roli and S. Vitulano (Eds.), Springer, Berlin Heidelberg New York, LNCS 3617, pp. 479–486, 2005
19. A. Senior, A Combination Fingerprint Classifier, *IEEE Trans. PAMI*, 23 (10), 1165–1174, 2001
20. R. Cappelli, D. Maio, and D. Maltoni, A Multi-Classifier Approach to Fingerprint Classification, *Patt. Anal. Appl.*, 5 (2), 136–144, 2002
21. G.L. Marcialis, F. Roli, and A. Serrau, Fusion of Statistical and Structural Fingerprint Classifiers, *Proc. of 4th Int. Conf. on Audio- and Video-Based Person Authentication AVBPA'03*, in: J. Kittler and M.S. Nixon (Eds.), Springer, Berlin Heidelberg New York, LNCS 2688, pp. 310–317, 2003
22. A. Serrau, G.L. Marcialis, H. Bunke, and F. Roli, An Experimental Comparison of Fingerprint Classification Methods Using Graphs, *5th Int. Work. on Graph-based Representations in Pattern Recognition GbR05*, in: L. Brun and M. Vento (Eds.), Springer, Berlin Heidelberg New York, LNCS 3434, pp. 281–290, 2005
23. R. Duda, P. Hart, and D. Stork, *Pattern Classification – Second Edition*, Wiley, New York, 2001
24. P. Frasconi, M. Gori, and A. Sperduti, A General Framework for Adaptive Processing of Data Structures, *IEEE Trans. Neural Networks*, 9 (5), 768–786, 1998
25. A. Sperduti and A. Starita, Supervised Neural Networks for the Classification of Structures, *IEEE Trans. Neural Networks*, 8 (3), 714–735, 1997
26. T.G. Dietterich and G. Bakiri, Solving Multiclass Learning Problems via Error-Correcting Output Codes, *J. Artif. Intell. Res.*, 2, 263–286, 1995
27. H. Bunke and G. Allermann, Inexact Graph Matching for Structural Pattern Recognition, *Patt. Recog. Lett.*, 1, 245–253, 1983
28. T. Windeatt and F. Roli (Eds.), *Multiple Classifier Systems*, LNCS 2709, Springer, Berlin Heidelberg New York, 2003
29. D. Maio and D. Maltoni, A Structural Approach to Fingerprint Classification, *Proc. 13th ICPR*, Vienna, pp. 578–585, 1996
30. D. Conte, P. Foggia, C. Sansone, and M. Vento, Thirty years of graph matching in pattern recognition, *Int. J. Patt. Recog. Artif. Intell.*, 18, 265–298, 2004
31. C.M. Bishop, *Neural Networks for Pattern Recognition*, Oxford University Press, 1995
32. C. Goller and A. Kuchler, Learning Task-Dependent Distributed Structure-Representations by Backpropagation Through Structure, *IEEE Int. Conf. on Neural Networks*, 347–352, 1996

33. P. Frasconi, M. Gori, and A. Sperduti, Special Section on Connectionist Models for Learning in Structured Domains, *IEEE Trans. Knowl. Data Eng.*, 13 (2), 2001
34. C.I. Watson and C.L. Wilson, NIST Special Database 4, Fingerprint Database, US National Institute of Standard and Technology, 1992
35. C.K. Chow, On Optimum Recognition Error and Reject Tradeoff, *IEEE Trans. Information Theory*, 16, 41–46, 1970

Graph Sequence Visualisation and its Application to Computer Network Monitoring and Abnormal Event Detection

H. Bunke, P. Dickinson, A. Humm, Ch. Irniger and M. Kraetzl

Summary. In this chapter, a new visualisation method for time series of graphs is introduced. This method is based on graph edit distance and multidimensional scaling. The proposed procedure maps each graph in a time series of graphs to a point on the two-dimensional real plane, such that graphs with a high (low) similarity are mapped to points with a small (large) Euclidean distance. In this way, similar graphs in the time series can be easily identified. As a potential application of this method, we consider the problem of computer network monitoring and abnormal event detection. A number of results on simulated data and graphs obtained from real computer networks are presented, highlighting the advantages of the proposed method over previous approaches.

1 Introduction

Graph representations play an important role in many disciplines of science and engineering. Particularly in pattern recognition and machine vision, graphs have been used in a variety of applications. For a recent survey see [1]. Measuring the similarity of objects is a fundamental task in pattern recognition and computer vision. When graphs are used for object representation, the problem of object recognition turns into that of measuring the similarity of two given graphs, one that represents an unknown object and another that represents a prototype stored in a database of known objects. The unknown object is then assigned to the class of its most similar prototype. The problem of measuring the similarity of two graphs is also known as graph matching. A number of graph matching problems have been addressed in the literature, including graph isomorphism [2], subgraph isomorphism [3], maximum common subgraph [4, 5] and graph edit distance computation [6, 7]. Solutions to these matching problems are based on a variety of computational paradigms, including combinatorial and heuristic search [8–10], constraint satisfaction [11], stochastic relaxation [12, 13], genetic algorithms [14, 15], neural networks [16, 17] eigenspace methods [18–20], random walks [21] and kernel methods [22–24].

Time series, or sequence, data are encountered in many application domains, such as financial engineering, audio and video databases, biological and medical research and weather forecast. Consequently, the analysis of time series has become an important area of research [25]. Particular attention has been paid to problems such as time series segmentation [26], retrieval of sequences or partial sequences [27], indexing [28], classification of time series [29], or detection of frequent subsequences [30]. Although a rather large effort has been devoted to both time series analysis and the processing and interpretation of graph based data, not much work in the intersection of the two fields has been reported. In [31, 32], graph based approaches to video sequence analysis have been proposed. An algorithm for comparing two sequences of graphs with each other has been proposed in [33]. However, this method has not been applied on real data until now.

In this chapter, we propose a new method for the analysis of time series of graphs. Although our motivation is in the field of computer network monitoring and abnormal event detection, the method is general and can be applied in other problem domains as well, wherever a system can be represented as a sequence of graphs and one is interested in detecting abnormal system behaviour. The aim of our method is to visualise the behaviour of a computer network over time for abnormal event detection. As computational tools we will use graph edit distance [6, 7] and multidimensional scaling [34, 35]. The proposed method is mainly to be understood as a visualisation tool suitable to assist a human network operator. However, as we will point out at the end of this chapter, it is rather straightforward to fully automate the method.

This chapter is organized as follows. In Sect. 2, we will introduce some basic concepts from both graph theory and multidimensional scaling. Next, in Sect. 3, we will describe how graph edit distance computation and multidimensional scaling can be applied to the problem of computer network monitoring. In Sect. 4, a number of experimental results will be presented, using synthetically generated data and graph sequences acquired from real computer networks. Finally, in Sect. 5, we will present a summary, conclusions and a number of possible extensions.

2 Preliminaries

In this section, we will review some fundamental concepts from graph theory and multidimensional scaling, as a basis of the method proposed in Sect. 3.

2.1 Graph Matching

Let L_V and L_E denote two sets of node and edge labels, respectively. A graph is a four-tuple $g = (V, E, \alpha, \beta)$, where V is the finite set of nodes, $E \subseteq V \times V$ is the set of edges, $\alpha : V \to L_V$ is the node labelling function, and $\beta : V \to L_E$ is the edge labelling function. Edges are directed, i.e., edge $(x, y) \in E$ originates at node $x \in V$ and terminates at node $y \in V$. The case of undirected edges is obtained if we define, for each edge $(x, y) \in E$, an edge (y, x) in the opposite direction. Unlabelled graphs are a special case of our definition if $|L_V| = |L_E| = 1$.

In many applications there is a need to compare graphs with each other. Graph comparison is also known as graph matching. It includes the computation of graph isomorphism, subgraph isomorphism and maximum common subgraph [2–5]. In the present chapter, we are concerned with a more general problem, namely, the computation of graph difference, or graph distance. One well-known distance measure for graphs, which has emerged in the domain of pattern recognition, is graph edit distance [6, 7]. In graph edit distance computation, one applies a sequence of edit operations on the two given graphs so as to make the first graph identical, or isomorphic, to the second one. The length of the shortest edit sequence of this kind is defined as the edit distance of the two graphs under consideration. Often a cost is assigned to each edit operation. In this case, edit distance is defined as the cost of the cheapest sequence of edit operations that make the two graphs identical to each other.

For general graphs, as introduced above, graph edit distance computation belongs to the class of NP-complete problems. Often there exist problem dependent heuristics that allow one to cut the search space. Nevertheless, graph edit distance computation is feasible for small graphs only. In the context of the application considered in Sect. 3, however, we deal with a special class of graphs. Graphs of this kind are characterised by the existence of unique node labels. That is, for any graph $g = (V, E, \alpha, \beta)$ and two nodes $x, y \in V$, if $x \neq y$ then $\alpha(x) \neq \alpha(y)$. Consequently, when computing the edit distance of two graphs g_i and g_j, there is a one-to-one correspondence between the nodes of g_i and g_j, which can be exploited to greatly reduce the computational complexity of the search process. As a matter of fact, it has been shown in [36] that common graph matching tasks, such as computation of graph isomorphism, subgraph isomorphism, maximum common subgraph and graph edit distance, can be accomplished in linear time with respect to the number of nodes plus the number of edges of the larger of the two graphs involved.[1]

The particular graph edit distance measure we use in this chapter is quite simple. Given two graphs $g_i = (V_i, E_i, \alpha_i, \beta_i)$ and $g_j = (V_j, E_j, \alpha_j, \beta_j)$, their distance is defined as:

$$d(g_i, g_j) = |V_i| + |V_j| - 2|V_i \cap V_j| + |E_i| + |E_j| - 2|E_i \cap E_j|. \qquad (1)$$

Because of the property of unique node labels, we identify each node with its unique label. Here, for the implementation of this equation, we only need a procedure for set intersection. In (1), $|V|$ denotes the number of nodes in set V, and $|E|$ the number of edges in E. Therefore this distance measure is equal to the number of nodes plus the number of edges that occur in only one of the two graphs, but not in both. In other words, if the set of edit operations consists of a node insertion, a node deletion, an edge insertion and an edge deletion, then (1) reflects the minimum number of edit operations needed in order to make g_i and g_j identical. More generally, the distance measure is equal to the minimum cost needed to make the two graphs identical to each other provided each edit operation has a cost equal to one. Note that $d(g_i, g_j)$ is small if g_i and g_j have many nodes and edges in common. In the extreme case, when g_i and g_j are identical, we get $d(g_i, g_j) = 0$. On the other hand, if both graphs have

---

[1] In the application considered in Sect. 3, nodes correspond to the servers, clients and routers of a computer network and node labels represent their unique IP addresses, for example.

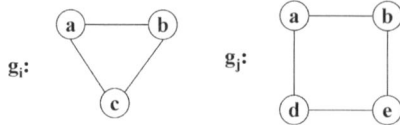

Fig. 1. Two graphs used to demonstrate a measure of graph distance

no node and no edge in common, then the distance assumes its maximum value, i.e., $d(g_i, g_j) = |V_i| + |V_j| + |E_i| + |E_j|$.

As an example, consider graphs g_i and g_j in Fig. 1. In order to make g_i and g_j identical, we have to remove node c and its two incident edges from g_i, and insert nodes d and e together with their incident edges in g_j. Assuming a cost equal to one for each edit operation, the total cost amounts to 8, i.e., $d(g_i, g_j) = 8$.

2.2 Multidimensional Scaling

Multidimensional Scaling (MDS) refers to a class of methods often used in the visualisation of high-dimensional data [34,35]. Consider n objects $o_1, ..., o_n$ in a metric space and assume that the only information we are given about these objects is their pairwise distances. Let d_{ij} denote the distance between objects o_i and o_j, where $d_{ii} = 0$ and $d_{ij} = d_{ji}; i, j = 1, ..., n; i \neq j$.

The starting point of MDS is an $n \times n$ distance matrix $\mathbf{D} = [d_{ij}]$. The goal of MDS is to reconstruct points $p_1, ..., p_n$ in the m-dimensional Euclidean space $\mathbb{R}^m$ such that the Euclidean distance between p_i and p_j approximates d_{ij} as closely as possible for all pairs i and j. In order to facilitate visualisation, the dimension m of the target space is usually chosen $m = 2$ or $m = 3$. In this chapter, we will exclusively consider the case $m = 2$.

There are several variations of MDS known from the literature. In this chapter we will focus on metric scaling [34,35]. Let d_{ij}^2 be the squared distance between objects o_i and o_j, and let $\hat{\mathbf{D}} = [d_{ij}^2]$ be the $n \times n$ matrix of pairwise squared distances. Note that the main diagonal of $\hat{\mathbf{D}}$ consists of zeros only. Define matrix

$$\mathbf{J} = \mathbf{I} - \mathbf{n}^{-1}\mathbf{1}\mathbf{1}', \qquad (2)$$

where $\mathbf{I}$ is the identity matrix, and let $\mathbf{1}$ be an n-dimensional column vector of 1's. We use $\mathbf{x}'$ and $\mathbf{X}'$ to denote the transpose of column vector $\mathbf{x}$ and matrix $\mathbf{X}$, respectively. From matrix $\hat{\mathbf{D}}$ we want to recover matrix

$$\mathbf{X} = \begin{pmatrix} x_{11} & \cdots & x_{1m} \\ \vdots & \ddots & \vdots \\ x_{n1} & \cdots & x_{nm} \end{pmatrix}, \qquad (3)$$

where $\mathbf{x_j} = (x_{j1}, ..., x_{jm})$ is the location of object o_j in $\mathbb{R}^m$. Because

$$d_{ij}^2 = (\mathbf{x_i} - \mathbf{x_j})'(\mathbf{x_i} - \mathbf{x_j}) = \mathbf{x_i'x_i} - 2\mathbf{x_i'x_j} + \mathbf{x_j'x_j}, \qquad (4)$$

matrices $\hat{\mathbf{D}}$ and $\mathbf{X}$ are related via the equation

$$\hat{\mathbf{D}} = \mathbf{c}\mathbf{1}' + \mathbf{1}\mathbf{c}' - 2\mathbf{X}\mathbf{X}', \qquad (5)$$

where $\mathbf{c} = (\mathbf{x}_1'\mathbf{x}_1, ..., \mathbf{x}_n'\mathbf{x}_n)'$. After multiplication of this equation with $\mathbf{J}$ from the left and from the right, and after some simplification, we obtain

$$\mathbf{B} = -\frac{1}{2}\mathbf{J}\hat{\mathbf{D}}\mathbf{J} = \mathbf{X}\mathbf{X}'. \qquad (6)$$

Now the term in the middle is factored by eigendecomposition, yielding

$$\mathbf{B} = \mathbf{Q}\mathbf{\Lambda}\mathbf{Q}' = (\mathbf{Q}\mathbf{\Lambda}^{1/2})(\mathbf{Q}\mathbf{\Lambda}^{1/2})' = \mathbf{X}\mathbf{X}', \qquad (7)$$

and

$$\mathbf{X} = \mathbf{Q}\mathbf{\Lambda}^{1/2}. \qquad (8)$$

Here, $\mathbf{\Lambda}$ is a matrix that contains the eigenvalues $\lambda_1, ..., \lambda_n$ of $\mathbf{B}$ in its diagonal and 0's elsewhere. By convention, we assume the eigenvalues being ordered such that $\lambda_1 \geq ... \geq \lambda_n \geq 0$. Matrix $\mathbf{Q}$ contains the eigenvectors of $\mathbf{B}$ as its columns. Now the coordinates $\mathbf{x_i} = (x_{i1}, x_{i2})$ of all objects o_i in the two-dimensional plane can be retrieved from the first two columns of matrix $\mathbf{X}$ (3).

3 Application to Computer Network Monitoring and Abnormal Event Detection

The application considered in this chapter is computer network monitoring and abnormal event detection. In particular, we will focus our attention on intranets. Intranets have been continuously growing in size and numbers because companies and organisations are becoming more and more information centred today. Consequently, intranet availability, reliability and security are becoming important issues. Ensuring a high degree of availability requires sophisticated tools for computer network monitoring and anomalous event detection. In the beginning, the identification of network anomalies has relied upon ad hoc methods developed by skilled network operators. Recently, however, anomalous event detection in computer networks has become an area of active research.

A number of principled methods for the detection of abnormal events in computer networks have been proposed in the literature. Some of these methods make use of signatures [37]. Signature based methods match current network patterns against abnormalities that have occurred in the past. Variants of signature based methods are rule-based methods [38], and case-based reasoning [39]. A further approach presents a data mining technique for discovering masquerader intrusion. User/system access data are used as a basis for deriving statistically significant event patterns. These patterns could be considered as a user/system access signature [40]. A shortcoming of signature based abnormal event detection is that anomalies that have not been observed in the past, and thus are not stored in the database of the system, remain undetected. Another approach to anomalous event detection is based on finite state machines [41]. However, a problem with this approach is that the number of states may grow very large when complex abnormalities need to be modelled. A number of other approaches make use of statistical methods [42, 43], including auto-regressive processes [43–45], hidden Markov models [46], wavelets [26] and Bayesian networks [27]. In [28] it is shown that network alarms produced by Intrusion Detection Systems (IDSs) attains a high-level description of threats. As the number of alarms is increasingly growing, automatic tools for alarm clustering have been proposed to provide such a high level description of the attack scenario. It has been shown that effective threat analysis requires the fusion of different sources of information, such as different IDSs, firewall logs, etc.

In our previous work, we used graph theoretic methods for network anomaly detection [29, 30, 47, 48]. The basic idea of this approach is to represent a computer network by a graph where the nodes represent servers, routers or clients and the edges represent physical or logical connections in the network. If the network is sampled at regular points in time $t_1, t_2, ..., t_i, ...$, a time series of graphs $g_1, g_2, ..., g_i, ...$ is obtained, which formally represents the network. Using the graph distance measure of (1), one can compute the amount of change, or distance, between consecutive graphs in such a time series. If the distance $d(g_{i-1}, g_i)$ between two consecutive graphs g_{i-1} and g_i is above a given threshold, it is assumed that an abnormal event has occurred in the network between time t_{i-1} and time t_i. Because all clients and servers can be uniquely identified in the application, the underlying graphs have the property of unique node labels. This property ensures that all required graph operations can be very efficiently computed and the method can deal with large graphs [36].

The approach proposed in [29, 30, 47] has proven effective in identifying abnormal network behaviour. Nevertheless, it is limited in that it can only classify network change as normal or abnormal, but cannot identify individual states, or clusters of states, of the network. In the context of this chapter, the state of a network is defined by a certain subset of the nodes and edges within a graph that are actually present in the graph at a certain point in time. In other words, the set of all possible networks states is defined by all subsets of the nodes and edges of the network. Here, when dealing with a time series of graphs, the same subset of graph nodes and edges will exist in adjacent graphs as long as the network remains in the same state. Now assume that $d(g_{i-1}, g_i) > \theta$ and $d(g_i, g_{i+1}) > \theta$ where θ is a threshold that indicates an abnormal event. Clearly, in this case we conclude that two abnormal events have occurred in the network, one between time t_{i-1} and t_i, and the other between time t_i and t_{i+1}. However, we do not know if at time t_{i+1} the network is in the same, or a similar state as it was at time t_{i-1}. That is, we do not know whether the changes that led from g_i to g_{i+1} are inverse to the changes that led from g_{i-1} to g_i, such that g_{i+1} is equal or similar to g_{i-1}. Information of this kind would be extremely valuable for a network operator. If it was known, for example, if the state at time t_{i-1} was a normal network state and the states at t_{i-1} and t_{i+1} were similar to each other, then one could conclude that after two abnormal events, between time t_{i-1} and t_i as well as t_i and t_{i+1}, the network has returned to a normal state again.

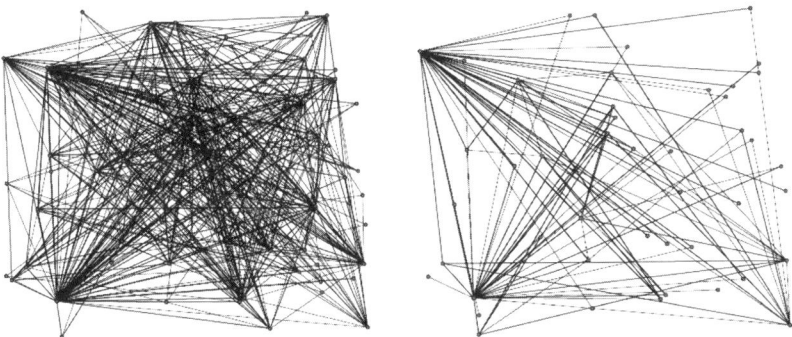

Fig. 2. Snapshots of computer network at two consecutive points in time

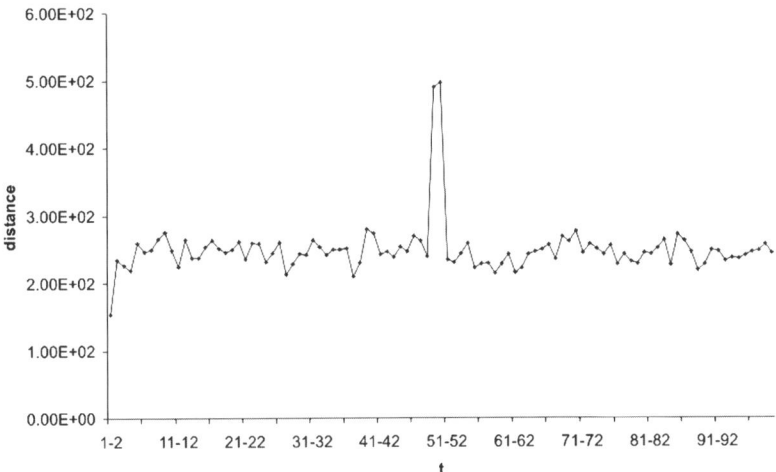

Fig. 3. Graph distance plot of the network over 102 consecutive points in time

To give an example, snapshots of a computer network at two consecutive points in time are given in Fig. 2. A plot of distances between pairs of consecutive graphs of a whole time series of graphs is shown in Fig. 3. There is one prominent peak in the distance plot of Fig. 3 at time $t = 50$, and this peak corresponds in fact the change between the first and the second graph shown in Fig. 2. A closer look at Fig. 3 reveals, however, that a large graph distance occurs not only at time $t = 50$, but also at $t = 51$. This leads to the conjecture that the network topologies at time $t = 49$ and $t = 51$ may be similar to one another, i.e., the changes that led to the topology at time $t = 51$ may be inverse to the changes that led to the topology at time $t = 50$. However this conjecture cannot be verified given only the information provided in Fig. 3.

In order to reveal similarities in network topology between pairs of graphs g_i and g_j that have a distance in time greater than one, i.e., $j \geqslant i + 1$, we propose to compute all pairwise distances $d(g_i, g_j)$ for $i, j = 1, ..., n; i \neq j$. This results in an

Fig. 4. MDS plot of the network

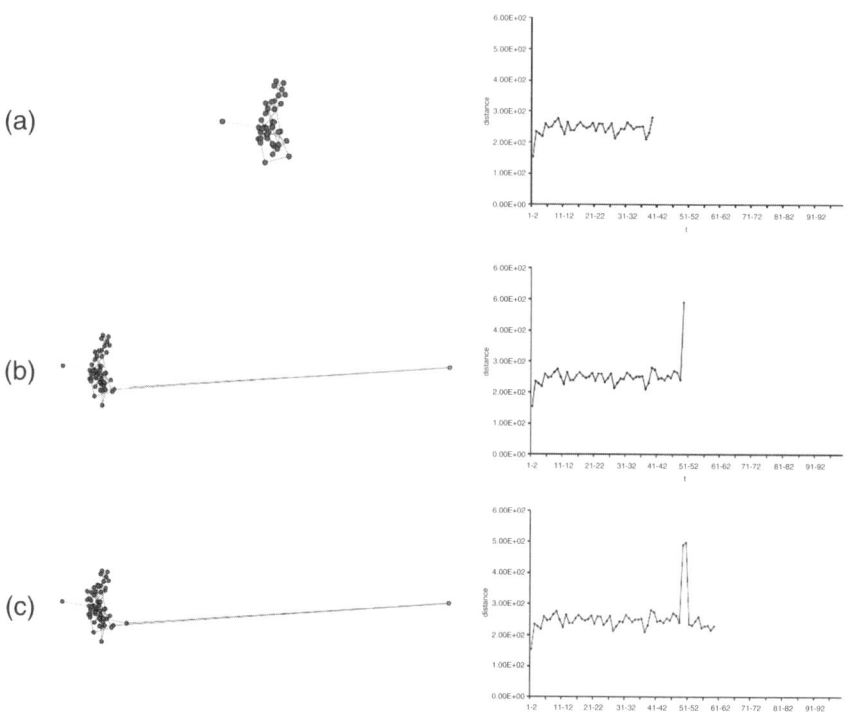

Fig. 5. Dynamic evolution of MDS and graph distance plots over time

$n \times n$ distance matrix $\mathbf{D} = [d_{ij}]$. As a matter of fact, from (1) it can be seen that $\mathbf{D}$ is a symmetric matrix with all elements in the diagonal equal to zero. Hence one actually needs to compute only $d(g_i, g_j)$ for $i > j$.

Mapping the graphs of the sequence underlying Fig. 3 into the two-dimensional plane by means of MDS yields the plot shown in Fig. 4. In addition to merely depicting the individual graphs, we show temporal relations by linking, through edges, pairs of points that belong to two consecutive graphs. In this figure one can identify one large cluster of points and one prominent outlier. As a matter of fact, the outlier corresponds to the network at time $t = 50$. Therefore, the conjecture that the network returns to its original state after the abnormal event can be verified by means of the MDS plot shown in Fig. 4, i.e., the network topologies at time $t = 49$ and $t = 51$ are similar to each other. To illustrate the behaviour of the network in greater detail, we show snapshots of the evolution of both the distance plot and the MDS plot in Fig. 5. Figure 5a shows the network at time $t = 40$ before the abnormal event occurred. Next, Fig. 5b illustrates the network at time $t = 50$ immediately after the abnormal event has happened, and Fig. 5c corresponds to time $t = 60$. In the MDS plot it can be clearly seen that the abnormal event causes a large distance between consecutive graphs (which can be seen in the graph distance plot as well). However, after the abnormal event has occurred, the network's topology becomes similar to the topology before the abnormal event as the corresponding points in the MDS plot belong to the same (i.e., the large) cluster. This phenomenon is only visible in the MDS plot, but not in the graph distance plot.

4 Experimental Results

In order to investigate the visualisation method proposed in Sect. 3 in a more systematic way, we generated a number of synthetic graph sequences with specific properties and applied the proposed method. In our first simulation, a sequence of 100 graphs was generated. All graphs had 150 nodes with randomly distributed edges. The sequence was divided into three subsequences s_1, s_2, and s_3, including graphs g_1 to g_{39}, g_{40} to g_{70}, and g_{71} to g_{100}, respectively. Sequences s_1 and s_3 had the same statistical properties, but for s_2 different parameters were used in the graph generation process.

In many real networks, there exist a number of nodes that communicate with each other frequently while others communicate only occasionally. Throughout this chapter we will refer to links arising from frequent communication as *group 1* edges of the network. By contrast, links between pairs of nodes that communicate infrequently will be called *group 2* edges. The two groups of edges are identified from the initial graph. The initial graph is generated in the following way. First, $N = 150$ nodes are generated. Out of the N^2 possible edges, 5% are randomly chosen as edges for the initial graph. The edges chosen are designated to be edges of group 1. Conversely, the edges not chosen are designated as edges of group 2. No self-loops, i.e., edges (x, x) are admitted in the graph generation process.

The two groups of edges then have different change probabilities applied to them. Given graph g_{i-1}, the edges of the next graph g_i are chosen according to the following conditional probabilities:

- $P($edge of group 1 exists in $g_i \mid$ edge of group 1 exists in $g_{i-1}) = 0.9$
- $P($edge of group 1 does not exist in $g_i \mid$ edge of group 1 does not exist in $g_{i-1}) = 0.3$
- $P($edge of group 2 exists in $g_i \mid$ edge of group 2 exists in $g_{i-1}) = 0.3$
- $P($edge of group 2 does not exist in $g_i \mid$ edge of group 2 does not exist in $g_{i-1}) = 0.99999$

In subsequence s_2, a subset of 75 nodes was randomly selected and all transition probabilities of edges between nodes from this subset were set equal to 0.5, i.e.,

- $P($edge exists in $g_i \mid$ edge exists in $g_{i-1}) = P($edge exists in $g_i \mid$ edge does not exist in $g_{i-1}) = 0.5$.

236   H. Bunke et al.

From the graph generation procedure we know that subsequences s_1 and s_3 are less dynamic than subsequence s_2, i.e., the distances between consecutive graphs in s_2 are expected to be higher than in s_1 and s_3. Figure 6 shows both the MDS and the graph distance plot. Our expectation of s_2 exhibiting larger graph distances than s_1 and s_3 is confirmed in the graph distance plot. In the MDS plot we see, in addition to some outliers, a compact cluster of points in the right-hand side, and a somewhat diffuse cluster in the left-hand side. Figure 7 shows three snapshots of the evolution

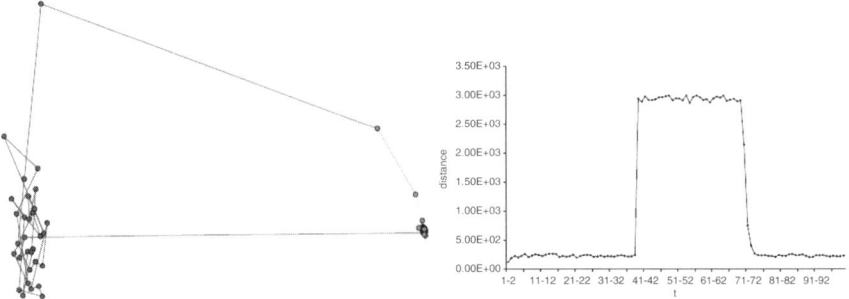

Fig. 6. MDS and graph distance plot of a simulated graph sequence

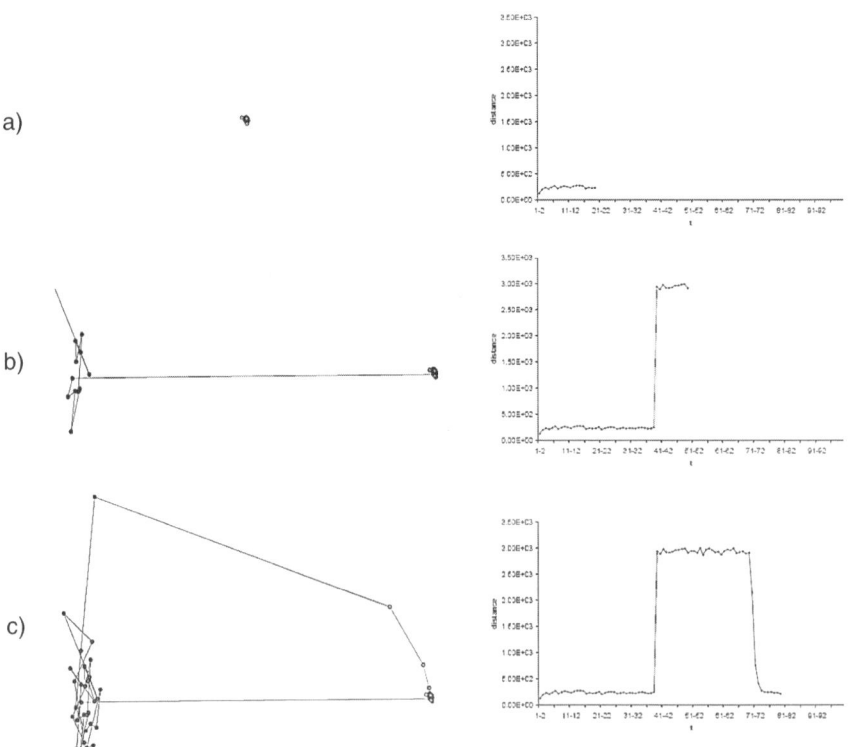

Fig. 7. Dynamic evolution of MDS and graph distance plots over time

of both plots over time. The three snapshots were taken at time 20, 50 and 80, i.e., during subsequence s_1, s_2 and s_3, respectively. From Fig. 7, we conclude that the compact cluster corresponds to subsequences s_1 and s_3, while the diffuse cluster represents the network during subsequence s_2. Note that in the compact cluster many points are printed on top of each other. Hence this cluster appears smaller than the diffuse cluster, although in fact it includes more points. We conclude that both, the distance and the MDS plot reflect our expectation and describe the behaviour of the network very well. The MDS plot, however, includes additional information that is not evident from the graph distance plot. First, it shows that there are two clusters of similar network states. Second, it indicates that the network states of subsequences s_1 and s_3 belong to one cluster, i.e., they are very similar.

This can also be seen in Fig. 8, where three graphs at consecutive points in time are shown from subsequence s_1 (Fig. 8a), s_2 (Fig. 8b) and s_3 (Fig. 8c), respectively. We observe that the three graphs in Fig. 8a are similar to the three graphs in Fig. 8c, while the graphs in Fig. 8b appear somewhat different.

In the second simulation, we generated a sequence of 100 graphs based on the same parameters that were used for the generation of subsequences s_1 and s_3 in the first experiment. Once the whole sequence was generated, a subset of 75 nodes were randomly selected, and each node of this subset that occurred in any of the graphs $g_{40}, \ldots, g_{70}$ was deleted together with all its incident edges. Due to this procedure one would expect distances between consecutive graphs to have similar

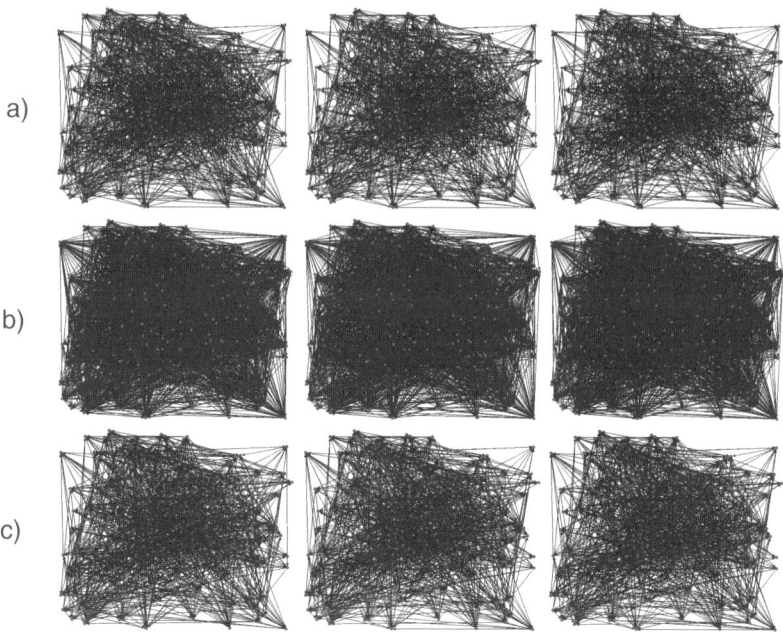

Fig. 8. Graphs from the first simulation: (**a**) three graphs from subsequence s_1; (**b**) three graphs from subsequence s_2 and (**c**) three graphs from subsequence s_3

values in subsequences $s_1 = g_1, ..., g_{39}$ and $s_3 = g_{71}, ..., g_{100}$, but be smaller in subsequence $s_2 = g_{40}, ..., g_{70}$, due to the reduced number of nodes and edges involved. This behaviour can be observed in the graph distance plot of Fig. 9, and can also be seen in Fig. 10, where similarly to Fig. 8 three consecutive graphs from subsequence s_1, s_2 and s_3 are shown. The two large peaks in Fig. 9 coincide with the points at which the subset of selected nodes, and their incident edges, were deleted and later re-inserted. In the MDS plot we identify two clusters and a few spurious points. The compact cluster in the left-hand side of the figure corresponds to sequence s_2 (smaller graph distances lead to smaller distances between points in the MDS plot), while the diffuse cluster in the right-hand side represents s_1 and s_3. The transition

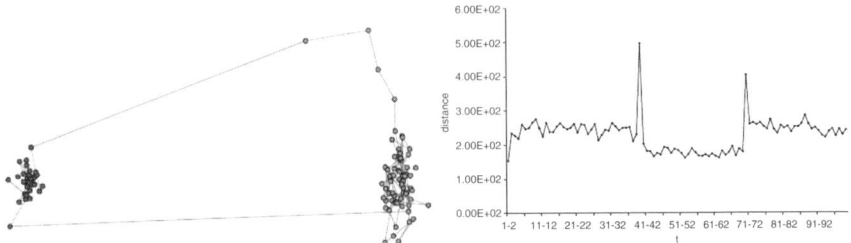

Fig. 9. MDS and graph distance plot of second simulated graph sequence

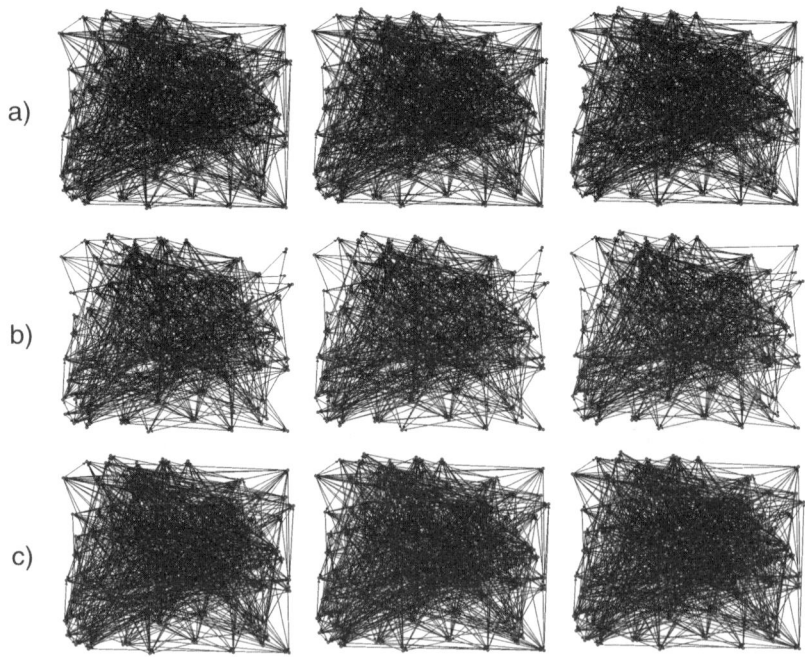

Fig. 10. Graphs from the second simulation: (**a**) three graphs from subsequence s_1; (**b**) three graphs from subsequence s_2 and (**c**) three graphs from subsequence s_3

between the two clusters occurs at points in the sequence corresponding to the large peaks in the graph distance plot. Similarly to the first experiment, we can clearly see from the MDS plot that there are two major states. Furthermore, it can be observed that the network returns to the first state after having changed from the first to the second state. Information of this kind is not evident from the graph distance plot.

In the third experiment, again a graph sequence of length 100 was generated using the same statistical parameters as for subsequences s_1 and s_3 in the first experiment. At time $t = 50$ the graph was significantly distorted by randomly selecting a subset V' of 75 nodes, deleting all edges existing between the nodes of V', and inserting an edge between any pair of nodes from V' that were not connected before. Such a graph would be considered an outlier with respect to the adjacent graphs in the sequence. In this experiment one would expect the graph distances $d(g_{49}, g_{50})$ and $d(g_{50}, g_{51})$ being significantly larger than all other graph distances. As a matter of fact, this experiment corresponds to Figs. 3–5. Our expectation is confirmed in the graph distance plot shown in Fig. 3. In the MDS plot we clearly identify the outlier that corresponds to the graph at time $t = 50$. One can also see that the topology of the network before and after time $t = 50$ is similar because the corresponding points are in the same cluster.

In our last experiment with synthetic data, a graph sequence of length 100 was generated with the same statistical properties as subsequences s_1 and s_3 in the first experiment. In this experiment no abnormal event was implanted into the graph sequence, i.e., the graph sequence was not altered. The MDS and graph distance plots obtained for this time series are shown in Fig. 11. As one would expect, all graph distances are of similar magnitude and no individual clusters emerge in the MDS plot. Note that the scaling of the MDS plot in Fig. 11 is different from the scaling used in previous figures. If the same scaling as in Fig. 9 was applied, the spread of the cluster in Fig. 11 would be about the same as the spread of the diffuse cluster in Fig. 9. Figure 12 shows three randomly selected consecutive graphs from this sequence. They all appear similar.

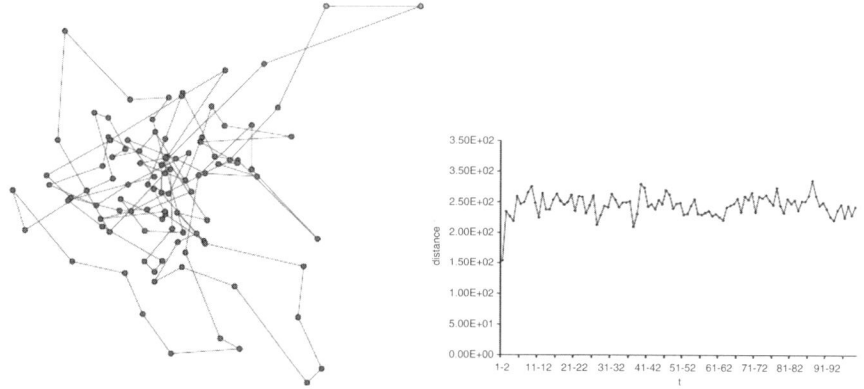

Fig. 11. MDS and graph distance plot of fourth simulated graph sequence

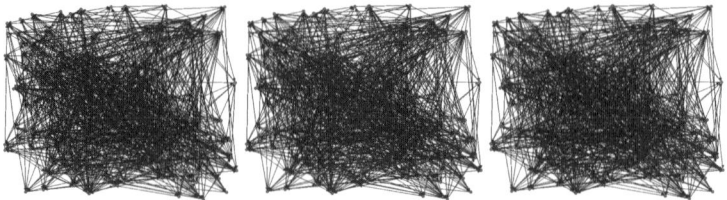

Fig. 12. Three graphs from the fourth simulation

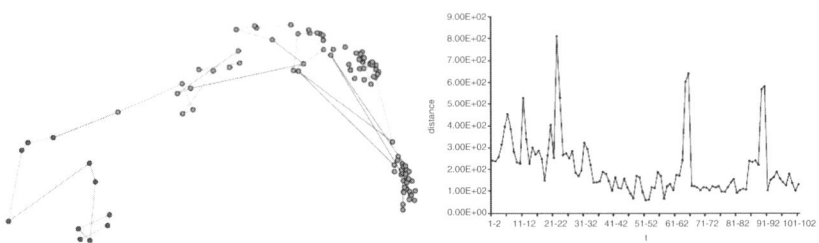

Fig. 13. MDS and graph distance plot of first sequence obtained from a real network

Finally, two experiments were conducted with time series of graphs obtained from real networks. The first network used in the study connects some 45,000 users around Australia. Origin–destination (OD) traffic statistics were collected using network monitoring tools, whereby five probes were placed on links in the core of an enterprise intranet. Probes were positioned on links in the network in such a way as to achieve wide coverage of traffic on the network. The number of nodes in the network was reduced to 150 by aggregating IP addresses to business domains. The OD traffic data for a single day was used to generate a graph representing the *logical* state of the network, in terms of topology and traffic, over a one day period. A time series of 102 graphs was derived using traffic data from 102 adjacent days of traffic. Average graph size was 70 nodes.

MDS and graph distance plots of this time series are shown in Fig. 13. Contrary to the synthetically generated sequences, only minimal 'ground truth' data exists for this time series, i.e., we do not have a description for many of the abnormal events that have occurred within the recorded period of time. In the graph distance plot we clearly observe three prominent peaks. The second peak coincides with the introduction of a new electronic pay system. Before the first peak, the plot looks rather dynamic, but between the first and second, the second and third, and after the third peak, graph distances are somewhat smaller. From the MDS plot we can draw a number of conclusions that cannot be inferred from the graph distance plot. There are two rather dense clusters of points in the MDS plot, one in the upper right and one in the lower right part. The upper cluster corresponds to the period between the first and second peak, while the lower one represents both the period between the

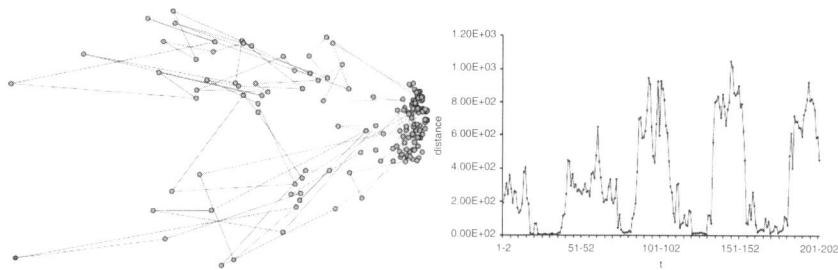

Fig. 14. MDS and graph distance plot of second sequence obtained from a real network

second and third, and after the third peak in the graph distance plot[2]. This means that the network has a different topology before and after the first peak. Likewise, the topology is different before and after the second peak. However, the network topology is similar before and after the third peak. Note that this kind of information cannot be concluded from the graph distance plot.

The second graph sequence based on real data was obtained from a wireless LAN used by delegates during the World Congress for Information Technology (WCIT) held in Adelaide, Australia, in 2002. The time series consists of 202 graphs with an average size of about 100 nodes each. Here each node represents an individual IP address. A graph was constructed from 30 min of traffic data. The sequence of graphs was therefore produced from traffic in adjacent time intervals. In the graph distance plot shown in Fig. 14, one can clearly observe a periodic behaviour of the network. There are five highly dynamic and four less dynamic periods, corresponding to day and night time, respectively.

In the MDS plot in Fig. 14, we observe one large and compact cluster in the right-hand side, and two rather diffuse clusters, one in the upper left and the other in the lower left part of the plot. The large compact cluster mainly corresponds to the network during the four less dynamic periods and to the first two dynamic periods. This cluster formed due to a reduced influence from traffic arising from user behaviour. The upper diffuse cluster represents the network during the third dynamic period and the lower diffuse cluster during the fourth and fifth dynamic periods. An even better visualisation is achieved through displaying the evolution as a movie. Obviously this kind of information cannot be inferred from the graph distance plot.

In Fig. 15a we show three graphs from the four less dynamic periods, while graphs from the third and fifth dynamic period are displayed in Figs. 15b and 15c, respectively. While all graphs in Fig. 15a look rather similar, the network is in different states in Figs. 15b and 15c. This again confirms that the MDS representation is very well suited to represent the different network states from the global point of view. The information visualised in the MDS plot cannot be represented by a graph distance plot, such as the one shown in the right-hand side of Fig. 14.

---

[2] This information is conveyed much clearer if we display the evolution of the graph distance and the MDS plot as a function of time, see Fig. 5. An even better visualisation is achieved through displaying the evolution as a movie.

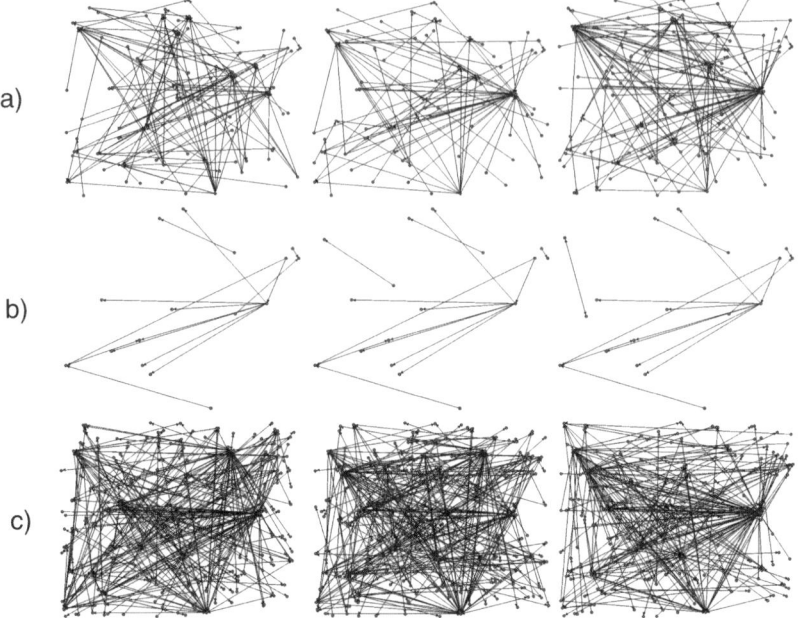

Fig. 15. A few graphs from the time series obtained from the second real network: (**a**) three graphs from one cluster; (**b**) and (**c**) graphs from different clusters

5 Conclusions, Discussion and Future Work

In this chapter, we propose a novel approach to the visualisation of computer network behaviour. We start by representing a given network as a time series of graphs, where the nodes represent either groups of users in common business domains or individual servers, routers or clients and the edges represent physical or logical links between nodes. A graph distance measure originally developed in the domain of pattern recognition is used to compare graphs that represent the network at different points in time. In our earlier work, only distances $d(g_i, g_{i+1})$ between graphs at consecutive points in time were computed and displayed as a plot showing graph distance over time. Abnormal events, or periods of abnormally high network activity, manifest themselves in such a plot through high values. In the present chapter we go one step further and compute distances between all pairs of graphs in a sequence. In this way not only local, but also global temporal network behaviour is taken into consideration. The pairwise graph distances are submitted to a multidimensional scaling procedure that renders a two-dimensional visualisation of the graph sequence. In this visualisation, each graph in the sequence is represented by a point in such a way that the distances between points in the two-dimensional plane resemble the distances between the underlying graphs as closely as possible. By means of this procedure, not only anomalous network change can be represented, but also clusters of network states and the transition between states can be visualised.

A number of open issues remain to be addressed in future research. For example, in the current chapter edge labels have been ignored. But it is a natural extension to include edge labels, or edge weights, in the underlying graphs so as to represent the amount of data transmitted over the links. As a matter of fact the considered graph edit distance measure can be easily extended such that edge labels are taken into account.

A limitation of the current method is imposed by the fact that a complete graph sequence must be given in order to apply the MDS procedure. This restricts the visualisation procedure to working exclusively in the 'off-line' mode. From the application oriented point of view, however, more flexibility would be achieved if an MDS plot could be built incrementally as new graphs of the time series are acquired. Such an approach could be applied in a streaming environment.

All steps required in the production of an MDS plot can be executed without user intervention, while the final interpretation of an MDS plot, i.e., the identification of clusters, abnormal events, etc. is left to a human operator. However, we argue that it takes just a small step to turn the method proposed in this chapter into a fully automatic procedure for the detection of abnormal network states. The essential addition required is an automatic clustering procedure that can identify clusters of points with a high proximity in the MDS plot. Alternatively one could apply the clustering procedure in a high-dimensional space before applying the projection into the two-dimensional space. That is, rather than using only the first two columns of matrix **X** in (3), one could use up to m columns. In principle, any of the known clustering techniques can be applied for this purpose [49].

References

1. Conte, D., Foggia, P., Sansone, C. and Vento, M.: Thirty years of graph matching in pattern recognition. International Journal of Pattern Recognition and Artificial Intelligence, vol. 18, no. 3, 2004, 265–298
2. McKay, B.: Practical graph isomorphism, Congressus Numerantium, vol. 30, 1981, 45–87
3. Ullman, J.: An algorithm for subgraph isomorphism, Journal of the Association for Computing Machinery, vol. 23, no. 1, 1976, 31–42
4. Levi, G.: A note on the derivation of maximal common subgraphs of two directed or undirected graphs. Calcolo, vol. 9, 1972, 341–354
5. McGregor: Backtrack search algorithms and the maximal common subgraph problem, Software-Practice and Experience, vol. 12, 1982, 23–13
6. Bunke, H. and Allermann, G.: Inexact graph matching for structural pattern recognition. Pattern Recognition Letters, vol. 1, 1983, 245–253
7. Sanfeliu, A. and Fu, K.: A distance measure between attributed relational graphs for pattern recognition. IEEE Transactions on Systems, Man and Cybernetics (Part B), vol. 13, no. 3, 1983, 353–363
8. Wong, A.K., You, M. and Chan, A.C.: An algorithm for graph optimal monomorphism, IEEE Transactions on SMC vol. 20, 1990, 628–636
9. Cordelle, L., Foggia, P., Sansone, C. and Vento, M.: A (sub)graph isomorphism algorithm for matching large graphs, IEEE Transactions on PAMI vol. 26, 2004, 1367–1372

10. Messmer, B.T. and Bunke, H.: A new algorithm for error-tolerant subgraph isomorphism detection, IEEE Transactions on PAMI, vol. 20, 1998, 493–505
11. Larossa, L. and Valiente, G.: Constraint satisfaction algorithms for graph pttern matching, Mathematical Structures in Computer Science, vol. 12, 2002, 403–422
12. Christmas, W., Kittler, J. and Petrou, M.: Structural matching in computer vision using probabilistic relaxation, IEEE Transactions on Pattern Analysis and Machine Intelligence, vol. 17, no. 8, 1995, 749–764
13. Wilson, R. and Hancock, E.: Structural matching by discrete relaxation, IEEE Transactions on Pattern Analysis and Machine Intelligence, vol. 19, no. 6, 1997, 634–648
14. Cross, A., Wilson, R. and Hancock, E.: Inexact graph matching using genetic search, Pattern Recognition, vol. 30, no. 6, 1997, 953–970
15. Wang, I., Fan, K.-C. and Horng, J.-T.: Genetic-based search for error correcting graph isomorphism, IEEE Transactions on Systems, Man and Cybernetics (Part B), vol. 27, no. 4, 1997, 558–597
16. Frasconi, P., Gori, M. and Sperduti, A.: A general framework for adaptive processing of data structures, IEEE Transactions on Neural Networks, vol. 9, no. 5, 1998, 768–786
17. Yao, Y., Marcialis, G., Pontil, M., Frasconi, P. and Roli, F.: Combining flat and structured representations for fingerprint classification with recursive neural networks and support vector machines, Pattern Recognition, vol. 36, no. 2, 2003, 397–406
18. Umeyama, S.: An eigendecomposition approach to weighted graph matching problems, IEEE Transactions on Pattern Analysis and Machine Intelligence, vol. 10, no. 5, 1988, 695–703
19. Luo, B., Wilson, R. and Hancock E.: Spectral embedding of graphs, Pattern Recogniction, vol. 36, no. 10, 2003, 2213–2223
20. Caelli, T. and Kosinov, S.: Inexact graph matching using eigen-subspace projection clustering, International Journal of Pattern Recognition and Artificial Intelligence, vol. 18, no. 3, 2004, 329–355
21. Gori, M., Maggini, M. and Sarti, L.: Exact and approxiamte graph matching using random walks, IEEE Transactions on Pattern Analysis and Machine Intelligence, vol. 27, no. 7, 2005, 1100–1111
22. Gärtner, T.: A survey of kernels for structured data, SIGKDD Explorations, vol. 5, no. 1, 2003, 49–58
23. Kashima, H., Tsuda, K. and Inokuchi, A.: Marginalized kernels between labeled graphs, in Proceedings of the 20th International Conference on Machine Learning, 2003, 321–328
24. Neuhaus, M. and Bunke, H.: Edit distance based kernel functions for structural pattern classification, Pattern Recognition vol. 39, 2006, 1852–1863
25. Last, M., Kandel, A. and Bunke, H. (Eds.): Data mining in time series databases, World Scientific, 2004
26. Magnaghi, A. and Hamada, T. and Katsuyama, T.: A wavelet-based framework for proactive detection of network misconfigurations. In SIGCOMM 2004, August 2004, 253–258
27. Hood, C.S. and Proactive, C.Ji.: Network-fault detection. IEEE Transactions on Reliability, vol. 46, no. 3, 1997, 333–341
28. Giacinto, G., Perdisci, R. and Roli, F.: Alarm clustering for intrusion detection systems in computer networks. In MLDM 2005: 4th International Conference, Leipzig, Germany, 2005, 184–193
29. Bunke, H., Kraetzl, M., Shoubridge, P. and Wallis, W.D.: Detection of abnormal change in time series of graphs, Journal of Interconnection Networks, vol. 3, nos. 1 and 2, 2002, 85–101

30. Dickinson, P., Bunke, H., Dadej, A. and Kraetzl, M.: Median graphs and anomalous change detection in communication networks, Proceedings of the International Conference on Information, Decision and Control, Adelaide, 2002, 59–64
31. Gomila, C. and Meyer, F.: Graph based object tracking, Proceedings of the ICIP 2003, vol. 3, 2003, 41–44
32. Conte, D., Foggia, P., Jolion, J.-M. and Vento, M.: A graph-based multi-resolution algorithm for tracking objects in presence of occlusions, Proceedings of the GbR 2005, Springer LNCS 3434, 2005, 193–202
33. Bunke, H., Dickinson, P. and Kraetzl, M.: Matching sequences of graphs with applications in computer network analysis, Proceedings of the 8th World Multi-Conference on Systemics, Cybernetics and Informatics, Orlando, 2004, 270–275
34. Cox, T.F. and Cox, M.A.A.: Multidimensional Scaling. Chapman & Hall, London, 1995
35. Borg, I. and Groenen, P.: Modern multidimensional scaling, Springer, Berlin Heidelberg New York, 1997
36. Dickinson, P., Bunke, H., Dadej, A. and Kraetzl, M.: Matching graphs with unique node labels, Pattern Analysis and Applications, vol. 7, no. 3, 2004, 243–254
37. Kruegel, C. and Toth, T: Using decision trees to improve signature-based intrusion detection. RAID, 2003
38. Mahoney, M. and Chan, P.: Learning rules for anomaly detection of hostile network traffic. In ICDM 2003: Third IEEE International Conference on Data Mining, 601–604, Washington, DC, USA, 2003. IEEE Computer Society
39. Lewis, L.: A case based reasoning approach to the managment of faults in communications networks. In IEEE INFOCOM, vol. 3, 1422–1429, San Francisco, CA, March 1993
40. Bon, K. S.: Signature-based approach for intrusion detection. In MLDM 2005: 4th International Conference, 526–536, Leipzig, Germany, 2005
41. Lazar, A., Wang, W. and Deng, R.: Models and algorithms for network fault detection and identification: A review. In ICC, Singapore, November 1992
42. Barford, P. and Plonka, D.: Characteristics of network traffic flow anomalies. In IMW '01: Proceedings of the 1st ACM SIGCOMM Workshop on Internet Measurement, 69–73, San Francisco, California, USA, 2001, ACM
43. Thottan, M. and Ji, C.: Proactive anomaly detection using distributed intelligent agents. IEEE Network, vol. 12, no. 5, September 1998, 21–27
44. Cabrera, J.B.D., Lewis, L., Qin, X., Lee, W., Prasanth, R.K., Ravichandran B. and Mehra, R.K.: Proactive detection of distributed denial of service attacks using MIB traffic variables – a feasibility study. In 2001 IEEE/IFIP International Symposium on Integrated Network Management Proceedings, May 2001, 609–622
45. Hellerstein, J. and Watson, T.J.: An approach to selecting metrics for detecting performance problems in information systems. Proceedings of Second IEEE International Workshop on Systems Management, 1996, 30–39
46. Hood, C.S. and Ji, C.: Intelligent network monitoring. In Proceedings of the 1995 IEEE Workshop on Neural Networks for Signal Processing, 1995, 521–530
47. Bunke, H. and Kraetzl, M.: Classification and detection of abnormal events in time series of graphs. In: Last M., Kandel, A. and Bunke, H. (Eds.): Data Mining in Time Series Databases, World Scientific, 2004, 127–148
48. Bunke, H., Dickinson, P.J., Kraetzl, M., Wallis, W.D.: A graph-theoretic approach to enterprise network dynamics, Birkhuser, Boston, Basel, Berlin, 2007
49. Jain, A., Murty, M. and Flynn, P.: Data clustering: a review. ACM Computing Surveys vol. 31, 1999, 264–323

Clustering of Web Documents Using Graph Representations

Adam Schenker, Horst Bunke, Mark Last and Abraham Kandel

Summary. In this paper we describe a clustering method that allows the use of graph-based representations of data instead of traditional vector-based representations. Using this new method we conduct content-based clustering of two web document collections. Clustering of web documents is performed to organize the documents with little or no human intervention. Benefits of clustering include easier browsing and improved retrieval speed. In order to measure the performance of our graph-matching approach, we compare it to the popular vector-based k-means method. We perform experiments using different graph distance measures as well as various document representations that utilize graphs. The results with the k-means clustering algorithm show that the graph-based approach can outperform traditional vector-based methods.

Key words: Graph distance, Graph representations, k-Means

1 Introduction

Clustering is the separation of a collection of objects into groups, called clusters, such that objects within the same cluster are similar to each other, yet dissimilar to the objects in other clusters. Clustering is an unsupervised method, meaning no labeled training examples are provided. Many different clustering algorithms have been proposed, such as k-means, fuzzy c-means, hierarchical agglomerative, and graph partitioning [1].

Clustering of natural language documents is an important research area for two major reasons. First, clustering a document collection into categories enables it to be more easily browsed and used. Second, clustering can improve the performance of search and retrieval on a document collection. Hierarchical clustering methods [1], for example, are used often for this purpose. When representing documents for clustering the vector model is typically used [2]. In this model, each meaningful term that can appear in a document becomes a feature (dimension).

The vector model is simple and allows the use of traditional clustering methods that deal with numerical feature vectors. However, it discards information such as the order in which the terms appear, where in the document the terms appear, how close the terms are to each other, and so forth. By keeping this kind of structural information we could possibly improve the performance of the clustering. The problem is that traditional clustering methods are often restricted to working on purely numeric feature vectors due to the need to compute distances between objects or to find some representative element of a cluster of objects, both of which are easily accomplished with numerical feature vectors. Thus either the original data needs to be converted to a vector of numeric values by discarding possibly useful structural information (that we do when using the vector model to represent documents) or we need to develop new, customized algorithms for a specific representation.

In order to overcome this problem, we have introduced an extension of classical clustering methods that allows us to work with graphs as fundamental data structures instead of being limited to vectors of numeric values [3, 4]. Our approach has two main benefits:

1. It allows us to keep the inherent structure of the original documents by modeling each document as a graph.
2. We can apply straightforward extensions to use existing clustering algorithms rather than needing to create new algorithms from scratch. In this paper we will address comparison of different graph similarity measures and document representations in the context of document clustering. We will use a graph-based k-means clustering algorithm to cluster two web document collections. We will use the cosine and Jaccard similarity measures [2] with the vector model representation as a baseline for comparison.

Recently, several papers have appeared in the literature that deal with graph representations of documents. Liang and Doermann represented the physical layout of document images as graphs [5]. In their *layout graphs* nodes represent elements on the page of a document, such as columns of text or headings, while edges indicate how these elements appear together on the page (i.e., spatial relationships). This method is based on the formatting and appearance of the documents when rendered, not the textual content (words) of a document as in our approach. Lopresti and Wilfong compared web documents using a graph representation that primarily utilizes HTML parse information, in addition to hyperlink and content order information [6]. In their approach they use *graph probing*, which extracts numerical feature information from the graphs, such as node degrees or edge label frequencies, rather than comparing the graphs themselves. In contrast, our representation uses graphs created solely from the content, and we use the graphs themselves rather than a set of extracted features. The *subject graphs* of Tomita et al. [7] are constructed using weights calculated from term occurrence frequencies; our method does not calculate any weights, and most of our models do not use any frequency information.

Other graph-based approaches to text or web representation that are well known in the literature include Sowa's *conceptual graphs* [8] and *directed acyclic word graphs* (DAWGs) [9]. Conceptual graphs provide for powerful knowledge representation capabilities, but are not widely used for web documents because "they are based on deep analysis, and so require well maintained dictionaries and an excessive amount of time to operate" [8]. DAWGs are used for compact representation and recognition of individual words in a text rather than representation of entire documents.

Clustering with graphs is well established in the literature. However, the paradigm in those methods is to treat the entire clustering problem as a graph: nodes represent the items to be clustered and weights on edges connecting two nodes indicate the distance (dissimilarity) between the objects the nodes represent. The usual procedure is to create a minimal spanning tree of the graph and then remove the remaining edges with the largest weight in the MST until the number of desired clusters (connected components) is achieved [10]. After applying the algorithm the edges indicate which objects belong to which clusters. In our method, by contrast, each object is represented by a graph (not a node), and we perform standard clustering methods on these graphs.

Lately there has been some progress with performing clustering directly on graph-based data. For example, an extension of self-organizing maps (SOMs) which allows the procedure to work with graphs has been proposed [11]; graph edit distance and weighted mean of a pair of graphs were introduced to deal with graph-based data under the SOM algorithm. Clustering of shock trees using tree edit distance has also been considered [12]. Both of these methods have in common that they use graph (or tree) edit distance for their graph distance measures. One drawback of this approach is that the edit cost functions must be specified for each application. Sanfeliu et al. have investigated clustering of attributed graphs using their own "function-described graphs" as cluster representatives [13]. However, their method is rather complicated and much more involved than our straightforward extension of a classical, simple clustering algorithm.

The remainder of this paper is organized as follows. In Sect. 2 we give the formal notations relating to graphs that will be used throughout the paper. We describe the graph-based extension of the k-means algorithm and the various graph distance measures in Sect. 3. The details of the different graph representations we utilize during clustering are provided in Sect. 4. In Sect. 5 we explain our experimental procedures and present the results. Finally, some concluding remarks are given in Sect. 6.

2 Formal Notation

In this section we will give the formal mathematical notation which pertains to graphs and their role in performing clustering with the k-means algorithm. *Graphs* are a mathematical formalism for dealing with structured entities and systems. In basic terms a graph consists of *vertices* (or *nodes*), which correspond to some objects or components. Graphs also contain *edges*, which indicate relationships between the vertices. The first definition we have is that of the graph itself. Each object (web document, in the context of this paper) in the data set we are clustering will be represented by such a graph:

Definition 1. *A graph G is defined by a four-tuple (quadruple): $G = (V, E, \alpha, \beta)$, where V is a set of vertices (also called nodes), $E \subseteq V \times V$ is a set of edges* connecting the vertices, $\alpha : V \to \Sigma_V$ is a function labeling the vertices, and $\beta : E \to \Sigma_E$ is a function labeling the edges (Σ_V and Σ_E being the sets of labels that can appear on the nodes and edges, respectively).

The graphs we will use in this paper are directed graphs with node and edge labels. The next definition we have is that of a *subgraph*. One graph is a subgraph of another graph if it exists as part of the larger graph:

Definition 2. *A graph $G_1 = (V_1, E_1, \alpha_1, \beta_1)$ is a subgraph of a graph $G_2 = (V_2, E_2, \alpha_2, \beta_2)$, denoted $G_1 \subseteq G_2$, if $V_1 \subseteq V_2$, $E_1 \subseteq E_2 \cap (V_1 \times V_1)$, $\alpha_1(x) = \alpha_2(x) \,\forall x \in V_1$, and $\beta_1((x,y)) = \beta_2((x,y)) \,\forall (x,y) \in E_1$. Conversely, graph G_2 is also called a supergraph of G_1.*

Next we have the important concept of the maximum common subgraph (mcs) for short, which is the largest subgraph a pair of graphs have in common:

Definition 3. *A graph G is a maximum common subgraph (mcs) of graphs G_1 and G_2, denoted $mcs(G_1, G_2)$, if: (1) $G \subseteq G_1$ (2) $G \subseteq G_2$ and (3) there is no other subgraph G' ($G' \subseteq G_1$, $G' \subseteq G_2$) such that $|G'| > |G|$.*

In the above definition, $|G|$ is intended to convey the "size" of the graph G; often it is taken to be $|V|$, i.e., the number of vertices in the graph. In most of the graph representations used in this paper we will define the size of a graph to be $|G| = |V| + |E|$, i.e., the sum of the number of nodes and edges in the graph. Complementary to Definition 3, we also have the concept of MCS:

Definition 4. *A graph G is a minimum common supergraph [14] (MCS) of graphs G_1 and G_2, denoted $MCS(G_1, G_2)$, if: (1) $G_1 \subseteq G$ (2) $G_2 \subseteq G$ and (3) there is no other supergraph G' ($G_1 \subseteq G'$, $G_2 \subseteq G'$) such that $|G'| < |G|$.*

Now that we have our formal notation, we are in a position to proceed to describing the k-means algorithm extended to cluster graphs instead of vectors.

3 Clustering with Graphs

3.1 Basic Clustering Algorithm

The k-means clustering algorithm is a simple and straightforward method for clustering data [15]. The basic algorithm is given in Fig. 1. Usually during clustering we represent each object, which consists of m numeric values, as a vector in the space $\Re^m$. When representing documents in this manner, each value is associated with a specific term (word) that may appear on a document, and the set of possible terms is shared across all documents; this is called the vector-space model of information retrieval. The values may be binary, indicating the presence or absence of the corresponding term. The values may also be nonnegative integers, which represent the number of times a term appears on a document (i.e., term frequency). Non-negative

Inputs:	the set of n data items and a parameter, k, defining the number of clusters to create
Outputs:	the centroids of the clusters and for each data item the cluster (an integer in $[1,k]$) it belongs to
Step 1.	Assign each data item randomly to a cluster (from 1 to k).
Step 2.	Using the initial assignment, determine the centroids of each cluster.
Step 3.	Given the new centroids, assign each data item to be in the cluster of its closest centroid.
Step 4.	Re-compute the centroids as in Step 2. Repeat Steps 3 and 4 until the centroids do not change.

Fig. 1. The k-means clustering algorithm

real numbers can also be used, in this case indicating the importance or weight of each term. These values are derived through a method such as the popular inverse document frequency model ($tf \cdot idf$) [2], which reduces the importance of terms that appear on many documents. Regardless of the method used, each series of values represents a document and corresponds to a point (i.e., vector) in a Euclidean feature space. This model is often used when applying data mining techniques to documents, as there is a strong mathematical foundation for performing distance measure and centroid calculations using vectors. However, this method of document representation does not capture important structural information, such as the order and proximity of term occurrence, or the location of term occurrence within the document. It is also common to restrict the number of dimensions by selecting some small set of discriminating or important terms, as the number of possible terms that can occur across a collection of documents can be quite large.

When representing data by vectors, the distances between two objects can be computed using the Euclidean distance in m dimensions:

$$dist_{EUCL}(x,y) = \sqrt{\sum_{i=1}^{m}(x_i - y_i)^2} \quad (1)$$

where x_i and y_i are the ith components of vectors $x = [x_1, x_2, \ldots, x_m]$ and $y = [y_1, y_2, \ldots, y_m]$, respectively. However, for applications in text and document clustering, the cosine similarity measure [2] is often used due to its length invariance property. We can convert this to a distance measure by the following:

$$dist_{COS}(x,y) = 1 - \frac{x \bullet y}{\|x\| \cdot \|y\|} \quad (2)$$

Here $\bullet$ indicates the dot product operation and $\|\ldots\|$ indicates the magnitude (length) of a vector. Another popular distance measure for determining document similarity is the extended Jaccard similarity [2], which is converted to a distance measure as follows:

$$dist_{JAC}(x,y) = 1 - \frac{\sum_{i=1}^{m} x_i y_i}{\sum_{i=1}^{m} x_i^2 + \sum_{i=1}^{m} y_i^2 - \sum_{i=1}^{m} x_i y_i} \quad (3)$$

We have determined that if methods of computing distance between graphs and constructing a representative of a set of graphs are available it is possible to extend many clustering and classification methods to work directly on graphs. First, any distance calculations between objects to be clustered, which are represented by graphs and not vectors, is accomplished with a graph-theoretical distance measure as we will discuss in Sect. 3.2. Second, since it is necessary to compute the distance between objects and cluster centers, it follows that the cluster centers (representatives) must also be graphs. Therefore, we compute the representative of a cluster as the median graph of the set of graphs in that cluster (as we will describe in Sect. 3.3).

3.2 Graph Distance Measures

As we mentioned above, we need a graph-theoretical distance measure in order to use graphs for clustering. We have implemented several distance measures and will compare their clustering performance. For brevity we will refer to the distance measures below as MCS, WGU, and MMCS.

The first distance measure MCS is a well-known graph distance measure based on the mcs [16]:

$$d_{MCS}(G_1, G_2) = 1 - \frac{|mcs(G_1, G_2)|}{\max(|G_1|, |G_2|)} \quad (4)$$

where G_1 and G_2 are the graphs to compare, $mcs(G_1, G_2)$ is their maximum common subgraph, $|\ldots|$ is the size of a graph, and $\max(\ldots)$ is the usual maximum operation. Here we define the size of a graph to be the sum of the number of nodes and edges in the graph. The concept behind this distance measure is that as the size of the maximum common subgraph of a pair of graphs becomes larger, the more similar the two graphs are (i.e., they have more in common). The larger the maximum common subgraph, the smaller $d_{MCS}(G_1, G_2)$ becomes, indicating more similarity and less distance. If the two graphs are in fact identical, their maximum common subgraph is the same as the graphs themselves and thus the size of all three graphs is equal: $|G_1| = |G_2| = |mcs(G_1, G_2)|$. This leads to the distance, $d_{MCS}(G_1, G_2)$, becoming 0. Conversely, if no maximum common subgraph exists, then $|mcs(G_1, G_2)| = 0$ and $d_{MCS}(G_1, G_2) = 1$. This distance measure has been shown to be a metric [16], and produces a value in $[0, 1]$.

A second distance measure WGU which has been proposed by Wallis et al. [17] is:

$$d_{WGU}(G_1, G_2) = 1 - \frac{|mcs(G_1, G_2)|}{|G_1| + |G_2| - |mcs(G_1, G_2)|} \quad (5)$$

This distance measure behaves similarly to MCS. If the maximum common subgraph does not exist (i.e., $|mcs(G_1, G_2)| = 0$), then $d_{WGU}(G_1, G_2) = 1$. If the maximum common subgraph is identical to the original graphs, $|G_1| = |G_2| = |mcs(G_1, G_2)|$, then the graphs G_1 and G_2 are identical and thus $d_{WGU}(G_1, G_2) = 0$. The denominator used in this method is based on the idea of "graph union." It represents the size of the union of the two graphs in the set theoretic sense; specifically adding the size of each graph ($|G_1| + |G_2|$) then subtracting the size of their intersection ($|mcs(G_1, G_2)|$) leads to the size of the union (the reader may easily verify this using a Venn diagram). The motivation for doing this is to allow for changes in the smaller graph to exert some influence over the distance measure, which does not happen with MCS [17]. This measure was also demonstrated to be a metric, and creates distance values in $[0, 1]$.

The third distance measure MMCS, proposed by Fernández and Valiente, is based on both the maximum common subgraph and the MCS [14]:

$$d_{MMCS}(G_1, G_2) = |MCS(G_1, G_2)| - |mcs(G_1, G_2)| \tag{6}$$

where $MCS(G_1, G_2)$ is the minimum common supergraph of graphs G_1 and G_2. The concept that drives this distance measure is that the maximum common subgraph provides a "lower bound" on the similarity of two graphs, while the MCS is an "upper bound." If two graphs are identical, then both their mcs and MCS are the same as the original graphs and $|G_1| = |G_2| = |MCS(G_1, G_2)| = |mcs(G_1, G_2)|$, which leads to $d_{MMCS}(G_1, G_2) = 0$. As the graphs become more dissimilar, the size of the maximum common subgraph decreases, while the size of the MCS increases. This in turn leads to increasing values of $d_{MMCS}(G_1, G_2)$. For two graphs with no maximum common subgraph, the distance will become $|MCS(G_1, G_2)| = |G_1| + |G_2|$. MMCS has also been shown to be a metric [14], but it does not produce values normalized to the interval $[0, 1]$, unlike the previously described distance measures. Note that if it holds that $|MCS(G_1, G_2)| = |G_1| + |G_2| - |mcs(G_1, G_2)| \, \forall G_1, G_2$, we can compute $d_{MMCS}(G_1, G_2)$ as $|G_1| + |G_2| - 2|mcs(G_1, G_2)|$. This is much less computationally intensive than computing the MCS.

We will describe our graph representation of documents in detail in Sect. 4. However, we wish to mention here an interesting feature our graph representation has on the time complexity of determining the distance using (4–6). For general graphs the computation of the mcs is NP-Complete. Methods for computing the mcs are presented in [18, 19]. However, for the graph representations of web documents presented in this paper, the computation of the maximum common subgraph is $O(n^2)$, with n being the number of nodes, due to the existence of unique node labels in the graph representations (i.e., we need only examine the intersection of the nodes, since each node has a unique label) [20]. Thus the maximum common subgraph, G_{mcs}, of a pair of graphs with unique node labels, G_1 and G_2, can be created by the following procedure:

1. Find the nodes V_{mcs} by determining the subset of node labels that the original graphs have in common with each other and create a node for each common label.
2. Find the edges E_{mcs} by examining all pairs of nodes from step 1 and introduce edges that connect pairs of nodes in both of the original graphs with identical edge labels.

Note that the calculation of the MCS can be reduced to the mcs problem [21]. Therefore the computation of the MCS can also be performed in $O(n^2)$ time.

3.3 Median of a Set of Graphs

The second ingredient required to apply clustering to graphs is that of a graph-theoretic cluster representative of a set of graphs. For this we have used the concept of the *median graph* [22], which is the graph which has the minimum average distance to all graphs in the cluster:

$$G = \arg \min_{\forall s \in S} \left(\frac{1}{n} \sum_{i=1}^{n} dist(s, G_i) \right) \quad (7)$$

Here $S = \{G_1, G_2, \ldots, G_n\}$ is a set of n graphs for which we want to compute the median (and thus $|S| = n$) and G is the median graph. The median is defined to be a graph in set S. Thus the median of a set of graphs is the graph from that set which has the minimum average distance to all the other graphs in the set. The distance $dist(\ldots)$ is computed using one of (4–6) above. There also exists the concepts of the generalized median and weighted mean [22], where we do not require that G be a member of S, but we will not consider them here because they are quite expensive to compute. In the case where the median is not unique (i.e., there is more than one graph that has the same minimum average distance) we select one of those graphs at random as the representative for the k-means algorithm. This variation of the k-means algorithm, where we use a median instead of a mean as cluster representatives, is also known as k-medoids [23].

4 Graph Representations of Web Documents

In this section we describe methods for representing web documents using graphs instead of the traditional vector representations. All representations are based on the adjacency of terms in a web document. These representations are named: *standard*, *simple*, *n-distance*, *n-simple distance*, *raw frequency* and *normalized frequency*.

Under the *standard* method each unique term (word) appearing in the document, except for *stop words* such as "the," "of," and "and" which convey little information, becomes a node in the graph representing that document. Each node is labeled with the term it represents. Note that we create only a single node for each word even if a word appears more than once in the text. Second, if word a immediately precedes word b somewhere in a "section" s of the document, then there is a directed edge from the node corresponding to term a to the node corresponding to term b with an edge label s. We take into account certain punctuation (such as periods) and do not create an edge when these are present between two words. Sections we have defined for web documents are: *title*, which contains the text related to the document's title and any provided keywords (meta-data); *link*, which is text that appears in hyperlinks on the document; and *text*, which comprises any of the readable text in the document (this includes link text but not title and keyword text). Next we remove the most infrequently occurring words on each document, leaving at most m nodes per graph (m being a user provided parameter). This is similar to the dimensionality reduction process for vector representations [2]. Finally we perform a simple stemming method and conflate terms to the most frequently occurring form by relabeling nodes and updating edges as needed. An example of this type of graph representation is given in Fig. 2. The ovals indicate nodes and their corresponding term labels. The edges are labeled according to title, link, or text. The document represented by the example has the title "YAHOO NEWS," a link whose text reads "MORE NEWS," and text containing "REUTERS NEWS SERVICE REPORTS." If a pair of terms appears together in more than one section, we create an edge for each section with the appropriate section label. Note there is no restriction on the form of the graph and that cycles are allowed. Also, disconnected components may occur in the graphs, which is not a problem with our approach. While this method of document representation appears superficially similar to the bigram, trigram, or N-gram methods, those are statistically oriented approaches based on word occurrence probability models [24]. The methods presented here, with the exception of the frequency representations described below, do not require or use the computation of term probability relationships.

The second type of graph representation we will look at is what we call the *simple* representation. It is basically the same as the standard representation, except that we look at only the visible text on the page (no title or meta-data is examined) and we do not label the edges between nodes. Thus we ignore the information about the "section" where the two respective words appear together. An example of this type of representation is given in Fig. 3.

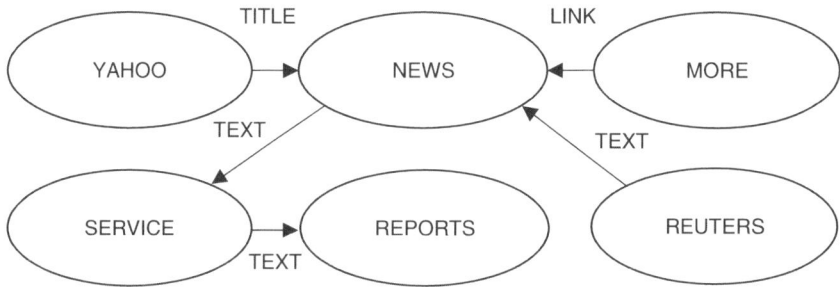

Fig. 2. Example of a *standard* graph representation of a document

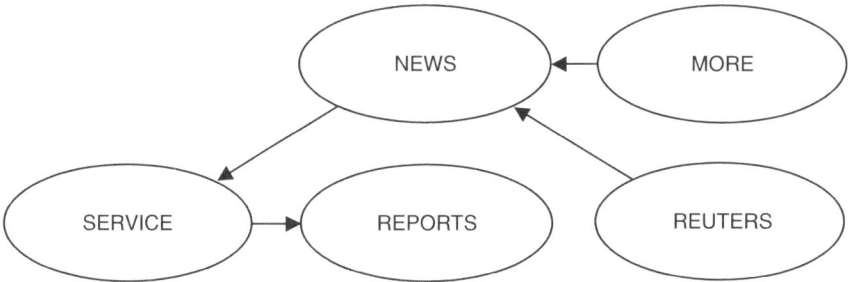

Fig. 3. Example of a *simple* graph representation of a document

The third type of representation is called the *n-distance* representation. Under this model, there is a user-provided parameter, n. Instead of considering only terms immediately following a given term in a web document, we look up to n terms ahead and connect the succeeding terms with an edge that is labeled with the distance between them (unless the words are separated by certain punctuation marks); here "distance" is related to the number of other terms which appear between the two terms in question. For example, if we had the following sequence of text on a web page, "AAA BBB CCC DDD," then we would have an edge from term AAA to term BBB labeled with a 1, an edge from term AAA to term CCC labeled 2, and so on. The complete graph for this example is shown in Fig. 4. The mcs for this representation is derived in the same manner as described previously, where we require the edge labels to be an exact match in both graphs.

Similar to n-distance, we also have the fourth graph representation, *n-simple distance*. This is identical to n-distance, but the edges are not labeled, which means we only know that the "distance" between two connected terms is not more than n.

The fifth graph representation is what we call the *raw frequency* representation. This is similar to the simple representation (adjacent words, no section-related information) but each node and edge is labeled with an additional frequency measure. For nodes this indicates how many times the associated term appeared in the web document; for edges, this indicates the number of times the two connected terms appeared adjacent to each other in the specified order. The raw frequency representation uses the total number of term occurrences (on the nodes) and co-occurrences (edges).

A problem with this representation is that large differences in document size could lead to skewed comparisons, similar to the problem encountered when using Euclidean distance with vector representations of documents. Under the *normalized frequency* representation, instead of associating each node with the total number of times the corresponding term appears in the document, a normalized value in $[0, 1]$ is assigned by dividing each node frequency value by the maximum node frequency value that occurs in the graph; a similar procedure is performed for the edges. Thus each node and edge has a value in $[0, 1]$ associated with it, which indicates the normalized frequency of the term (for nodes) or co-occurrence of terms (for edges).

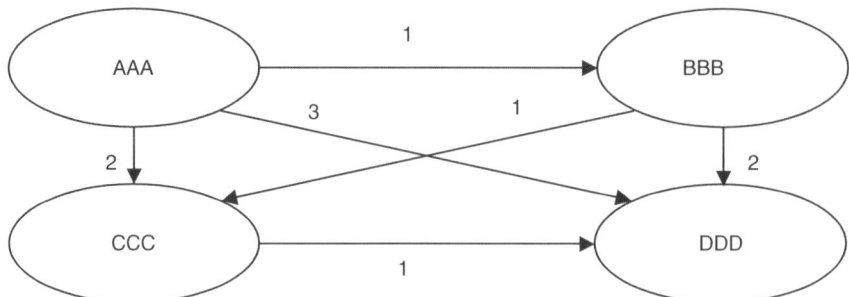

Fig. 4. Example of a *n-distance* graph representation of a document

For the raw frequency and normalized frequency representations the graph size is defined as the total of the node frequencies added to the total of the edge frequencies, rather than the previous definition of $|G| = |V| + |E|$. We need this modification to reflect the frequency information in the graph size. As an example, consider two raw frequency graphs each with a node "A"; however, term "A" appears twice in one document and 300 in the other. This difference in frequency information is not captured under the previous definition. Further, when we compute the mcs for these representations we take the minimum frequency element (either node or edge) as the value for the mcs. To continue the above example, node "A" in the mcs would have a frequency of 2, which is $\min(2, 300)$.

5 Experiments and Results

5.1 Web Document Data Sets

In order to evaluate the performance of the graph-based k-means algorithm as compared with the traditional vector methods, we performed experiments on three different collections of web documents, called the *F-series*, the *J-series*, and the *K-series* [25]; the data sets are available under these names at `ftp://ftp.cs.umn.edu/dept/users/boley/PDDPdata/`. These data sets were selected because of two major reasons. First, all of the original HTML documents are available, which is necessary if we are to represent the documents as graphs; many other document collections only provide a preprocessed vector representation, which is unsuitable for use with our method. Second, ground truth assignments are provided for each data set, and there are multiple classes representing easily understandable groupings that relate to the content of the documents. Some web document collections are not labeled or are presented with some other task in mind than content-related clustering (e.g., building a predictive model based on user preferences).

The F-series originally contained 98 documents belonging to one or more of 17 subcategories of four major category areas: *manufacturing*, *labor*, *business and finance*, and *electronic communication and networking*. Because there are multiple subcategory classifications for many of these documents, we have reduced the categories to just the four major categories mentioned above in order to simplify the problem. There were five documents that had conflicting classifications (i.e., they were classified to belong to two or more of the four major categories) which we removed, leaving 93 total documents. The J-series contains 185 documents and ten classes: *affirmative action*, *business capital*, *information systems*, *electronic commerce*, *intellectual property*, *employee rights*, *materials processing*, *personnel management*, *manufacturing systems*, and *industrial partnership*. We have not modified this data set. The K-series consists of 2,340 documents and 20 categories: *business*, *health*, *politics*, *sports*, *technology*, *entertainment*, *art*, *cable*, *culture*, *film*, *industry*, *media*, *multimedia*, *music*, *online*, *people*, *review*, *stage*, *television*, and *variety*. The last 14 categories are subcategories related to entertainment, while the entertainment category refers to entertainment in general. These were originally news pages hosted at Yahoo (http://www.yahoo.com). Experiments on this data set are presented in [26].

For the vector-model representation experiments there were already several term-document matrices available for our experiments at the same location where we obtained the document collections. We selected the matrices with the smallest number of dimensions. For the F-series documents there are 332 dimensions (terms) used, while the J-series has 474 dimensions; the K-series used 1,458 dimensions. We performed some preliminary experiments and observed that other term-weighting schemes (i.e., $tf \cdot idf$, see [2]) improved the accuracy of the vector-model representation for these data sets either only very slightly or in many cases not at all. Thus we have left the data in its original format.

5.2 Clustering Performance Measures

We use the following three clustering performance measures to evaluate the performance of each clustering. The first two indices measure the matching of obtained clusters to the "ground truth" clusters, while the third index measures the quality of clustering in general.

The first index is the *Rand index* [27]. To compute the Rand index, we perform a comparison of all pairs of objects in the data set after clustering. If both objects in a pair are in the same cluster in both the ground truth clustering and the clustering we wish to measure, this counts as an "agreement." If both objects in the pair are in different clusters in both the ground truth clustering and the clustering we wish to investigate, this is also an agreement. Otherwise, this is a "disagreement." The Rand index is computed as:

$$R_I = \frac{A}{A+D} \quad (8)$$

where A is the number of agreements and D is the number of disagreements, as described above. Thus the Rand index is a measure of how closely the clustering created by some procedure matches ground truth. It produces a value in the interval $[0, 1]$, with 1 representing a clustering that perfectly matches ground truth.

The second performance measure we use is *mutual information* [26, 28], which is defined as:

$$\Lambda^M = \frac{1}{n} \sum_{l=1}^{k} \sum_{h=1}^{g} n_l^{(h)} \log_{k \cdot g} \left(\frac{n_l^{(h)} \cdot n}{\sum_{i=1}^{k} n_i^{(h)} \sum_{i=1}^{g} n_l^{(i)}} \right) \quad (9)$$

where n is the number of objects, k is the number of clusters produced by our clustering algorithm, g is the actual number of ground truth clusters, and $n_i^{(j)}$ is the number of items in cluster i (in the created clustering) associated with cluster j (in the ground truth clustering). Note that k and g may not necessarily be equal, which would indicate we are attempting to create more (or fewer) clusters than exist in the ground truth clustering. However, for the experiments described in this paper we will create an identical number of clusters as is present in ground truth. Mutual information represents the overall degree of agreement between the clustering created by some method and the categorization provided by the ground truth clustering with a preference for clusters that have high purity (i.e., are homogeneous with respect to the objects clustered, as given by the clusters they belong to in ground truth). Higher numbers indicate clusters that are homogeneous (i.e., created clusters which contain objects mostly belonging to a single ground truth cluster). Lower numbers indicate less similarity between the clustering that was created and ground truth; a value of zero signifies no statistical correlation between the two clusterings (i.e., they are independent).

The third performance measure we use is the *Dunn index* [29], which is defined as:

$$D_I = \frac{d_{min}}{d_{max}} \qquad (10)$$

where d_{min} is the minimum distance between any two objects in different clusters and d_{max} is the maximum distance between any two items in the same cluster. The numerator captures the worst-case amount of separation between clusters, while the denominator captures the worst-case compactness of the clusters. Thus the Dunn index is an amalgam of the overall worst-case compactness and separation of a clustering, with higher values being better. It does not, however, measure clustering accuracy compared to ground truth as the other two methods do. Rather it is based on the basic underlying assumption of any clustering technique: items in the same cluster should be similar (i.e., have small distance, thus creating compact clusters) and items in separate clusters should be dissimilar (i.e., have large distance, thus creating clusters that are well separated from each other).

5.3 Results

In Tables 1–3 we show the clustering performance for the F-series, J-series, and K-series when using different graph distance measures (Sect. 3.2). The performance of the traditional vector-based approach using distances based on cosine and Jaccard similarity is also given for comparison. Because of the random initialization of the k-means algorithm, each number indicates the average performance taken over ten experiments. We used a maximum of 50 nodes per graph (i.e., $m = 50$, see Sect. 4) for the F and J data sets, while we used 70 nodes per graph for K, due to the higher number of classes and documents. The standard representation was used for the distance measure comparison experiments. The value of k used in the experiments matches the number of clusters present in the ground truth clustering for each data set; thus $k = 4$ for the F-series, $k = 10$ for the J-series, and $k = 20$ for the K-series.

We see that the graph-based methods that use normalized distance measures (MCS and WGU) generally performed similarly to or better than vector-based methods using cosine or Jaccard. MMCS, which is not normalized to the interval $[0, 1]$, performed poorly for all data sets. To see why this occurs, we have provided the following example. Let $|G_1| = 10, |G_2| = 10, |mcs(G_1, G_2)| = 0, |MCS(G_1, G_2)| = 20, |G_3| = 20, |G_4| = 20, |mcs(G_3, G_4)| = 5$, and $|MCS(G_1, G_2)| = 35$. Clearly

Table 1. Distance measure comparison for the F-series data set using the standard representation and 50 nodes per graph maximum

distance measure	Rand index	mutual information	Dunn index
Cosine (vector-based)	0.6788	0.1101	0.4168
Jaccard (vector-based)	0.6899	0.1020	0.6188
MCS	0.7748	0.2138	0.7202
WGU	0.7434	0.1744	0.7967
MMCS	0.6594	0.1120	0.3132

Table 2. Distance measure comparison for the J-series data set using the standard representation and 50 nodes per graph maximum

distance measure	Rand index	mutual information	Dunn index
Cosine (vector-based)	0.8648	0.2205	0.3146
Jaccard (vector-based)	0.8717	0.2316	0.5703
MCS	0.8618	0.2240	0.6476
WGU	0.8757	0.2598	0.7691
MMCS	0.1809	0.0273	0.1381

Table 3. Distance measure comparison for the K-series data set using the standard representation and 70 nodes per graph maximum

distance measure	Rand index	mutual information	Dunn index
Cosine (vector-based)	0.8537	0.2266	0.0348
Jaccard (vector-based)	0.8998	0.2441	0.0730
MCS	0.8957	0.1174	0.0284
WGU	0.8377	0.1019	0.0385
MMCS	0.1692	0.0127	0.0649

graphs G_3 and G_4 are more similar to each other than graphs G_1 and G_2 since G_1 and G_2 have no common subgraph whereas G_3 and G_4 do. However, the distances computed for these graphs are $d_{MCS}(G_1, G_2) = 1.0$, $d_{MCS}(G_3, G_4) = 0.75$, $d_{MMCS}(G_1, G_2) = 20$, and $d_{MMCS}(G_3, G_4) = 30$. So we have the case that the unnormalized distance is actually greater for the pair of graphs that are more similar. This is both counter-intuitive and the opposite of what happens in the cases of the normalized distance measures. Thus this phenomenon leads to the poor clustering performance for MMCS.

In Tables 4–6 we show the clustering performance for the F-series, J-series, and K-series for the different graph representations presented in Sect. 4. For these experiments we use the MCS distance measure (4). For the representations n-distance and

Table 4. Representation comparison for the F-series data set using MCS distance and 50 nodes per graph maximum

representation	Rand index	mutual information	Dunn index
Cosine (vector-based)	0.6788	0.1101	0.4168
Jaccard (vector-based)	0.6899	0.1020	0.6188
Standard	0.7748	0.2138	0.7202
Simple	0.6823	0.1314	0.7364
2-distance	0.6924	0.1275	0.7985
5-distance	0.6731	0.1044	0.8319
2-simple distance	0.7051	0.1414	0.7874
5-simple distance	0.7209	0.1615	0.8211
Raw frequency	0.7070	0.1374	0.7525
Normalized frequency	0.7242	0.1525	0.7077

Table 5. Representation comparison for the J-series data set using MCS distance and 50 nodes per graph maximum

representation	Rand index	mutual information	Dunn index
Cosine (vector-based)	0.8648	0.2205	0.3146
Jaccard (vector-based)	0.8717	0.2316	0.5703
Standard	0.8618	0.2240	0.6476
Simple	0.8562	0.2078	0.5444
2-distance	0.8674	0.2365	0.6531
5-distance	0.8598	0.2183	0.7374
2-simple distance	0.8655	0.2285	0.7056
5-simple distance	0.8571	0.2132	0.6874
Raw frequency	0.8650	0.2141	0.6453
Normalized frequency	0.8812	0.2734	0.6119

n-simple distance, we use values of $n = 2$ and $n = 5$ (i.e., 2-distance, 2-simple distance, 5-distance, and 5-simple distance) in these experiments.

For the F-series, standard was the best performing representation, achieving the best value for Rand index and mutual information, while for the Dunn index, 5-distance was the best representation. For the J-series, normalized frequency was the best for Rand index and mutual information, with 5-distance again being best for the Dunn index. It is not a surprising result that Rand and mutual information should perform similarly to each other and differently than Dunn, as both Rand and mutual information are based on comparison with ground truth while Dunn is a measure of compactness and separation of the clusters with no regard to "accuracy."

For the K-series, the best performing graph representation was standard. However, the graph-based method in this case did not outperform the Jaccard distance-based vector approach. The K-series is a highly homogeneous data set; all the pages

Table 6. Representation comparison for the K-series data set using MCS distance and 70 nodes per graph maximum

representation	Rand index	mutual information	Dunn index
Cosine (vector-based)	0.8537	0.2266	0.0348
Jaccard (vector-based)	0.8998	0.2441	0.0730
Standard	0.8957	0.1174	0.0284
Simple	0.8870	0.0972	0.0274
2-distance	0.8753	0.0832	0.0229
5-distance	0.8813	0.1013	0.0206
2-simple distance	0.8813	0.0947	0.0218
5-simple distance	0.8663	0.0773	0.0234
Raw frequency	0.8770	0.0957	0.0335
Normalized frequency	0.8707	0.0992	0.0283

Table 7. Statistical analysis of experimental results

data set	performance measure	confidence $(1 - P)$	significant?
F-series	Rand	0.9998	yes (better)
F-series	MI	1.0000	yes (better)
J-series	Rand	0.9255	no (same)
J-series	MI	0.4767	no (same)
K-series	Rand	0.3597	no (same)
K-series	MI	1.0000	yes (worse)

have a similar format and some of the same terms appear on every document. To improve the performance of the graph method in this case, we should look at either removing the common terms (nodes) from all graphs (which is often done with the vector model and can also be applied to our approach), or greatly increase the size of the graphs to capture more terms. In our experiments, Rand increases to 0.9053 and mutual information to 0.1618 for the standard representation and MCS distance when using 200 nodes maximum per graph.

In Table 7 we give a statistical analysis of some of the experimental results. Six comparisons are listed in the table, which represent comparing the Jaccard and graph methods for Rand index and mutual information for all three data sets. The graph experiments represented in Table 7 use the standard graph representation, MCS distance, and either 50 nodes per graph (F and J) or 70 nodes per graph (K). The *Confidence* column in the table represents the probability that the means of the results for the vector and graph methods are statistically different, as determined by a two-tailed t-test. Values higher than 0.95 are considered significant, as shown in the last column of the table; we also show whether the graph method was considered better, the same, or worse than the vector method.

6 Conclusions

In this paper we have examined the problem of clustering data which is represented by graphs instead of simpler feature vectors. To perform the clustering we have developed a graph-based version of the k-means clustering algorithm, substituting a suitable graph-theoretical distance measure in the place of the usual vector-related distance and median graphs in place of centroids.

The application we presented here was clustering of web documents. We implemented six different methods of representing web documents by graphs and three different graph distance measures. Our experiments compared the clustering performance of the various proposed methods with the usual vector model approach using cosine and Jaccard-based distance measures. Experimental results showed that the graph-based methods can outperform the traditional vector methods in terms of clustering performance under three different clustering performance measures. We saw that graph distance measures that were not normalized performed poorly, while those that were normalized to the interval $[0, 1]$ yielded good results. The standard representation produced the best results for one data set in terms of comparison with ground truth, while normalized frequency was better for another.

For future work we intend to extend our graph-based method to other classification and clustering methods, such as hierarchical agglomerative clustering and distance weighted k-nearest neighbors. We also wish to look for the optimal graph size and associated terms to represent each specific document. Further, we only examined using two values of n for the n-distance and n-simple distance representations in this paper. Finding the optimal value of n is another subject of ongoing research. Given the good results for the normalized frequency representation for one of the document collections, we will explore similar representations that incorporate more explicit term weighting components (i.e., a model similar to $tf \cdot idf$ but for graphs). However, such an extension is not immediately obvious, since we must deal with adjusting the weights of edges as well as terms (nodes). Finally, we can look at incorporating specific domain knowledge in the distance measure definitions.

Acknowledgement. This work was supported in part by the National Institute for Systems Test and Productivity at the University of South Florida under the US Space and Naval Warfare Systems Command Contract No. N00039–02–C–3244.

References

1. A. K. Jain, M. N. Murty, and P. J. Flynn. Data clustering: a review. *ACM Computing Surveys*, 31(3):264–323, 1999
2. G. Salton. *Automatic Text Processing: The Transformation, Analysis, and Retrieval of Information by Computer*. Addison–Wesley, Reading, 1989
3. A. Schenker, M. Last, H. Bunke, and A. Kandel. Comparison of distance measures for graph-based clustering of documents. In *Proceedings of the Fourth IAPR-TC15 Workshop on Graph-based Representations*, pp. 202–213, 2003

4. A. Schenker, M. Last, H. Bunke, and A. Kandel. Graph representations for web document clustering. In *Proceedings of the First Iberian Conference on Pattern Recognition and Image Analysis*, pp. 935–942, 2003
5. J. Liang and D. Doermann. Logical labeling of document images using layout graph matching with adaptive learning. In D. Lopresti, J. Hu, and R. Kashi, editors, *Document Analysis Systems V*, Volume 2423 of *Lecture Notes in Computer Science*, pp. 224–235. Springer, Berlin Heidelberg New York, 2002
6. D. Lopresti and G. Wilfong. A fast technique for comparing graph representations with applications to performance evaluation. *International Journal on Document Analysis and Recognition*, 6(4):219–229, 2004
7. J. Tomita, H. Nakawatase, and M. Ishii. Graph-based text database for knowledge discovery. In *World Wide Web Conference Series*, pp. 454–455, 2004
8. J. F. Sowa. Conceptual graphs for a database interface. *IBM Journal of Research and Development*, 20(4):336–357, 1976
9. M. Crochemore and R. Vérin. Direct construction of compact directed acyclic word graphs. In A. Apostolico and J. Hein, editors, *CPM97*, Volume 1264 of *Lecture Notes in Computer Science*, pp. 116–129. Springer, Berlin Heidelberg New York, 1997
10. C. T. Zahn. Graph-theoretical methods for detecting and describing gestalt structures. *IEEE Transactions on Computers*, C-20:68–86, 1971
11. S. Günter and H. Bunke. Self-organizing map for clustering in the graph domain. *Pattern Recognition Letters*, 23:405–417, 2002
12. B. Luo, A. Robles-Kelly, A. Torsello, R. C. Wilson, and E. R. Hancock. Clustering shock trees. In *Proceedings of the Third IAPR-TC15 Workshop on Graph-based Representations in Pattern Recognition*, pp. 217–228, 2001
13. A. Sanfeliu, F. Serratosa, and R. Alquézar. Clustering of attributed graphs and unsupervised synthesis of function-described graphs. In *Proceedings of the 15th International Conference on Pattern Recognition*, Volume 2, pp. 1026–1029, 2000
14. M.-L. Fernández and G. Valiente. A graph distance metric combining maximum common subgraph and minimum common supergraph. *Pattern Recognition Letters*, 22:753–758, 2001
15. T. M. Mitchell. *Machine Learning*. McGraw–Hill, Boston, 1997
16. H. Bunke and K. Shearer. A graph distance metric based on the maximal common subgraph. *Pattern Recognition Letters*, 19:225–259, 1998
17. W. D. Wallis, P. Shoubridge, M. Kraetz, and D. Ray. Graph distances using graph union. *Pattern Recognition Letters*, 22:701–704, 2001
18. G. Levi. A note on the derivation of maximal common subgraphs of two directed or undirected graphs. *Calcolo*, 9:341–354, 1972
19. J. J. McGregor. Backtrack search algorithms and the maximal common subgraph problem. *Software Practice and Experience*, 12:23–34, 1982
20. P. Dickinson, H. Bunke, A. Dadej, and M. Kraetzl. Matching graphs with unique node labels. *Pattern Analysis and Applications*, 7(3):243–254, 2004
21. H. Bunke, X. Jiang, and A. Kandel. On the minimum common supergraph of two graphs. *Computing*, 65:13–25, 2000
22. X. Jiang, A. Muenger, and H. Bunke. On median graphs: properties, algorithms, and applications. *IEEE Transactions on Pattern Analysis and Machine Intelligence*, 23(10):1144–1151, 2001
23. L. Kaufman and P. J. Rousseeuw. *Finding Groups in Data: An Introduction to Cluster Analysis*. Wiley, New York, 1990
24. C.-M. Tan, Y.-F. Wang, and C.-D. Lee. The use of bigrams to enhance text categorization. *Information Processing and Management*, 38:529–546, 2002

25. D. L. Boley. Principal direction divisive partitioning. *Data Mining and Knowledge Discovery*, 2(4):325–344, 1998
26. A. Strehl, J. Ghosh, and R. Mooney. Impact of similarity measures on web-page clustering. In *AAAI-2000: Workshop of Artificial Intelligence for Web Search*, pp. 58–64, 2000
27. W. M. Rand. Objective criteria for the evaluation of clustering methods. *Journal of the American Statistical Association*, 66:846–850, 1971
28. T. M. Cover and J. A. Thomas. *Elements of Information Theory*. Wiley, New York, 1991
29. J. Dunn. Well separated clusters and optimal fuzzy partitions. *Journal of Cybernetics*, 4:95–104, 1974

Printing: Krips bv, Meppel
Binding: Stürtz, Würzburg